Learning Materials in Biosciences

Learning Materials in Biosciences textbooks compactly and concisely discuss a specific biological, biomedical, biochemical, bioengineering or cell biologic topic. The textbooks in this series are based on lectures for upper-level undergraduates, master's and graduate students, presented and written by authoritative figures in the field at leading universities around the globe.

The titles are organized to guide the reader to a deeper understanding of the concepts covered.

Each textbook provides readers with fundamental insights into the subject and prepares them to independently pursue further thinking and research on the topic. Colored figures, step-by-step protocols and take-home messages offer an accessible approach to learning and understanding.

In addition to being designed to benefit students, Learning Materials textbooks represent a valuable tool for lecturers and teachers, helping them to prepare their own respective coursework.

Karl-Erich Jaeger • Andreas Liese •
Christoph Syldatk
Editors

Introduction to Enzyme Technology

 Springer

Editors
Karl-Erich Jaeger
Institute of Molecular Enzyme Technology,
Heinrich Heine University Düsseldorf and
Forschungszentrum Jülich GmbH
Jülich, Germany

Andreas Liese
Institute of Technical Biocatalysis, Hamburg
University of Technology
Hamburg, Germany

Christoph Syldatk
Institute of Process Engineering in Life
Sciences II - Electro Biotechnology
KIT - Karlsruhe Institute of Technology
Karlsruhe, Germany

ISSN 2509-6125   ISSN 2509-6133   (electronic)
Learning Materials in Biosciences
ISBN 978-3-031-42998-9   ISBN 978-3-031-42999-6   (eBook)
https://doi.org/10.1007/978-3-031-42999-6

Translation from the German language edition: "Einführung in die Enzymtechnologie" by Karl-Erich Jaeger et al.,
© Springer-Verlag GmbH Deutschland, ein Teil von Springer Nature 2018. Published by Springer Spektrum. All
Rights Reserved.

The translation was done with the help of artificial intelligence (machine translation by the service DeepL.com). A
subsequent human revision was done primarily in terms of content.

This Springer imprint is published by the registered company Springer Nature Switzerland AG
The registered company address is: Gewerbestrasse 11, 6330 Cham, Switzerland

Paper in this product is recyclable.

Contents

Introduction to Enzyme Technology

1

Karl-Erich Jaeger, Andreas Liese, and Christoph Syldatk

What You Will Learn in This Chapter
Enzyme technology is one of the most important areas of modern biotechnology and involves numerous disciplines of natural sciences. It is a highly interdisciplinary field, which from the beginning until today could only be further developed through integrated interdisciplinary research. This chapter provides an introduction, starting from the fields of work, through the historical development, a classification of biocatalysts to industrial applications.

K.-E. Jaeger (✉)
Institute for Molecular Enzyme Technology, Heinrich Heine University Düsseldorf and Forschungszentrum Jülich GmbH, Jülich, Germany
e-mail: karl-erich.jaeger@fz-juelich.de

A. Liese
Institute for Technical Biocatalysis, Hamburg University of Technology, Hamburg, Germany
e-mail: liese@tuhh.de

C. Syldatk
Institute of Process Engineering in Life Sciences II - Electro Biotechnology, KIT - Karlsruhe Institute of Technology, Karlsruhe, Germany
e-mail: christoph.syldatk@kit.edu

© The Author(s), under exclusive license to Springer Nature Switzerland AG 2024
K.-E. Jaeger et al. (eds.), *Introduction to Enzyme Technology*, Learning Materials in Biosciences,
https://doi.org/10.1007/978-3-031-42999-6_1

1.1   The Field of Enzyme Technology

Enzymes and enzyme-catalyzed reactions have been used in the production, processing, and preservation of food and feed since the beginning of human history, long before the term "enzyme" was coined; these include the use or production of yeast and sourdough, alcoholic beverages such as beer and wine, vinegar, cheese and other dairy products, sauerkraut, silage, and compost. In more recent times, enzymes as products of biotechnology have become an integral part of everyday life; their areas of application include household applications (e.g., as detergent enzymes), environmental analysis (e.g., cholesterol esterase for the detection of organophosphates), medical diagnostics (e.g., glucose oxidase for the determination of blood sugar), the production of basic and fine chemicals (e.g. acrylamide and so-called chiralica), of pharmaceuticals (e.g. semisynthetic penicillins and cephalosporins as well as insulin) and direct use as therapeutics (e.g. urokinase to dissolve blood clots or thrombin to promote blood coagulation). Likewise, all molecular, biological, and genetic engineering work cannot be carried out without enzymes. The climate and energy problems combined with a steadily increasing demand for renewable resources in the chemical industry will result in an even further increasing importance of enzymes and enzymatic processes in the near future.

Enzymes are natural catalysts that, like their conventional chemical counterparts, accelerate chemical reactions without being consumed. Accordingly, they are referred to as "biocatalysts" by analogy with chemical ones. Enzymes differ significantly from chemical catalysts in their composition, structure, and kinetics (Bisswanger 2015). Therefore, with the wide use of enzymes in all areas of life, all methods related to their identification, characterization, and use are of great importance (Aehle 2007; Buchholz et al. 2012) and precisely describe the field of enzyme technology.

Major contributions to enzyme technology are made by biology, in particular microbiology; molecular biology and genetics; biochemistry; analytical, organic, and technical chemistry; structural biology and computer-based simulations, bioinformatics, and process engineering (Table 1.1). The establishment of processes involving multienzyme reactions has resulted in increasing importance also of systems biology and synthetic biology.

Table 1.1 clearly shows that the successful cooperation of biologists, chemists, engineers, and mathematicians is indispensable for the technical production of enzymes and the establishment of new enzyme processes. Here, the mutual understanding of the respective disciplines is a prerequisite for success. In industry, the research and development of enzyme processes is almost exclusively carried out by interdisciplinary teams. It is therefore all the more important that everyone is a specialist in his or her own field but, at the same time, has learned during training to communicate in an interdisciplinary manner and to understand the technical language of other disciplines.

Table 1.1 Fields of expertise contributing to enzyme technology

Microbiology	Isolation of microorganisms, strain identification and maintenance
Biochemistry	Assay development, purification, and characterization of enzymes
Genetics	Construction of new strains and expression of recombinant enzymes
Molecular biology	Enzyme optimization, construction of enzyme libraries
Organic chemistry	Synthesis of enzyme substrates and enzyme inhibitors, biocatalysis
Analytical chemistry	Identification of proteins, substrates, and products
Structural biology	Elucidation of enzyme 3D structures
Computer simulation	Modeling of enzyme structures and catalytic mechanisms
Bioinformatics	Enzyme identification in protein and gene databases
Technical chemistry	Enzyme immobilization, reaction control, product processing
(Bio)process engineering	Production, process modeling and management, scaling, downstream processing
Systems biology	Construction of whole-cell catalysts, optimization of metabolic pathways
Synthetic biology	Establishment of also nonnatural multienzyme reactions in vivo and in vitro

1.2   The Development of Enzyme Research

Processes based on the action of enzymes, such as alcoholic fermentation, the use of yeast for baking, or the production of cheese, have been used for several thousand years. The first evidence of the existence of enzymes came in 1833 from experiments by the French chemist Anselme Payen with extracts from barley seedlings, which could be used to produce large quantities of sugar from starch. The existence of a so-called ferment (from Latin: *fermentum* = fermentation or leaven) was postulated. The Swedish chemist Jöns Jakob Berzelius suggested in 1830–1840 that this ferment had a catalytic effect, on which the process of fermentation was based. The term "enzyme" in use today was introduced by Friedrich Wilhelm Kühne in 1878 (from Greek: *énzymon* = leaven or yeast). Emil Fischer, who was awarded the Nobel Prize in Chemistry in 1902 for his work on sugar chemistry, postulated in 1890 the so-called lock-and-key principle which is still in use today, according to which a substrate fits into the active center of an enzyme like a key into a lock. The founder of modern enzymology is Eduard Buchner, who proved that a fermentation process can also take place in the absence of living cells and is catalyzed by a protein-containing substance, which he called zymase. For this discovery, Buchner was awarded the Nobel Prize in Chemistry in 1907. Figure 1.1 shows that since that time, numerous other Nobel Prizes have been awarded for work on or involving enzymes. Another milestone in enzyme research in 1926 was the proof by James B. Sumner that the enzyme urease is a pure protein that can be crystallized; this was also confirmed by John

The Nobel Prize in Chemistry

Laureate(s)	Justification	Year
E. Charpentier, J. A. Doudna	"for the development of a method for genome editing"	2020
F. H. Arnold	"for the **directed evolution of enzymes**"	2018
G. P. Smith, G. P. Winter	"for the **phage display of peptides** and antibodies"	
R. J. Lefkowitz, B. K. Kobilka	"for studies of G-protein-coupled receptors"	2012
V. Ramakrishnan, T. A. Steitz, A. E. Yonath	"for studies of the structure and function of the ribosome"	2009
O. Shimomura, M. Chalfie, R. Y. Tsien	"for the discovery and development of the green fluorescent protein, GFP"	2008
P. D. Boyer, J. E. Walker	"for their elucidation of the enzymatic mechanism underlying the synthesis of adenosine triphosphate (ATP)"	1997
J. C. Skou	"for the first discovery of an ion-transporting enzyme, *Na⁺, K⁺ -ATPase*"	
K. B. Mullis	"for his invention of the **polymerase** chain reaction (PCR) method"	1993
M. Smith	"for his fundamental contributions to the establishment of oligonucleotide-based, site-directed mutagenesis and its development for protein studies"	
J. Deisenhofer, R. Huber, H. Michel	"for the determination of the three-dimensional structure of a photosynthetic reaction centre"	1988
J. W. Cornforth	"for his work on the stereochemistry of **enzyme-catalyzed reactions**"	1975
C. B. Anfinsen	"for his work on **ribonuclease**, especially concerning the connection between the amino acid sequence and the biologically active conformation"	1972
S. Moore, W. H. Stein	"for their contribution to the understanding of the connection between chemical structure and catalytic activity of the active centre of the **ribonuclease** molecule"	
M. F. Perutz, J. C. Kendrew	"for their studies of the structures of globular proteins"	1962
F. Sanger	"for his work on the structure of proteins, especially that of insulin"	1958
A. R. Todd	"for his work on nucleotides and nucleotide co-enzymes"	1957
J. B. Sumner	"for his discovery that **enzymes can be crystallized**"	1946
J. H. Northrop, W. M. Stanley	"for their **preparation of enzymes** and virus proteins in a pure form"	
A. Harden, H. K. v. Euler-Chelpin	"for their investigations on the fermentation of sugar and **fermentative enzymes**"	1929
E. Buchner	"for his biochemical researches and his discovery of cell-free fermentation" **Buchner is considered the father of enzymology.**	1907

The Nobel Prize in Physiology or Medicine

Year	Laureate(s)	Justification
2009	E. H. Blackburn, C. W. Greider, J. W. Szostak	"for the discovery of how chromosomes are protected by telomeres and the enzyme **telomerase**"
1999	G. Blobel	"for the discovery that proteins have intrinsic signals that govern their transport and localization in the cell"
1997	S. B. Prusiner	"for his discovery of Prions – a new biological principle of infection"
1994	A. G. Gilman, M. Rodbell	"for their discovery of G-proteins and the role of these proteins in signal transduction in cells"
1978	W. Arber, D. Nathans, H. O. Smith	"for the discovery of **restriction enzymes** and their application to problems of molecular genetics"
1968	R. W. Holley, H. G. Khorana, M. W. Nirenberg	"for their interpretation of the genetic code and its function in protein synthesis"
1965	F. Jacob, A. Lwoff, J. Monod	"for their discoveries concerning genetic control of enzyme and virus synthesis"
1958	G. W. Beadle, E. L. Tatum	"for their discovery that genes act by regulating definite chemical events" ("**one-gene-one-enzyme-hypothesis**")
1955	A. H. T. Theorell	"for his discoveries concerning the nature and mode of action of **oxidation enzymes**"
1931	O. H. Warburg	"for his discovery of the nature and mode of action of the **respiratory enzyme**"
1923	F. G. Banting, J. J. R. Macleod	"for the discovery of insulin"
1910	A. Kossel	"in recognition of the contributions to our knowledge of cell chemistry made through his work on proteins, including the nucleic substances"

1901

Fig. 1.1 Scientists who have received the Nobel Prize for their research on proteins or enzymes (selection). If enzymes or enzyme classes are specifically named in the Nobel Committee's justification for the prize, these and the corresponding year are shown in red

H. Northrop for the enzymes pepsin, trypsin, and chymotrypsin; both received the Nobel Prize in Chemistry for this in 1946 together with Wendell M. Stanley.

The first milestone in process engineering for enzyme processes was set by Nelson and Griffin in 1916, when they immobilized the enzyme invertase on activated carbon by adsorption for the first time. They used this immobilisate packed into a chromatography column as the first fixed-bed reactor for the continuous cleavage of sucrose to α-D-glucose and β-D-fructose, which can be used as a sweetener. Later, during World War II, the company Tate and Lyle (Decatur, IL, USA) used invertase to produce the so-called golden syrup.

Industrial processes for the production of L-amino acids laid the technological foundations for the efficient large-scale use of isolated enzymes. Since 1954, homogeneously dissolved aminoacylase has been used in batch reactors in an aqueous medium to obtain enantiomerically pure L-amino acids by racemic cleavage. Due to the lack of possibility to recycle the enzymes, very high process costs were incurred. The first large-scale application of an immobilized enzyme was established in 1969 by the company Tanabe Seiyaku (Tokyo, Japan), which used aminoacylase immobilized on DEAE-Sephadex ion exchange resins in continuously operated fixed-bed reactors for the synthesis of ʟ-methionine. In order to circumvent limitations due to diffusion on the heterogeneously bound enzymes, in 1980 Degussa, now Evonik Industries (Essen, Germany), established a continuously operated membrane reactor in large-scale production, with which the aminoacylase can be used homogeneously dissolved and recycled (enzyme membrane reactor, EMR) via membrane filtration. This circumvents possible diffusion limitations on the immobilisates.

Since the 1970s, it has been known from the work of Sidney Altmann and Thomas Cech (Nobel Prize 1989) that RNA can also have an enzymatic function, either together with an enzyme (in RNase P) or as a so-called ribozyme (for more details on the history of enzyme research, see Buchholz et al. (2012), Panagiotopoulos and Fasouilakis (2017), and https://de.wikipedia.org/wiki/Enzym).

The historical development of enzyme research shows that from the very beginning, it has always been interdisciplinary research, the success of which could only be achieved through the active cooperation of different scientific disciplines.

Questions

1. What is the meaning of the word "enzyme"?
2. Which technology represents the first milestone for enzyme processes?
3. Which two fundamental process technologies were applied in the first industrial biotransformations to produce ʟ-amino acids?
4. Does enzymatic activity always require a protein?

1.3 Modern Enzyme Research

The development of modern methods of molecular biology, structural biology, and computer-based modeling has revolutionized research and biotechnological applications of enzymes. Today, it is possible to identify an almost unlimited number of new enzyme genes. This is done with the help of high-throughput sequencing methods for DNA (*next-generation sequencing* or *NGS methods*) and RNA (so-called RNA-seq methods) but also by applying bioinformatic methods when searching databases (sequence-based screening; Chap. 7). The nucleic acids originate from culturable (micro)organisms that are either isolated from environmental habitats or kept in strain collections, or the DNA is isolated directly from environmental habitats (metagenomics). The enzyme-encoding DNA can then be amplified by PCR, cloned into expression vectors, and expressed in suitable prokaryotic or eukaryotic host organisms such as *Escherichia coli*, *Bacillus subtilis*, *Saccharomyces cerevisiae*, or *Aspergillus niger* (Chap. 9). Subsequently, (often high-throughput) screening methods are used to identify the enzyme activity (activity-based screening; Chap. 6). Optimization of the properties of the enzymes found (for industrial applications, e.g., stability in organic solvents, at high or low pH values and temperatures) is then carried out by applying methods of rational design or *directed evolution* (Chap. 8).

After purification and biochemical characterization of the corresponding enzymes, their production is carried out on a larger scale. This often requires complex and cost-intensive process development, which above all must also include a suitable *downstream processing* method (dsp) (Chap. 12), before the enzymes are available for various biotechnological applications. A future vision, for which, however, there are already first experimental examples, is the computer-based prediction of an amino acid sequence that codes for an enzyme with predictable properties. However, many problems still need to be solved here, ranging from the functional expression of a theoretically predicted enzyme (a so-called theozyme) to the precise simulation of the catalytic mechanism (Bornscheuer et al. 2012).

Enzymes are classified into seven classes according to the recommendations of the International Union of Biochemistry and Molecular Biology, IUBMB (http://www.sbcs. qmul.ac.uk/iubmb/enzyme/). Each enzyme is assigned a classification number consisting of four numbers, beginning with EC (for Enzyme Commission)—EC A.B.C.D—whose first one-digit number A stands for the type of reaction from a chemical point of view that the respective enzyme catalyzes (Table 1.2). In each case, the reversible reaction is considered, i.e., in enzyme class 1 of the oxidoreductases, both oxidation and reduction. As a rule, each enzyme described has a trivial name and an EC number.

Detailed information on the individual representatives of an enzyme class can be researched on the Internet via the freely accessible database BRENDA-Braunschweig Enzyme Database, www.brenda-enzymes.org, founded in 1987 by Dietmar Schomburg, TU Braunschweig. Here one can find detailed information on the microbiology of the host organisms, integration into metabolic pathways, molecular biology, kinetics, stability data, substrate spectra, enzyme inhibitors, and the most important literature references.

In the following, a model enzyme from the biotechnologically important group of hydrolases will be presented as an example of the methods used in enzyme research today.

The Gram-positive bacterium *Bacillus subtilis* produces and secretes a number of hydrolases that are important for biotechnological purposes; in addition to several proteases and amylases, these also include the lipase LipA (EC 3.1.1.3). This enzyme was first described in 1992 by a Belgian research group and has been the subject of intensive work ever since. In general, the identification of hydrolases (and other enzymes) can be carried out in so-called metagenome libraries. For this purpose, research projects that are financially supported by the European Union, for example, are carried out. Environmental samples are obtained; DNA is isolated from these samples, cloned into specially constructed expression vectors, and expressed in expression host organisms, e.g., *B. subtilis* or *E. coli*. For activity-based identification, high-throughput screening methods are being developed that allow many 100,000 clones to be tested to determine whether they contain a DNA fragment that codes for an enzymatically active lipase, i.e., a lipase gene. The lipase produced by *B. subtilis* was expressed, purified, and biochemically characterized, and the purified enzyme was crystallized and its three-dimensional structure determined by X-ray crystallography. This structure made it possible, with the aid of computer simulations, to predict which amino acids are important for the temperature stability of this enzyme and which ones need to be exchanged by targeted mutations in

Table 1.2 Enzyme classification by EC number—EC A.B.C.D (EC: Enzyme Commission; (**a**) describes the main type of reaction (1–6); (**b**) describes the chemical structure of the substrate or the transferred molecule; (**c**) describes the cosubstrate or the substrate specificity; (**d**) individual counting number in class A.B.C.)

Enzyme class A: Name	Reaction type	Cofactors	Examples
EC 1: oxidoreductases	Oxidation/reduction (electron transfer)	NAD(P)$^+$, FAD, FMN, lipoic acid	Alcohol dehydrogenase (EC 1.1.1.x) Carbonyl reductase (EC 1.1.1.x) Glucose oxidase (EC 1.1.3.4) D-amino acid oxidase (EC 1.4.3.3) Catalase (EC 1.11.1.16)
EC 2: transferases	Transfer of functional groups from one molecule (donor) to another (acceptor)	*S-adenosylmethionine*, ATP, cAMP, biotin, thiamine diphosphate (ThDP, TPP), tetrahydrofolic acid	Hexokinase (HK) (EC 2.7.1.1) Polymerases (EC 2.7.7. x) Acetyl, amino, methyl transferases (EC 2.6.1.x)
EC 3: hydrolases	Hydrolytic/nucleophilic cleavage/formation of C–O, C–N, C–S, C–C, P–O, S–O bonds		Esterases (EC 3.1.1.x) Lipases (EC 3.1.1.3) Glucosidases (EC 3.2.1. x) Proteases (EC 3.4.2x.x)
EC 4: lyases	Non-hydrolytic/ oxidative cleavage/ formation of C=O, C=N, and C=C bonds via elimination/addition	Thiamine diphosphate (ThDP, TPP), pyridoxal phosphate	Decarboxylase (EC 4.1.1.X) Oxynitrilase (EC 4.1.2. X) Hydratase (EC 4.2.1.X) Carbonic anhydrase (EC 4.2.1.1)
EC 5: isomerases	Intramolecular transformation, e.g., epimerization, racemization, rearrangement	Cobalamin, glucose-1,6-bisphosphate	Racemases (EC 5.1.x.x) Epimerases (EC 5.1.3.x)
EC 6: ligases	Cleavage/formation of C–O, C–N, C–S, C–C, and P–O bonds of two substrates under ATP consumption	ATP, NAD(P)$^+$, biotin	Synthetases (EC 6.x.x. x) DNA ligases (EC 6.5.1. x)
EC 7: translocases	Moving an ion or molecule across a cell membrane	ATP, diverse	TOM complex (EC 7.x. x.x)

order to increase the temperature stability. Computer simulation was also used to optimize another property of this lipase that is at least as important from a biotechnological point of view: enantioselectivity. Using a complete saturation mutagenesis, each of the 181 amino acids of the wild-type lipase was exchanged for all 19 remaining amino acids that do not occur at this position in the wild-type lipase. The lipase library thus constructed consists of exactly 3439 *E. coli* clones, each producing a lipase variant that differs in one amino acid from the other lipases. This library makes it possible to identify amino acids that are important for biotechnologically relevant properties, such as stability to detergents or organic solvents. A larger amount of this lipase was produced by fermentation, and its biotechnological application was demonstrated for the production of an enantiomerically pure alcohol, which is used as an intermediate for the synthesis of pharmaceuticals. These research approaches and results, summarized in Fig. 1.2, demonstrate that modern enzyme research is a methodologically very diverse field that requires multidisciplinary work involving numerous scientists from very different disciplines.

1.4   Enzymes as Biocatalysts

Enzymes are composed of L-α-amino acids linked by peptide bonds. Therefore, they are usually nontoxic and readily biodegradable. Compared to chemical catalysts, these biocatalysts are active at comparatively mild reaction conditions in terms of temperature, pH, and pressure, and this is already the case at much lower catalyst concentrations than is usual for chemical catalysts (Faber 2018; Jeromin and Bertau 2005).

In general, six major advantages are mentioned that can make enzymes attractive for use in the abovementioned areas:

– *Activity under mild reaction conditions*. In contrast to chemical syntheses, reactions catalyzed by enzymes generally take place in an aqueous environment, avoiding extreme pH values and temperatures.
– *Reaction specificity*. In general, an enzyme catalyzes only one type of reaction.
– *Substrate specificity*. Unlike a chemical catalyst, enzymes are often highly specific to a particular substrate and can therefore catalyze its targeted conversion in a highly specific manner even in a mixture of different compounds (e.g., the oxidation of glucose to gluconic acid in blood serum).
– *Regiospecificity*. Enzymes can distinguish between functional groups of the same reactivity due to their active site and will only convert very specific ones (e.g., specific reaction at the C-11 carbon atom in the hydroxylation of steroids).
– *Enantioselectivity*. Due to the structure of their active site, most enzymes recognize and distinguish not only between different regions of a substrate molecule but also between different enantiomers of a substrate and usually selectively convert only one of them. Since natural enzymes are themselves composed of L-amino acids, they are themselves chiral catalysts.

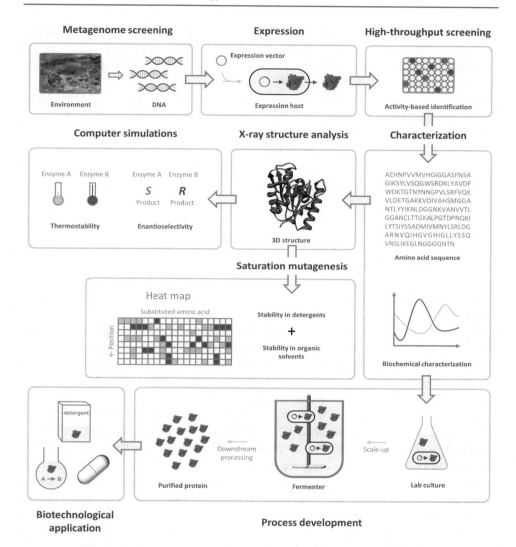

Fig. 1.2 Modern methods of enzyme research exemplified with the model enzyme lipase A from *Bacillus subtilis*. This enzyme was isolated, purified, and characterized in the laboratory of K.-E. Jaeger in collaboration with the working groups mentioned below: metagenome libraries (Prof. Peter Golyshin, Bangor University, England, and Prof. Wolfgang Streit, University of Hamburg, Germany), expression optimization (Prof. Jochen Büchs, RWTH Aachen, Germany), activity-based screening (Prof. Manuel Ferrer, CSIC Madrid, Spain), elucidation of the 3D structure (Prof. Bauke Dijkstra, University of Groningen, The Netherlands), computer simulations to predict mutations for increased temperature stability (Prof. Holger Gohlke, Heinrich Heine University Düsseldorf, Germany) and enantioselectivity (Prof. Walter Thiel and Prof. Manfred Reetz, Max-Planck-Institute for Coal Research, Mülheim an der Ruhr, Germany), complete saturation mutagenesis (Prof. Ulrich Schwaneberg, RWTH Aachen, Germany), process development (Dr. Thorsten Eggert, evoxx technologies, Monheim, Germany), biotechnological applications (Prof. Michael Müller, University of Freiburg and Prof. Stephan Lütz, TU Dortmund, Germany)

- *Not bound to their natural role*. Many enzymes exhibit high substrate tolerance in that they also accept nonnatural substrates. Furthermore, they can be used in nonaqueous media, which makes it possible, for example, to use hydrolases for synthesis.

In addition to these potential advantages, however, there are also a number of limitations and disadvantages of enzymes compared to chemical catalysts, which must be taken into account in any technical application (Buchholz et al. 2012):

Enzymes are usually sensitive to the following:

- High temperatures
- Extreme pH values
- Metal ions
- Aggressive chemicals
- Many solvents
- Susceptibility to degradation by proteases, since they are proteins themselves

Possibilities for dealing with these limitations or even overcoming them can be found in Chaps. 8, 11, 12, and 13.

Enzymes are produced either microbially or from fresh biological material, which can be of plant or animal nature, and, depending on the purity required for the corresponding subsequent application, require a greater or lesser degree of effort for purification, which can be associated with considerable costs if, for example, expensive chromatography processes have to be used for this purpose (Aehle 2007).

Since enzymes are usually dissolved in an aqueous medium, their reuse requires prior separation from the reaction medium while retaining the activity. This can be done either by retention of dissolved enzymes by means of membranes or alternatively after their previous binding to an insoluble carrier, i.e., immobilization, by filtration, centrifugation, or use in a fixed-bed reactor analogous to a chromatography column. Both procedures are associated with additional effort and costs. Whether a biocatalytic or a chemocatalytic process is ultimately implemented on an industrial scale depends on the overall economic considerations of the company in question. If there are already established processes, the use of chemical catalysts may be more advantageous and cost-effective for many reactions.

Nevertheless, the use of enzymes is also increasingly observed in the chemical industry (Liese et al. 2006; Grunwald 2015). This is mainly the case for applications involving the functionalization of a specific, nonactivated C atom in complex molecules, the selective conversion of a specific functional group of a molecule among several groups with the same reactivity, the introduction of chiral centers, or the racemate cleavage of chiral compounds (Faber 2018). Often, immobilized enzymes are used because of the easier separation of catalyst and product from the reaction medium and the possibility to reuse the catalyst (Buchholz et al. 2012); increasingly, this is also done in unusual nonaqueous reaction media.

Questions

5. What makes the use of enzymes as catalysts particularly interesting?
6. What are the disadvantages of enzymes compared to chemical catalysts?

1.5 Industrial Applications of Enzymes

Since the end of the nineteenth century, there have been successful examples of the use of immobilized whole biocatalysts to carry out enzyme reactions on a large scale in industry with the technical production of vinegar essence using the generator process. Chemical conversion reactions, so-called biotransformations, can be carried out not only with enzymes in free or immobilized form but also with dead microbial cells, e.g., after treatment with glutaraldehyde, living growing or living immobilised microbial cells. The use of living cells has the advantage that any necessary coenzyme regeneration, as in the case of nicotinamide adenine dinucleotide (NAD) or flavin adenine dinucleotide (FAD), can be carried out in a simple manner by the cells themselves in the presence of an inexpensive carbon source such as glucose. However, side and degradation reactions occurring in this process can be problematic. Dead microbial cells are used, for example, to stabilize complex enzymes consisting of several subunits, such as the nitrile hydratase or tryptophan synthetase. Figure 1.3 shows the main areas of application for industrial enzymes and the sales achieved by the world's largest enzyme producer, the company Novozymes (Bagsværd, Denmark), in the various areas of application. The requirements placed on enzymes for industrial use vary greatly depending on the application.

1.5.1 Enzymes in the Food, Feed, and Textile Industries

When used in the food industry, the enzymes often remain in the end product. Therefore, it is important to note that the enzymes themselves must be suitable *as* food, which means that they are either of animal or plant origin or originate from so-called GRAS (*generally regarded* as *safe*) microorganisms. Furthermore, the enzymes should be nontoxic and nonallergenic, and they should be neutral in taste, simple, and inexpensive to produce and should not represent a significant cost factor in the corresponding food production process (Aehle 2007).

Examples of plant and animal enzymes that have long been used in the food sector are enzyme-containing cereal malt extracts produced in a gentle process to promote starch degradation, e.g., in beer production, pepsin from pig stomach mucosa as a digestive enzyme in pepsin wine for better protein digestion in the stomach, rennet enzyme from calf stomachs for protease-catalyzed protein precipitation in cheese production, papain and chymopapain from the melon tree as meat tenderizers, or bromelain from pineapple and ficin from the fig tree also as meat tenderizers or to promote digestion in diets.

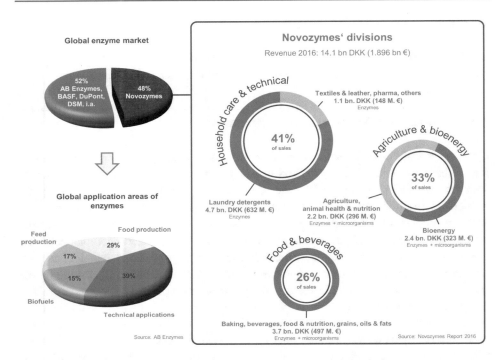

Fig. 1.3 Global enzyme market, divided into the application areas of technical application (including textile, leather, paper/pulp, detergent and cosmetics production, agriculture, wastewater management, pharmaceutical and chemical industry), food production (including food, food supplement, and beverage production), feed production (animal feed, animal feed additive production), and biofuels (including biomass degradation and conversion), as well as the share of the largest enzyme producer Novozymes in this. Enzymes produced on the largest scale worldwide include proteases (approx. 2000 t a-1), amylases (approx. 1200 t a-1), phytases (approx. 50 t a-1), and lipases (approx. 20 t a-1). DKK: Danish kroner (Source: Novozymes Report 2016)

However, microbial enzymes such as amylases, glucanases, and proteases are also used in the food sector, e.g., for starch saccharification and for breaking down the "gluten" in cereal flours, proteases are used as a substitute for animal calf rennet in cheese production, and pectinases are used for clarification, degumming, and reducing the viscosity of fruit juices. The use of microbial lactase or β-galactosidase in the production of lactose-free dairy products is also now a common practice. Since microbial enzymes are classified as "vegan," their use in the production of corresponding foods is also possible. The enzymes mentioned generally remain in the end product. An overview for this area is given in Chap. 16.

In the feed sector, the enzyme phytase is used in nonruminant animals as a feed additive for phosphate mobilization in plant feeds.

One of the world's largest enzyme processes by volume is the glucose isomerase reaction for the production of the sweetener glucose–fructose syrup (*high fructose corn syrup*, HFCS) used in the beverage and baking industries. As a rule, this reaction is preceded by an enzymatic saccharification of starch consisting of several steps with a first enzymatic reaction step at about 100 °C.

Successful examples of the use of hydrolytic enzymes for synthesis reactions with addition or in organic solvents are the lipase-catalyzed transesterification of sunflower oil to cocoa butter or the protease-catalyzed synthesis of the sweetener aspartame (L-aspartyl-L-phenylalanine methyl ester); in both cases, the regioselectivity or stereoselectivity of the enzymes used is exploited.

In the textile industry, microbial amylases are used industrially on a large scale to remove the "sizing agent" starch in textile processing, microbial cellulases to break down cotton fibers to improve the softness and visual appearance of laundry, and microbial proteases to remove hair and meat residues in leather processing (Aehle 2007).

1.5.2 Enzymes in the Chemical and Pharmaceutical Industry

When processes in the chemical and pharmaceutical industry that were carried out with the aid of chemical catalysts are replaced by enzyme processes, the aspects of cost reduction, quality improvement, raw material utilization, and reduction of environmental pollution play an important role (Buchholz et al. 2012). In the industrial use of enzymes in the chemical and pharmaceutical industry to date, however, it is noticeable that enzymes of enzyme class EC 3 "hydrolases" are predominantly used on a large scale in free or immobilized form. The main reasons for this are that many of these enzymes are extracellular in microbial production, are easy to process, and are generally quite robust and stable. In addition, they do not require expensive cofactors or coenzymes such as enzymes from class EC 1 "oxidoreductases." Corresponding reactions with enzymes from this enzyme class are therefore often carried out with wild-type or recombinant whole-cell catalysts.

In almost all industrial reactions carried out in the chemical and pharmaceutical industries to date, the properties of reaction regio- and stereoselectivity are used. Important examples for the use of enzymes in this sector are the production of vitamin C and enantiomerically pure non-proteinogenic L- and D-amino acids, e.g., with the enzymes L-*N*-aminoacylase or D-hydantoinase, the production of 6-aminopenicillanic acid (6-APA) and 7-aminocephalosporanic acid (7-ACA) and derived thereof semisynthetic penicillins and cephalosporins with the corresponding penicillin and cephalosporin acylases and the production of enantiomerically pure alcohols and amines as building blocks for organic synthesis. In the cosmetics industry, for example, cosmetic esters, in some cases even on a large scale, are produced without the use of solvents in the pure reactant mixtures of fatty acid and fatty alcohol. An overview of this field is given in Chap. 14.

Most of the enzymes used for this purpose are now produced recombinantly using genetic engineering processes. Compared with the use of animal, plant, or microbial enzymes from wildtype strains, this has in many cases led to a significant increase in economic efficiency (Buchholz et al. 2012).

High-cell-density cultivation can provide large amounts of enzyme within a short time. The recombinant enzyme can account for more than 20% of the total cell protein.

A deliberately incorrectly carried out polymerase chain reaction (*error-prone* or *ep-PCR*) allows the production of numerous gene variants of an enzyme and thus rapid optimization. After expression of these variants, so-called enzyme libraries are obtained, which can be searched for better enzyme variants that do not occur in nature using suitable *high-throughput screening* (HTS) methods (Chap. 8).

So-called tagging technologies enable fast and simple enzyme purification. In this process, either the N- or the C-terminus of a recombinant protein is tagged with a specific sequence of amino acids (e.g., with the hexa-histidine sequence His6), which enables highly selective separation of the protein from foreign protein.

In the use of living cells for biotransformation reactions, there are examples of the successful design of recombinant whole-cell biocatalysts (*designer bugs*) that can be cultivated to high cell densities, e.g., based on *Escherichia coli* cells, for carrying out multienzyme reactions with recombinant enzymes contained therein. The recombinant enzymes can originate from different wildtype strains, they can already be molecular biologically optimized, and the concentrations in the cell can be optimally matched. One challenge with living whole-cell biocatalysts is to overcome possible transport limitations caused by the cell membrane and cell wall, which may require permeabilization of the cells. This in turn can lead to a "bleeding" of the cells and the release of undesired by-products.

Worldwide, the conversion of the chemical industry from petroleum-based to renewable raw materials has begun and is being driven forward as part of the establishment of a bio-based economy. Therefore, there is no doubt that enzymatic processes will play an increasingly important role in the future, both in the pretreatment and supply of renewable raw materials and in the production of basic chemicals (Hilterhaus et al. 2016).

Questions

7. What is the prerequisite for the use of enzymes remaining in food?
8. Which is one of the world's largest enzyme processes by volume?
9. Which factors play a role when processes in the chemical and pharmaceutical industry are replaced by enzyme processes?
10. What have been great advances in recombinant enzyme production for the chemical and pharmaceutical industry?

Take-Home Message

- Enzymes are biocatalysts which have found a large number of biotechnological applications.
- Enzyme technology is an interdisciplinary field comprising expertise from (micro)biology, (bio)chemistry, (bio)physics, and (bio)engineering.
- Enzymes are classified into seven classes numbered EC 1–7 with this first digit indicating the type of catalyzed reaction.
- Enzymes are usually active under mild reaction conditions, i.e., in aqueous solution, at ambient temperature, and at neutral pH.
- Many enzymes show exquisite substrate specificity and enantioselectivity.
- Enzymes are used, among others, in food and feed, chemical, and pharmaceutical industries.

Answers

1. The term "enzyme" introduced by Friedrich Wilhelm Kühne in 1878 originates from the Greek word *énzymon* meaning leaven or yeast.
2. The immobilization of the enzyme invertase on activated carbon by adsorption is carried out by Nelson and Griffin in 1916.
3. First, the application of a fixed-bed reactor with immobilized enzymes and, second, the use of homogeneously solubilized enzymes being retained by an ultrafiltration membrane enable continuous production of L-amino acids.
4. No; ribozymes discovered by Sidney Altman and Thomas Cech in the late 1970s are RNAs which catalyze the site-specific cleavage of RNA molecules.
5. Enzymes are active under mild reaction conditions. In contrast to many chemical catalysts, enzymes show substrate and reaction specificity as well as regio- and enantioselectivity.
6. In contrast to chemical catalysts, enzymes are usually sensitive to high temperatures, extreme pH values, heavy metal ions, aggressive chemicals, and many solvents and susceptible to degradation by proteases, since they are proteins themselves.
7. When used in the food industry, enzymes themselves must be suitable as food, which means that they are either of animal or plant origin or originate from so-called GRAS (*generally regarded as safe*) microorganisms. They should be nontoxic, nonallergenic, and neutral in taste and not represent a significant cost factor in the corresponding food production process.
8. One of the world's largest enzyme processes by volume is the glucose isomerase reaction for the production of the sweetener glucose-fructose syrup (*high fructose corn syrup*, HFCS).

9. When processes in the chemical and pharmaceutical industry are replaced by enzyme processes, the aspects of cost reduction, quality improvement, raw material utilization, and reduction of environmental pollution play an important role.

10. *High-cell-density cultivation* can provide large amounts of enzyme within a short time. The *error-prone* or *ep-PCR* in combination with *high-throughput screening* (HTS) methods allows the production of numerous variants of an enzyme and thus its rapid optimization. *Tagging technologies* enable fast and simple enzyme purification.

References

Aehle W. Enzymes in industry. 3rd ed. Wiley-VCH; 2007.

Bisswanger H. Enzymes – structure, kinetics and applications. Wiley-VCH; 2015.

Bornscheuer UT, Huisman GW, Kazlauskas RJ, Lutz S, Moore JC, Robins K. Engineering the third wave of biocatalysis. Nature. 2012;485:185–94.

Buchholz K, Kasche V, Bornscheuer U. Biocatalysts and enzyme technology. 2nd ed. Wiley-VCH; 2012.

Faber K. Biotransformations in organic chemistry. 7th revised and corrected ed. Springer; 2018.

Grunwald P. Industrial biocatalysis. Pan Stanford Publishing; 2015.

Hilterhaus L, Kettling U, Antranikian G, Liese A. Applied biocatalysis: from fundamental science to industrial applications. Wiley-VCH; 2016.

Jeromin GE, Bertau M. Bioorganikum – practical course in biocatalysis. Wiley-VCH; 2005.

Liese A, Seelbach K, Wandrey C. Industrial biotransformations. 2nd revised ed. Wiley-VCH; 2006.

Panagiotopoulos AA, Fasouilakis EG. Introduction to enzymes and biotechnology. Lab Lambert Academic Publishing; 2017.

Part I

Fundamentals

Enzyme Structure and Function

2

Vlada B. Urlacher and Katja Koschorreck

What You Will Learn in This Chapter

This chapter provides an overview on the general architecture of enzymes, describes protein structure and the types of enzyme cofactors, and discusses the advantages of enzymes as efficient and selective biocatalysts. A short overview of the thermodynamics of enzyme-catalyzed reactions is followed by the description of different catalytic mechanisms and strategies that enzymes use to speed up reaction rates. All aspects of enzyme structure, function, and catalysis presented in this subchapter are illustrated by respective examples.

2.1 Structure of Enzymes

Enzymes are biocatalysts that accelerate biochemical reactions up to 10^{17}-fold and thus maintain the metabolism of all living organisms. They do so by reducing the energetic barriers that have to be overcome in the conversion of a substrate to a product. Enzymes are mainly proteins.

V. B. Urlacher (✉) · K. Koschorreck
Institute of Biochemistry, Heinrich Heine University Düsseldorf, Düsseldorf, Germany
e-mail: vlada.urlacher@uni-duesseldorf.de; katja.koschorreck@uni-duesseldorf.de

2.1.1 General Architecture

Enzymes are biocatalysts that accelerate all biochemical reactions in living organisms and thereby maintain their metabolism and reproduction. Apart from a small group of catalytic RNA and DNA, most biocatalysts are proteins. Proteins are biopolymers that are mainly composed of 22 proteinogenic amino acids. Proteinogenic amino acids consist of a central α-carbon atom to which an amino group, a carboxyl group, a hydrogen atom, and a variable side chain are attached. Twenty proteinogenic amino acids are canonical, which means that they are directly encoded by the codons of the universal genetic code. According to the chemical properties of the side chain, canonical amino acids can be classified into four groups:

– Amino acids with a nonpolar side chain: glycine, alanine, valine, leucine, isoleucine, methionine, proline, phenylalanine, tryptophan
– Amino acids with a polar, uncharged side chain: serine, threonine, asparagine, glutamine, tyrosine, cysteine
– Amino acids with a polar, positively charged side chain: lysine, arginine, histidine
– Amino acids with a polar, negatively charged side chain: aspartic acid, glutamic acid

Amino acids build linear polymers by forming peptide bonds. The peptide bond is a type of amide bond that occurs between the carboxyl group of one amino acid and the amino group of another amino acid, upon releasing a water molecule. Peptide bonds have a rigid, planar structure stabilized by resonance of amides. In the polypeptide chain, the atoms of the peptide bonds form the backbone of the chain, while the side chains, also known as amino acid residues, represent the variable part. Due to the large number of possible combinations of proteinogenic amino acids, each polypeptide chain has an individual character. The polypeptide end, which carries a free amino group, is called the N-terminus and defines the beginning of the polypeptide chain. The end of the polypeptide chain carries a free carboxyl group and is called the C-terminus.

2.1.2 Primary, Secondary, Tertiary, and Quaternary Structure

The structure of proteins is described on four levels (Berg et al. 2013, p. 25). The first level is the linear sequence of amino acids, which are covalently linked via peptide bonds. It is referred to as primary structure.

The primary structure of a protein determines further structural levels. The secondary structure is formed by the spatial arrangement of amino acids that are located close to each other in the primary structure. This specific arrangement is stabilized by hydrogen bonds between the amino and carbonyl groups of the peptide backbone. Local structural elements are α-helices, β-sheets, and loops. In some proteins, the linear polypeptide chain is linked by disulfide bridges, which are formed by oxidation of two cysteine residues.

The α-helix is a right-handed helix that contains an average of 3.6 amino acid residues per helical turn, with the side chains pointing outward. Stabilization occurs through hydrogen bonds formed between every backbone amino group and the backbone carbonyl group of the amino acid located four amino acid residues before this in the sequence. β-Sheets have a pleated shape and are stabilized by hydrogen bonds between the carbonyl and amino groups of neighboring polypeptide extended strands, so-called β-strands. In parallel β-sheets, the β-strands run in the same direction, whereas in antiparallel β-sheets, the N-terminus of one strand is adjacent to the C-terminus of the next strand. On average, β-sheets in proteins are formed from six β-strands containing an average of six amino acid residues. Two regular secondary structures are often connected by so-called loops. Loops are often located on the surface of proteins. Different combinations of α-helices, β-sheets, and loops result in different protein structures. Proteins can be grouped into two subtypes: fibrillar (or fibrous) and globular (or spherical). Fibrillar proteins such as collagen or elastin are composed of helical elements, that do not build α-helices. These non- or poorly water-soluble proteins have a stabilizing function. Globular proteins such as enzymes are formed by the connection of α-helices and/or β-sheets via loops. They are mostly water-soluble and have multiple functions (Voet et al. 2010, p. 144).

Superordinate to the secondary structure are the tertiary structure and quaternary structure. The tertiary structure contains several secondary structure elements and represents the three-dimensional form of a protein. In a compact tertiary structure of globular proteins, nonpolar amino acid residues are usually hidden inside the folded molecule, while amino acids with polar, charged side chains are arranged on the protein surface. In addition to hydrophobic interactions, ionic bonds, hydrogen bonds, as well as disulfide bonds play an important role in stabilizing the tertiary structure. The tertiary structure of a protein is essential for its biological function (Voet et al. 2010, p. 158).

Some proteins consist of multiple polypeptide chains, so-called subunits. Their arrangement in a three-dimensional form is referred to as quaternary structure. It is stabilized by electrostatic interactions, hydrogen bonds, and van der Waals forces. In the simplest case, two identical polypeptide chains form a dimer. There are many proteins with a more complex quaternary structure. For example, catalase from bovine liver, which cleaves hydrogen peroxide, is a tetramer consisting of four identical subunits. Quite often, different polypeptide chains cluster together to form a multimer. Pyruvate decarboxylase, which decarboxylates pyruvate to acetaldehyde and CO_2, is a tetramer of two different subunits. The eukaryotic cytochrome c oxidase, a large enzyme complex of the respiratory chain, consists of a total of 13 different subunits (Voet et al. 2010, p. 173).

2.1.3 Posttranslational Modifications

Enzymes need to acquire their three-dimensional native conformation to catalyze chemical reactions. However, the amino acid sequence alone does not necessarily determine the activity of an enzyme. Many enzymes only obtain their active form by so-called posttranslational protein modifications. Posttranslational protein modifications are changes in proteins that take place after translation. In addition to proteolytic cleavage, these changes include chemical modification of amino acid residues via addition of inorganic groups, e.g., by phosphorylation or hydroxylation, or of organic groups, e.g., by acetylation or glycosylation. Proteolytic cleavage leads to removal of a part of the protein by cleaving one or more peptide bonds. This converts, for example, digestive enzymes into their active form or activates blood clotting factors. Phosphorylation of serine, threonine, or tyrosine residues often alters the activity of proteins. Hydroxylation of proline and lysine residues is an important step in collagen synthesis. Acetylation of the N-terminus impedes the degradation of proteins. During glycosylation, oligosaccharides are covalently bound to specific amino acid residues on the protein surface, rendering the proteins more hydrophilic. Thereby, glycosylation affects the activity, structure, and stability of proteins and is mainly found in secreted and membrane-bound eukaryotic proteins.

2.1.4 Cofactors

Many enzymes require cofactors for their activity. Cofactors are nonproteinogenic low-molecular components of enzymes that participate in catalysis (Voet et al. 2010, p. 357).

Cofactors are either small organic molecules, metal complexes, or metal ions. They enable/assist reactions that cannot be catalyzed by the protein component of the enzyme alone. This catalytically inactive protein component is referred to as apoenzyme. The catalytically active enzyme form in which both the cofactor and the apoenzyme are present is called holoenzyme. One type of cofactors are coenzymes. Coenzymes are complex organic molecules that are usually loosely and temporarily bound to the enzyme. Since coenzymes are chemically modified and not regenerated during the reactions in which they participate, they are often referred to as cosubstrates. Cosubstrates often serve as diffusible electron donors or acceptors in redox reactions like nicotinamide adenine dinucleotide ($NADH/NAD^+$) and nicotinamide adenine dinucleotide phosphate ($NADPH/NADP^+$) (Table 2.1). During oxidation of a substrate molecule, two electrons and a proton (a hydride ion) are transferred to $NAD(P)^+$, forming $NAD(P)H$. For example, alcohol dehydrogenase catalyzes the oxidation of ethanol to acetaldehyde in the presence of NAD^+, which is reduced to NADH. NADH can be regenerated to NAD^+ in an independent enzymatic reaction during reduction of another substrate, like during the reduction of pyruvate to lactate, catalyzed by lactate dehydrogenase. Another cosubstrate adenosine triphosphate (ATP) is used as an energy supplier in many enzymatic reactions. Hydrolysis of ATP to adenosine diphosphate (ADP) and orthophosphate or to adenosine monophosphate (AMP) and diphosphate releases energy that can drive endergonic reactions. For example, energy to form glutamine from glutamate and ammonia catalyzed by glutamine synthetase (also called glutamate-ammonium ligase) is provided by the hydrolysis of ATP to ADP and orthophosphate.

Table 2.1 Cosubstrates and prosthetic groups in enzymes

Cofactor	Enzyme
Cosubstrate	
Nicotinamide adenine dinucleotide (NADH/NAD$^+$)	Alcohol dehydrogenase
Nicotinamide adenine dinucleotide phosphate (NADPH/NADP$^+$)	Glucose-6-phosphate dehydrogenase
Adenosine triphosphate (ATP)	Glutamine synthetase
Nonmetal-containing prosthetic group	
Flavin adenine dinucleotide (FAD)	Glucose oxidase
Flavin mononucleotide (FMN)	Glycolate oxidase
Biotin	Pyruvate carboxylase
Pyridoxal phosphate (PLP)	Aspartate aminotransferase
Thiamine pyrophosphate (TPP)	Pyruvate decarboxylase
Metal-containing prosthetic group	
Iron porphyrin	Cytochrome P450 monooxygenase
Iron sulfur cluster	Ferredoxin
Molybdenum cofactor	Xanthine oxidase
Molybdenum iron cofactor	Nitrogenase
Vanadium cofactor	Vanadium-dependent haloperoxidase

Cofactors that are tightly bound to the enzyme are called prosthetic groups. The binding of the prosthetic group to the enzyme is usually covalent and often supported by hydrophobic interactions and hydrogen bonding. Prosthetic groups can be divided into metal-containing and nonmetal-containing. Flavin adenine dinucleotide (FAD), biotin, pyridoxal phosphate (PLP), and thiamine pyrophosphate (TPP) are examples of nonmetal-containing prosthetic groups (Table 2.1). FAD functions as an electron acceptor similar to NAD$^+$ and NADP$^+$; however, unlike the latter, it can accept two electrons and two protons. This results in the formation of FADH$_2$. Glucose oxidase uses FAD to catalyze the oxidation of glucose to gluconolactone. The FADH$_2$ formed in this process is regenerated back to FAD by the subsequent reduction of oxygen to hydrogen peroxide catalyzed by the same enzyme glucose oxidase.

Many prosthetic groups in enzymes are vitamins. Biotin is a water-soluble vitamin (B$_7$ or vitamin H). Chemically, it is a bicyclic ring formed by imidazolidone and thiophane with a valeric acid side chain. Biotin is covalently bound to enzymes via an amide bond between the valeric acid chain and the ε-amino group of a specific lysine residue of the enzyme. Biotin-dependent enzymes catalyze carboxylation reactions, for example, the carboxylation of pyruvate to oxaloacetate is catalyzed by pyruvate carboxylase.

Pyridoxal phosphate is a derivative of pyridoxine (vitamin B$_6$). It is found as a prosthetic group in many enzymes involved in amino acid metabolism. In this process, pyridoxal phosphate forms a Schiff base with the α-amino group of an amino acid substrate. Depending on the site at which the Schiff base is cleaved, a transamination, decarboxylation, or elimination reaction occurs. The type of the catalyzed reaction is controlled by the apoenzyme.

Thiamine pyrophosphate is a derivative of thiamine (vitamin B_1). As a prosthetic group, it is present in various enzymes including pyruvate decarboxylases, pyruvate dehydrogenase complex, and transketolases. In these reactions, the thiazolium ring represents the catalytically active species. The thiazolium ring can be readily deprotonated at C2 to become a zwitterionic ylide with a formally positively charged nitrogen and a formally negatively charged carbanion. The carbanion of the thiamine pyrophosphate ylide can then react as a nucleophile with the carbonyl carbon atom of a substrate like the keto acid pyruvate. This leads to decarboxylation of pyruvate in the further course of this reaction.

About 30% of all enzymes contain metal ions which are essential for their function, structure, or stability. In these so-called metalloenzymes, Fe^{2+}, Fe^{3+}, Cu^{2+}, Co^{2+}, Zn^{2+}, Mg^{2+}, Mn^{2+}, Ni^{2+}, Mo^{4+}, or V^{5+} are found as cofactors. The catalytic activity of a metalloenzyme is defined by the nature of the metal ion and in particular by its electronic structure and oxidation state. The protein environment has a significant influence on the binding, but also on the structure and properties of the metal cofactor. In enzymes, metal ions can either be coordinated in macrocyclic prosthetic groups (Table 2.1) or are directly bound to the protein via amino acid residues such as aspartic acid, glutamic acid, histidine, or cysteine (Table 2.2).

Among the ubiquitous metal-containing prosthetic groups are cyclic tetrapyrroles also known as porphyrins. They consist of four pyrrole rings cyclically linked by four methine bridges, optimal for the coordination of metal ions like Fe^{2+} and Fe^{3+} (Fig. 2.1). Out of six coordination sites of iron, four are occupied by the nitrogen atoms of the porphyrin ring. The fifth coordination site is occupied, for example, by the imidazole group of a conserved histidine residue of the protein, as in hemoglobin, or by the thiol group of a conserved cysteine residue, as in cytochrome P450 monooxygenases. The sixth coordination site is coordinated with either oxygen, water, or another amino acid residue. Iron porphyrin is also known as heme. Heme-containing proteins, the so-called heme proteins, have a variety of functions. They are involved, for example, in oxygen transport in the blood (hemoglobin), in oxygen storage (myoglobin), in the reduction of peroxides (catalases and heme peroxidases), in electron transfer in redox chains (cytochromes), and in electron transfer to oxygen coupled with substrate oxidation (cytochrome P450 monooxygenases).

Porphyrins can also build complexes with nickel ions (cofactor F430 in methyl-coenzyme M reductases) or cobalt ions (cobalamin in methylmalonyl-CoA mutases) (Fig. 2.1).

Another type of metal-containing prosthetic groups are iron-sulfur clusters (Fe-S). They are used for electron transfer in enzymes and contain iron and sulfide ions coordinated via the thiol group of cysteine residues. The most common forms are 2Fe-2S and 4Fe-4S clusters, such as in electron-transferring ferredoxins (Fig. 2.1). For example, in the 2Fe-2S cluster of ferredoxins, two iron ions are coordinated by two inorganic sulfides and sulfurs provided by four cysteine residues, whereas the 4Fe-4S cluster contains four iron ions, four sulfide ions, and four cysteine residues.

Table 2.2 Metal ions as cofactors in enzymes

Metal ion	Enzyme
Fe^{2+}	Prolyl-4-hydroxylase
Fe^{3+}	Catechol 1,2-dioxygenase
$Cu^{+/2+}$	Laccase
Co^{2+}	3-Dehydroquinate synthase
Zn^{2+}	Carboxypeptidase A
Mg^{2+}	Hexokinase
Mn^{2+}	Manganese containing catalase
Ni^{2+}	Urease

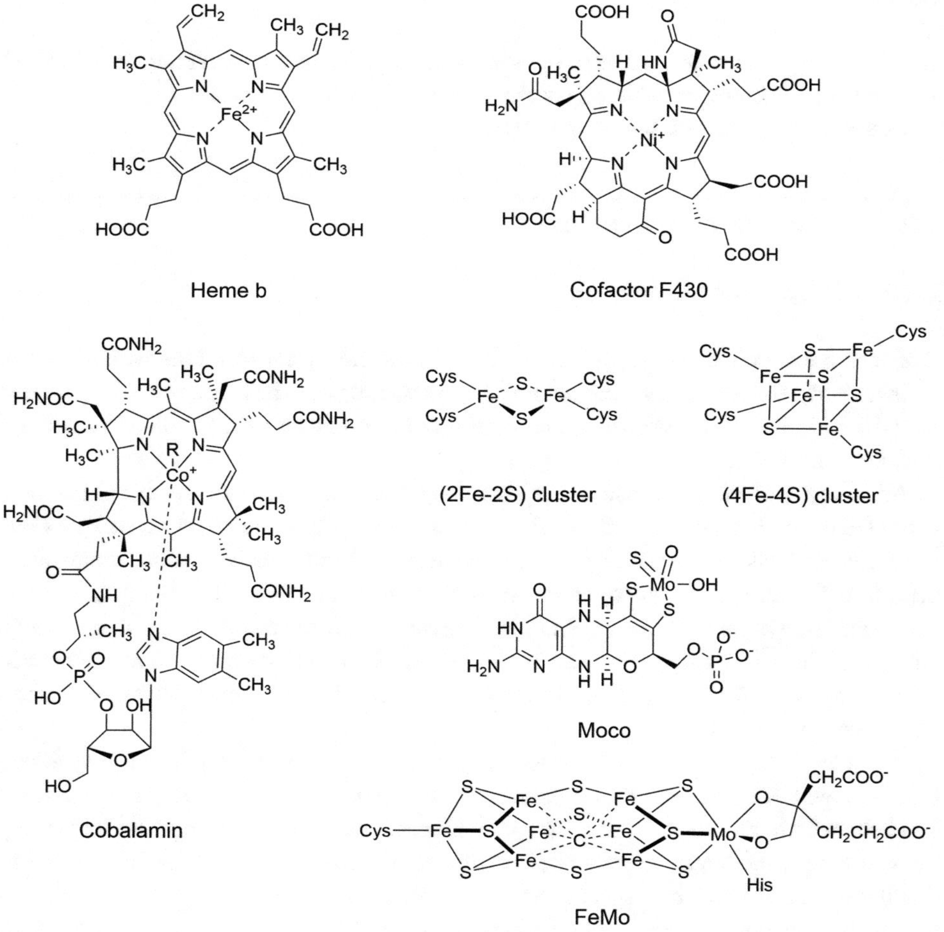

Heme b

Cofactor F430

Cobalamin

(2Fe-2S) cluster

(4Fe-4S) cluster

Moco

FeMo

Fig. 2.1 Examples of metal-containing prosthetic groups in enzymes

Among the more complex metal-containing prosthetic groups are the molybdenum cofactor (Moco) in molybdenum enzymes such as xanthine oxidase and the molybdenum-iron cofactor (FeMo) in nitrogenase (Fig. 2.1; Mendel 2013; MacLeod and Holland 2013).

Vanadium is found in vanadium-dependent haloperoxidases, which oxidize halides to hypohalites at the expense of hydrogen peroxide. In these enzymes, vanadium is bound in a trigonal-bipyramidal coordination geometry. It is coordinated to the imidazole ring of a conserved histidine residue at the apical position, to three nonproteinous oxygen atoms in the equatorial plane, and to an OH group at the other apical position (Crans et al. 2004).

Questions

1. Which interactions contribute to the stabilization of the tertiary structure of enzymes?
2. What is the difference between cosubstrates and prosthetic groups?
3. How can metal ions be bound in proteins?

2.2   Function of Enzymes

2.2.1   Enzymes as Catalysts

Enzymes are excellent biocatalysts. In nature, most biological reactions would proceed unimaginably slowly without the help of enzymes. Enzyme-catalyzed reactions can proceed 10^6- to 10^{17}-fold faster than the corresponding non-catalyzed reactions (Table 2.3; Radzicka and Wolfenden 1995).

Advantages of enzymes become apparent by comparing enzyme-catalyzed reactions with chemically catalyzed reactions. Enzymes are generally more effective catalysts than chemical catalysts and work under mild reaction conditions such as room temperature, neutral pH value, and atmospheric pressure. Chemical catalysts often require high temperatures, extreme pH values, and high pressures. Enzymes are also far more specific than chemical catalysts, as the precise arrangement of catalytically active amino acid residues enables them to specifically select and convert one or more substrates from a variety of compounds.

One can distinguish between absolute, moderate, and relative substrate specificity. Enzymes with an absolute substrate specificity accept and convert only one substrate. Maltase, which cleaves maltose into two glucose molecules, belongs to this group. Enzymes with moderate substrate specificity convert several substrates that contain a specific chemical bond in a specific environment, such as the digestive enzyme chymotrypsin. This enzyme cleaves peptide bonds in proteins after aromatic amino acid residues. Enzymes with relative substrate specificity convert substrates that carry specific chemical bonds, such as ester-cleaving esterases. Some enzymes are able to accept and convert a

Table 2.3 Reaction rates of non-catalyzed and enzyme-catalyzed reactions (adapted from Radzicka and Wolfenden 1995)

Enzyme	non-catalyzed reaction rate (converted substrate molecules/sec)	Enzyme-catalyzed reaction rate (converted substrate molecules/ enzyme molecule/sec)	Increase in reaction rate
Staphylococcus nuclease	$1.7 \cdot 10^{-13}$	95	$5.6 \cdot 10^{14}$
Carboxypeptidase A	$3.0 \cdot 10^{-9}$	578	$1.9 \cdot 10^{11}$
Triosephosphate isomerase	$4.3 \cdot 10^{-6}$	4,300	$1.0 \cdot 10^{9}$
Ketosteroid isomerase	$1.7 \cdot 10^{-7}$	66,000	$3.9 \cdot 10^{11}$
Carbonic anhydrase	$1.3 \cdot 10^{-1}$	$1 \cdot 10^{6}$	$7.7 \cdot 10^{6}$

large number of structurally quite different substrates, such as cytochrome P450 monooxygenases in the liver. This is referred to as substrate promiscuity.

Generally, enzymes demonstrate high product selectivity. They often convert a substrate into only one of several possible products. Due to the high selectivity of enzymes, often no by-products are formed. A distinction is made between chemo-, regio-, and stereoselectivity of an enzyme. In a chemoselective reaction, the enzyme acts on only one out of two or more functional groups of a substrate. This holds true for, e.g., ene reductases, which exhibit high chemoselectivity for the reduction of C=C double bonds relative to other unsaturated compounds such as those containing C=O bonds. In a regioselective reaction, the enzyme attacks only one of several chemically equivalent positions or groups of a substrate, such as aspartate-4-decarboxylase in the decarboxylation of aspartic acid to L-alanine. Stereoselective enzymes prefer the conversion or formation of one stereoisomer over the other(s). In a diastereoselective reaction, one diastereomer is selectively formed or reacted, as in the diastereoselective reduction of α-alkyl-1,3-diketones by NADPH-dependent ketoreductases. Enantioselectivity of an enzyme means that one enantiomer is preferentially formed or reacted. Lipases, for example, catalyze the hydrolysis of only one (either (*R*)- or (*S*)-) enantiomer of a racemic ester. The selective reaction with only one enantiomer is used in the so-called kinetic resolution of racemates. Depending on the enzyme's enantioselectivity, both the product and the unreacted substrate can be obtained in a very high enantiomeric purity.

If alkenes are used as substrate, enzymes can exhibit *cis*–/*trans*- or *E*/*Z*-selectivity and preferably convert either the *cis*- (*E*-) isomer or *trans*- (*Z*-) isomer of the substrate. For example, a nitrilase from *Arabidopsis thaliana* selectively hydrolyzes the *E*-isomer of α,β-unsaturated nitriles to the *E*-carboxylic acid (Effenberger and Oßwald 2001).

2.2.2   Active Site

Enzyme-catalyzed reactions take place in a strictly defined and spatially limited region of the enzyme called active center or active site. The active site is usually located in a cavity of the enzyme and has the shape of a cleft. The active site is formed by amino acid side chains that are involved in the binding and conversion of the substrate and, if present, the binding of the cofactor (Berg et al. 2013, p. 228). These amino acid side chains are often located far away from each other in the primary sequence. The precise spatial arrangement of the amino acid side chains is responsible for the specific positioning and highly selective binding of the substrate in the active site. Upon this binding, a temporary enzyme-substrate complex is formed. The substrate is thereby bound to the enzyme by covalent bonds and many non-covalent interactions such as hydrogen bonds, electrostatic interactions, hydrophobic interactions, ion-dipole and dipole-dipole interactions, and van der Waals forces.

Substrate and enzyme have to match perfectly for substrate binding. To explain why enzymes exhibit a high degree of substrate specificity, in 1890, Emil Fischer postulated the lock-and-key theory. The substrate fits exactly into the rigid active site of the enzyme like a "key" to a "lock," which suggests that the active site of the enzyme and the substrate are equally shaped. However, this theory does not explain the fact that many enzymes can convert several substrates and undergo conformational changes upon substrate binding. According to the theory of *induced fit* postulated in 1958 by Daniel E. Koshland Jr., the enzyme changes its form during the formation of the enzyme-substrate complex, thereby attaining a conformation complementary to the substrate. Only this conformation allows the enzyme to catalyze the reaction. Thus, an enzyme is preformed for a substrate (which corresponds to the lock-and-key theory) but also flexible (which corresponds to the *induced-fit* theory).

2.2.3   Activation Energy

The conversion of a substrate into a product occurs via the formation of a transition state $\ddagger$ which has the highest Gibbs free energy G (Gibbs energy) compared to the substrate and product (Fig. 2.2). The transition state is a transient structure which is no longer a substrate but not yet a product. The Gibbs free energy is a measure of the maximum amount of work that can be performed in a closed thermodynamic system at constant pressure and temperature.

ΔG of a chemical reaction ($\Delta G_{reaction}$) corresponds to the difference between the Gibbs free energies of the substrates and the products and indicates the direction in which the reaction can proceed under certain conditions. If $\Delta G < 0$, the reaction is exergonic and runs spontaneously in the direction of products. If $\Delta G > 0$, the reaction is endergonic and does not proceed spontaneously in the direction of products but runs spontaneously in the reverse direction. If $\Delta G = 0$, the forward and backward reactions are at thermodynamic equilibrium.

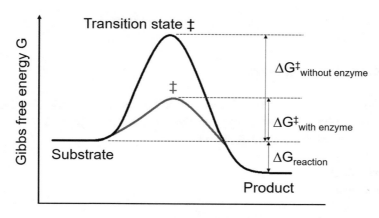

Fig. 2.2 Transition state diagram of a catalyzed (red) and an uncatalyzed reaction

The difference between the free energy of the transition state ‡ and that of the substrate is $\Delta G^{\ddagger}$ (Berg et al. 2013, p. 226). The change in free energy, ΔG, is calculated from the change in enthalpy (ΔH) and the change in entropy (ΔS) via the Gibbs-Helmholtz equation: $\Delta G = \Delta H - T\Delta S$ (Berg et al. 2013, p. 11). The enthalpy of activation (ΔH) is approximately equal to the energy of activation (Ea) in the Arrhenius equation, in accordance to which the reaction rate k is proportional to $e^{-Ea/RT}$ with R as the universal gas constant and T as the absolute temperature. Thus, the higher the activation energy, the lower the reaction rate. When the activation energy decreases from, for example, 30 kJ mol^{-1} to 20 kJ mol^{-1}, the reaction rate increases by a factor of 55, and when the activation energy decreases to 10 kJ mol^{-1}, the reaction rate increases by a factor of 3,000. Thus, enzymes accelerate reactions mainly by lowering the activation energy Ea due to formation of a lower-energy transition state. During the enzyme-catalyzed reaction, energy is released through the formation of a variety of weak interactions between substrate and enzyme. It is believed that this released energy decreases the activation energy of the reaction. For example, the formation of a hydrogen bond between the enzyme and substrate alone can reduce the activation energy and markedly accelerate the reaction rate. However, it is important to note that the entropy of activation (ΔS) may also play a role in determining the efficiency of an enzyme.

In reversible reactions, enzymes similar to any other catalysts accelerate the rate of both the forward and backward reactions but do not alter the free energy difference ($\Delta G_{reaction}$). They cannot alter the thermodynamic equilibrium constant but increase the rate of approach to this equilibrium.

Questions

4. Which types of enzyme selectivity can be distinguished?

5. How do enzymes accelerate the rates of catalyzed reactions?

2.2.4 Catalytic Mechanisms

In enzymes, different molecular processes contribute to the reduction of $\Delta G^{\ddagger}$ of the transition state. Five different catalytic mechanisms are distinguished, which are used individually or in combination to catalyze reactions:

1. Acid-base catalysis
2. Covalent catalysis
3. Metal ion catalysis
4. Proximity and orientation effects
5. Transition state binding

2.2.4.1 Acid–Base Catalysis

In acid catalysis, the substrate is protonated by the acid catalyst, while in base catalysis, the substrate is deprotonated by the base catalyst. This proton transfer lowers the Gibbs free energy of the transition state, thereby accelerating the reaction rate. If both processes occur during the reaction, it is referred to as acid–base catalysis (Voet et al. 2010, p. 361).

A distinction is made between general and specific acid catalysis and base catalysis. Specific acid-base catalysis occurs frequently in chemical catalysis. Here, the reaction rate is proportional to the concentration of hydronium ions (H_3O^+) or hydroxide ions (OH^-) and thus depends on the pH value and not on the concentration of the buffering solution.

General acid-base catalysis is dependent on the concentration of the buffer, and occurs in enzymes, where certain amino acid side chains are able to donate or accept a proton under physiological conditions. In the pH range including their pK_a values, aspartate, glutamate, lysine, cysteine, histidine, and tyrosine act as proton donors or proton acceptors. Since the pH of the environment affects the protonation state of the amino acid side chains, the catalytic activity of enzymes is considered also pH dependent. However, the pK_a values of free amino acids often differ from those of protein-bound amino acids. This is due to the altered microenvironment of amino acid residues in proteins. Usually, the differences between the pK_a values are very small (e.g., pK_a of ≈ 10 for protein-bound lysine instead of ≈ 10.5 for free lysine) but can also be relatively large (e.g., pK_a of ≈ 12 for protein-bound tyrosine instead of ≈ 10 for free tyrosine).

Due to the three-dimensional structure of enzymes, several catalytic groups, capable of transferring a proton, can be located in close proximity to a substrate and jointly contribute to increase the reaction rate. Usually, both a general acid and a general base are present in the active site of an enzyme, which are used to facilitate proton transfers in the reaction. For example, bovine pancreatic ribonuclease A (RNase A), a digestive enzyme that is secreted by the pancreas into the duodenum, catalyzes the hydrolysis of RNA to mono-, di-, or oligonucleotides. Biochemical and X-ray structural analyses have shown that two histidine residues, His12 and His119, act as general acid and general base catalysts, respectively. The imidazole ring of histidine, with a pK_a value of about 7, can act as both an acid and a base at pH around 7. His12, as general base catalyst, deprotonates the 2′-OH group of the 3′-nucleotide, promoting a nucleophilic attack of the 2′-O on the phosphorus atom (Fig. 2.3). This results in breaking of the P-5′O bond and formation of a 2′,3′-cyclic phosphodiester intermediate. His119, as general acid catalyst, protonates the oxygen of the leaving group. The cyclic phosphodiester intermediate is subsequently hydrolyzed by the addition of a water molecule. Thereby, His119 acts as general base catalyst, abstracting a hydrogen atom from water. The resulting hydroxide reacts as nucleophile at the phosphorus atom, which leads to the break of the P-2′O bond. His12 acts as general acid catalyst and protonates the 2′O atom, leading to enzyme regeneration (Voet et al. 2010, p. 362).

Other examples of general acid-base catalysis can be found in hydratases, dehydratases, epimerases, and isomerases such as ribulose-5-phosphate isomerase, which catalyzes the keto-enol tautomerization of ribulose-5-phosphate to ribose-5-phosphate.

2.2.4.2 Covalent Catalysis

In covalent catalysis, a nucleophilic group in the active site of the enzyme attacks an electrophilic group of the substrate. This results in a temporary formed covalent bond between the enzyme and substrate (Voet et al. 2010, p. 364). Carbonyl groups, which have a partial positive charge on the carbonyl carbon atom, often act as electrophiles. Especially in nonpolar enzyme pockets, a nucleophilic group can be very active. To ensure that the enzyme remains unchanged after the reaction, the covalent bond must be broken again. To achieve this, the catalytic group must be easily polarizable to act as both a nucleophile and a good leaving group. Table 2.4 shows some enzymes and their functional groups that meet these requirements and are involved in covalent catalysis.

Serine proteases like chymotrypsin use covalent catalysis to hydrolytically cleave peptide bonds in proteins. Chymotrypsin contains a reactive serine (Ser195) in its active site. Ser195 is part of a catalytic triad formed by the three amino acids aspartate (Asp102), histidine (His57), and serine (Ser195) (Fig. 2.4).

In the presence of a substrate, His57 facilitates the deprotonation of the hydroxyl group of Ser195, allowing the nucleophilic addition to the carbonyl carbon of the scissile peptide bond of the substrate. The carbonyl group of Asp102 thereby facilitates the deprotonation of His57, making the histidine nitrogen more electronegative. An electron pair from the double bond of the peptide carbonyl oxygen moves to the oxygen, and a tetrahedral intermediate is generated. The electrons of the peptide bond move toward the hydrogen of His57, breaking the bond between the nitrogen and the carbon in this bond. The electrons from the negative oxygen move back, yielding the acyl enzyme. The N-terminus of the cleaved polypeptide chain detaches and is replaced by water. The formation of a hydrogen bond between His57 and the water molecule enables its nucleophilic addition to the carbonyl group of the acyl enzyme. The second tetrahedral intermediate is formed. In

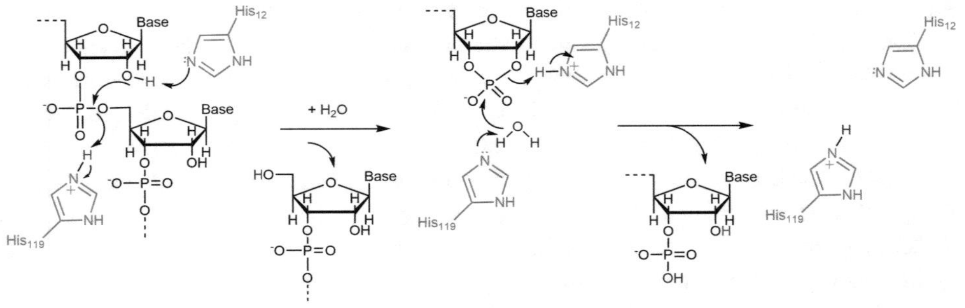

Fig. 2.3 Catalytic mechanism of RNase A. RNA is hydrolyzed by RNase A by acid–base catalysis via a 2′,3′-cyclic intermediate

Table 2.4 Functional groups in enzymes with covalent catalysis

Functional groups	Enzyme
–OH (serine)	Serine proteases, lipases
–COOH (aspartic acid)	Pseudouridine synthase
–SH (cysteine)	Acyltransferase
–Imidazole-NH (histidine)	Phosphotransferase
–NH$_2$ (lysine)	Aldolase

the last step, the original hydrogen bond between His57 and Ser195 is reestablished, and the covalent bond between enzyme and peptide residue is cleaved. Due to the tetrahedral form of both unstable covalent intermediates (transition states), the carbonyl carbon atom achieves the so-called oxyanion hole, where it is stabilized by two additional hydrogen bonds.

In type I aldolases and transaldolases, a Schiff base is formed in the active site between the amino group of the catalytic lysine residue and the C-atom of the reacting ketone. Catalysis of Schiff base forming enzymes involves a series of intermediates covalently bound to the catalytic lysine. Following this mechanism, the class I fructose-1,6-bisphosphate aldolase catalyzes a reversible reaction in which fructose-1,6-bisphosphate is splitted into glyceraldehyde-3-phosphate and dihydroxyacetone phosphate.

Some enzymes use cofactors like biotin, thiamine pyrophosphate (TPP), or pyridoxal phosphate (PLP) (Sect. 2.1.4) for covalent catalysis. For instance, in aspartate aminotransferase (also known as glutamate-oxalacetate-transaminase) that catalyzes a reversible transfer of an α-amino group between aspartate and glutamate, the aldehyde carbon of pyridoxal phosphate is bound to the ε-amino group of a lysine residue in the active site to form a Schiff base (internal aldimine). In the reaction, the amino group of the aspartate substrate replaces the ε-amino group of the lysine. A tetrahedral covalent intermediate, external aldimine, is formed. After protonation and deprotonation, ketimine is formed and subsequently hydrolyzed to release the first product, oxaloacetate. The cofactor in form of pyridoxamine phosphate binds the second substrate α-ketoglutarate yielding the second

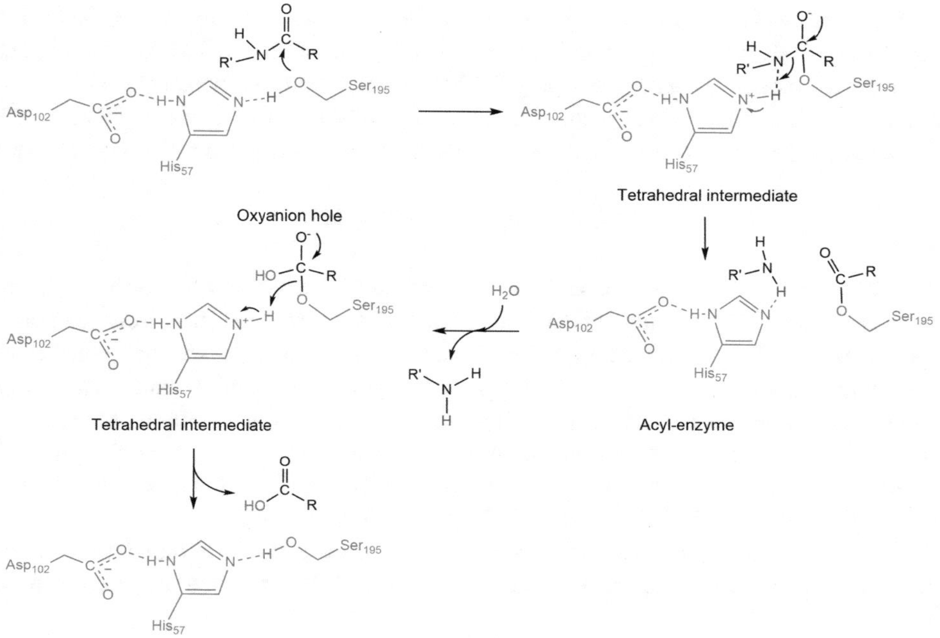

Fig. 2.4 Catalytic mechanism of the serine protease chymotrypsin. Acid-base catalysis and covalent catalysis lead to the cleavage of the peptide bond

covalent intermediate. The amino group of this intermediate is subsequently replaced by ε-amino group of the lysine in the active site of the enzyme. The second product glutamate is released, and the regenerated enzyme is available for the next catalytic cycle.

Questions

6. Which amino acid residues can act as general acid and general base catalysts?
7. Which amino acids are involved in cleavage of peptide bonds by serine proteases?

2.2.4.3 Metal Ion Catalysis

As mentioned in Sect. 2.1.4, about one-third of all known enzymes require tightly bound metal ions such as Fe^{2+}, Fe^{3+}, Cu^{2+}, Zn^{2+}, Mn^{2+}, or Co^{2+} for their catalytic activity. In contrast, metal ions such as Na^+, K^+, Ca^{2+}, and Mg^{2+} stabilize the protein structure in the so-called metal-activated proteins but have no catalytic function. These metal ions are only weakly bound (Voet et al. 2010, p. 366).

The tasks of metal ions in enzymes include the binding of substrates in the correct orientation, the electrostatic balancing of negative charges, and catalysis of redox reactions. Metal ions stabilize the negative charge of a leaving group that is released, thereby improving the properties of the leaving group as such.

The positive charge of metal ions also lowers the pK_a value of bound water molecules. As a result, deprotonation of a water molecule and the formation of a hydroxide ion bound to the metal ion easily occur at neutral pH. The hydroxide ion can then act as nucleophile and attack a substrate bound at the active site. A well-known example of this is Zn^{2+}-containing carbonic anhydrase. This ubiquitous enzyme catalyzes the following reaction:

$$CO_2 + H_2O \leftrightarrow HCO_3^- + H^+$$

The zinc ion is coordinated in carbonic anhydrases via three histidine residues and one water molecule. The Zn^{2+} ion acts as Lewis acid and withdraws electron from the bound water molecule. This lowers the pK_a value of the water molecule from 15.7 to 7 even before the actual reaction. The nearby His64 supports the deprotonation of the zinc-bound water molecule and transfers a proton to the protein surface. Nucleophilic addition of the formed zinc-bound hydroxide ion to the enzyme-bound carbon dioxide leads to the formation of the hydrogen carbonate ion. The latter is released from the active site via binding of a new water molecule to the Zn^{2+} ion (Fig. 2.5; Berg et al. 2013, p. 270).

Metalloproteases such as carboxypeptidase A or thermolysin also exploit the activation of a water molecule by an enzyme-bound metal ion, usually a zinc ion, to cleave peptide bonds.

An example of redox catalysis with metal ions is represented by the heme-containing cytochrome P450 monooxygenases (Fig. 2.6). In the initial state, a water molecule is bound as an axial ligand to the heme iron in the active site of the enzyme (1). The catalytic cycle begins with the binding of the substrate (RH) near the heme group in the active site with displacement of the water molecule (2). The accompanying change in the redox potential of the iron complex (2) facilitates the subsequent reduction of heme iron(III) to heme iron (II) (3). The first electron required is provided by the cosubstrate NAD(P)H and transferred to the heme iron via redox partner proteins. In the next step, molecular oxygen is bound, leading to the formation of the so-called iron(III)-superoxo complex (4). The second electron reduces the complex to an iron(III)-peroxo complex (5). After protonation, a heme-iron(III) hydroperoxo complex (6) is formed, which is also called *compound 0*. Further protonation leads to heterolytic cleavage of the oxygen–oxygen bond with release of a water molecule. A reactive intermediate is formed, which is a complex of a heme-iron (IV)-oxo porphyrin radical species (7) known as *compound I*. In the next step, the introduction of an oxygen atom into the substrate takes place (8). The hydroxylated product (ROH) is released, and the enzyme is available for the next catalytic cycle (Denisov et al. 2005).

2.2.4.4 Proximity and Orientation Effects

It is well known that intramolecular reactions with one substrate molecule generally proceed much faster than intermolecular reactions. Enzymes can catalyze reactions extremely efficiently by bringing the substrates into direct spatial proximity to the active site and to each other. This also increases the effective concentration of substrates. Thus,

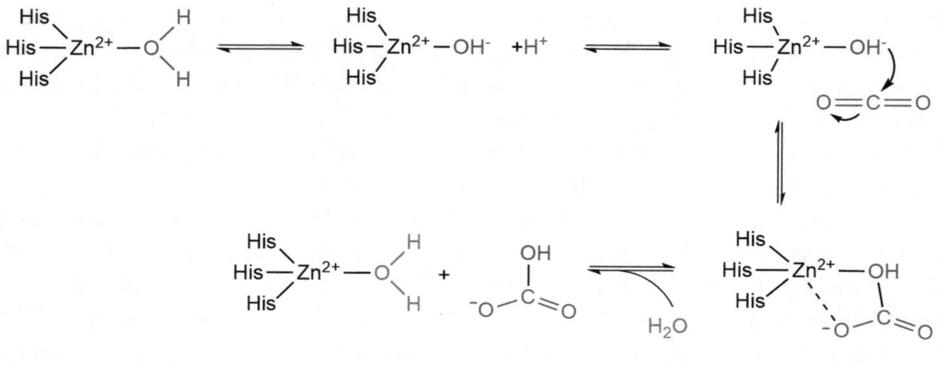

Fig. 2.5 Catalytic mechanism of the metalloenzyme carbonic anhydrase. The bound zinc ion deprotonates the bound water molecule so that the hydroxide ion formed can nucleophilically attack and hydrate a carbon dioxide molecule

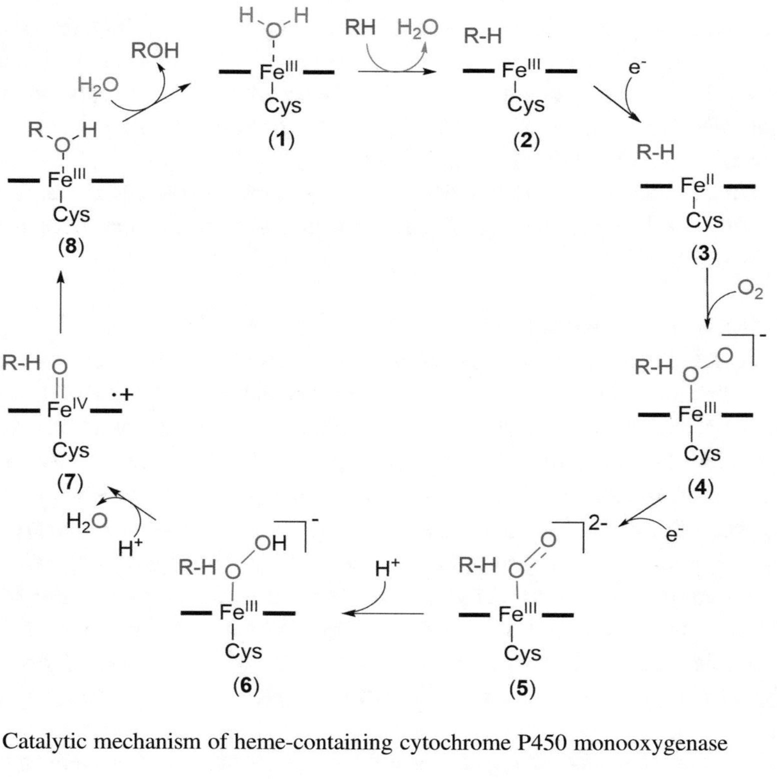

Fig. 2.6 Catalytic mechanism of heme-containing cytochrome P450 monooxygenase

binding of substrates to the enzyme imparts an intramolecular character to the reaction. Furthermore, enzymes bind their substrates in the active site in the correct orientation favoring the reaction (Horton et al. 2008, p. 238). For example, enzymes that catalyze S_N2

reactions bind their substrate in a way that the nucleophile can attack from the direction opposite the leaving group. Restriction of the translational and rotational movements of substrates in enzymes leads to a loss of entropy, which should be largely compensated by the substrate binding energy, resulting in an overall increased reaction rate.

Proximity effect can be illustrated on example of nucleoside monophosphate kinases. These kinases catalyze the transfer of a terminal phosphoryl group from a nucleoside triphosphate (usually ATP) to a nucleoside monophosphate. Mg^{2+} and Mn^{2+} are essential for the activity of these enzymes. The metal ion coordinated by water molecules first binds ATP, resulting in the formation of a metal ion-nucleotide complex. Supported by the interactions between the metal ion and the oxygen atom of the phosphoryl group, ATP is bound in a specific orientation in the active site of the enzyme. This triggers a movement of the so-called P-loop, the Gly-X-X-X-Gly-Lys sequence, which interacts with the phosphoryl groups of the bound ATP. This leads to a strong conformational change of the enzyme. Binding of the second substrate nucleoside monophosphate induces further conformational changes. The ATP molecule is bound in such way that its terminal phosphoryl group is positioned in close proximity to the nucleoside monophosphate. Thus, the catalytically active conformation of the enzyme is only formed with both substrates bound in the active site, which prevents the competitive reaction—the transfer of the phosphoryl group to a water molecule—and enables a direct transfer of the phosphoryl group from ATP to the nucleoside monophosphate (Matte et al. 1998).

This example shows that in enzymes, several processes often contribute to effective catalysis: effect of the metal ion, conformational changes of the enzyme structure (*induced fit*), and proximity effect.

2.2.4.5 Transition State Binding

The significantly higher total binding energy of the substrate than the binding energy of the transition state leads to a higher activation energy and slower reaction rate. Enzymes can significantly accelerate the reaction rate by binding the reaction transition state more specifically than the substrate or product (Horton et al. 2008, p. 243). "Good" substrates that are converted quickly are therefore sometimes bound by an enzyme with comparable or even lower affinity than "poor" substrates that are bound stronger but converted slower.

Serine proteases such as chymotrypsin better bind the tetrahedral intermediates. After nucleophilic addition to the carbonyl carbon atom, it is transformed from trigonal to tetrahedral. In the tetrahedral intermediates, the carbonyl oxygen atom reaches the "oxyanion hole" (Fig. 2.4). This leads to the formation of an additional hydrogen bond with Gly193 and thus to further reduction of the activation energy (Berg et al. 2013, p. 260).

Transition states are short-living and exist for approximately 10^{-13} s. Studies on natural transition states help to synthesize transition state analogs. These are stable molecules that are used in medicine as antibiotics or enzyme inhibitors. For example, 2-phosphoglycolate, a transition state analog, binds much tighter to triose phosphate isomerase than the actual substrate dihydroxyacetone phosphate.

Questions

8. Which metal ion is used by carbonic anhydrase for catalysis, and which reaction is catalyzed by this enzyme?
9. What is the cofactor of cytochrome P450 monooxygenases, and which type of reaction do these enzymes catalyze?
10. What does proximity effect in enzyme catalysis mean?

Take-Home Message
- Protein structure can have up to four levels of complexity.
- Cofactors are small organic molecules, metal complexes, or metal ions that participate in enzyme catalysis and are either loosely or tightly bound to the enzyme.
- Enzyme function can be influenced by posttranslational modifications of the protein molecule.
- Enzymes accelerate reaction rates by lowering the activation energy.
- Various catalytic mechanisms are used by enzymes individually or in combination to catalyze a reaction.
- Proximity and orientation effects contribute to efficient enzyme catalysis.

Answers

1. Hydrophobic interactions, ionic bonds, hydrogen bonds, as well as disulfide bonds contribute to the stabilization of the enzyme's tertiary structure.
2. Cosubstrates are diffusible compounds that are loosely bound to the enzyme and are changed during the enzymatic reaction, while prosthetic groups are tightly bound to the enzyme and remain unchanged after the reaction.
3. Metal ions can be directly bound to the enzyme via amino acid residues such as aspartic or glutamic acid, histidine, or cysteine, or they can be coordinated in macrocyclic prosthetic groups.
4. Enzymes can be chemo-, regio-, and/or- stereoselective. If an alkene acts as substrate, enzymes can demonstrate $cis-/trans$ (E/Z) selectivity.
5. Enzymes accelerate reactions by lowering the activation energy.
6. Amino acid residues of aspartic acid, glutamic acid, lysine, cysteine, histidine, and tyrosine can act as general acid and general base catalysts.
7. Serine proteases commonly use a catalytic triad usually consisting of serine, histidine, and aspartate for cleavage of peptide bonds.
8. Carbonic anhydrase is a Zn^{2+}-containing enzyme that forms hydrogen carbonate from carbon dioxide and water.
9. Cytochrome P450 monooxygenases are heme-containing redox enzymes that catalyze the oxidation of a broad range of substrates at the expense of molecular oxygen.

10. Proximity effect means the arrangement/binding of two or more substrate molecules within the active site close to each other so that the reaction can easily take place.

References

Berg JM, Tymoczko JL, Stryer L. Stryer biochemistry. 7th ed. Berlin, Heidelberg: Springer Spectrum; 2013.

Crans DC, Smee JJ, Gaidamauskas E, Yang L. The chemistry and biochemistry of vanadium and the biological activities exerted by vanadium compounds. Chem Rev. 2004;104:849–902.

Denisov IG, Makris TM, Sligar SG, Schlichting I. Structure and chemistry of cytochrome P 450. Chem Rev. 2005;105:2253–2277.

Effenberger F, Oßwald S. (E)-Selective hydrolysis of (E,Z)-α,β-unsaturated nitriles by the recombinant nitrilase AtNIT1 from *Arabidopsis thaliana*. Tetrahedron Asymmetry. 2001;12:2581–2587.

Horton HR, Moran LA, Scrimgeour KG, Perry MD, Rawn JD. Biochemistry. 4th ed. Pearson Education; 2008.

MacLeod KC, Holland PL. Recent developments in the homogeneous reduction of dinitrogen by molybdenum and iron. Nat Chem. 2013;5:559–565.

Matte A, Tari LW, Delbaere LTJ. How do kinases transfer phosphoryl groups? Structure. 1998;6: 413–419.

Mendel RR. The molybdenum cofactor. J Biol Chem. 2013;288:13165–13172.

Radzicka A, Wolfenden R. A proficient enzyme. Science. 1995;267:90–93.

Voet D, Voet JG, Pratt CW. Textbook of biochemistry. 2nd ed. Weinheim: Wiley-VCH; 2010.

Enzyme Modeling: From the Sequence to the Substrate Complex

3

Silvia Fademrecht and Jürgen Pleiss

What Will You Learn in This Chapter?

This chapter gives an overview on computational methods to study the relationship between sequence, structure, and function of proteins. Computational methods are in widespread use to search for novel enzymes with the desired properties in sequence and structure databases and to guide enzyme engineering toward variants with improved properties. The structure and function of enzymes can be modeled by two complementary methods, namely, data-driven modeling and mechanistic modeling. The integration of both methods allows the discovery of novel enzymes and the improvement of known enzymes by amino acid exchanges. Because the costs of DNA sequencing of whole genomes and metagenomes are exponentially decreasing, predicting structure and function of proteins from their sequence is a promising route toward novel bioprocesses.

Enzymes are complex nanomachines with high structural stability, rich dynamics, and a finely tuned active site. All the properties of an enzyme—its catalytic function, substrate recognition, enzyme dynamics, three-dimensional structure, and folding pathway—are

S. Fademrecht · J. Pleiss (✉)

Institute of Biochemistry and Technical Biochemistry, University of Stuttgart, Stuttgart, Germany

e-mail: Juergen.Pleiss@itb.uni-stuttgart.de

© The Author(s), under exclusive license to Springer Nature Switzerland AG 2024

K.-E. Jaeger et al. (eds.), *Introduction to Enzyme Technology*, Learning Materials in Biosciences,

https://doi.org/10.1007/978-3-031-42999-6_3

Fig. 3.1 Increase in the amount of data in GenBank (http://www.ncbi.nlm.nih.gov/genbank). The number of nucleotides at the end of the respective year and the trend (increase by a factor of 100 per decade) are indicated

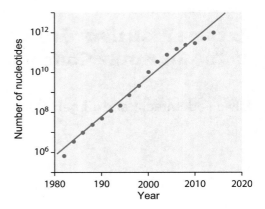

encoded in the protein sequence. The life sciences are only at the beginning of reading, understanding, and eventually writing this code.

Rational protein engineering is based on two complementary modeling methods: data-driven and mechanistic modeling. In data-driven modeling, the rapidly growing amount of data on sequence, structure, and function of proteins is statistically analyzed to derive correlations between individual amino acid positions and interesting biochemical properties. On this basis, novel enzymes with desired properties can be found in databases, or the properties of a particular enzyme can be changed in a desired direction by mutation. In the mechanistic modeling of structure and function of enzymes, an atomistic model of the enzyme, its substrates, and its environment is established to understand experimentally determined biochemical data at a molecular level and to predict the effect of mutations. Data-driven modeling is fueled by the rapidly growing data stream generated by genomic and metagenomic sequencing projects. Over the past 30 years, the amount of DNA sequence data has increased by a factor of 100 every decade (Fig. 3.1), and there is no end to this growth, yet. Mechanistic modeling is fueled by an increasingly available computing power and enables realistic simulations of structure and dynamics of large molecular systems. During the last 40 years, microprocessor performance has increased by a factor of 50 every decade (Moore's law), and this trend will continue in the near future due to massive parallelization.

While the sequences of more than 200 million proteins are now known (Protein Information Resource PIR; http://pir.georgetown.edu/, August 2022), there are a total of only 170,000 protein structure entries in the Protein Data Bank (http://www.rcsb.org, August 2022). The small number of experimentally determined structures and the much slower growth severely limit the possibility of mechanistic modeling. Therefore, in most cases, the structure of the enzyme under investigation must be modeled. The basis of template-based methods is the observation that proteins with a similar sequence also have a similar structure. For example, proteins with a sequence identity of 60% show a mean protein backbone deviation of 1 Å, and with a lower sequence identity of 30%, the structural difference increases to 1.8 Å. This remarkable conservation of protein structure between homologous proteins provides a reliable basis for template-based methods of structure prediction (Sect. 3.2). In addition, template-free methods are now available to predict the structure of proteins that have less than 25% sequence identity to a protein of known structure (Sect. 3.2.5). The predicted structure of an enzyme is the basis for docking methods that model the binding of a substrate molecule into the binding pocket (Sect. 3.3). Biochemical properties such as substrate specificity and regio- and stereoselectivity can then be derived from the enzyme-substrate complex, and mutations can be predicted to optimize enzyme properties (Sect. 3.4).

3.1   Investigation of the Protein Sequence Space

As a rule, proteins or protein domains with a similar sequence also have a similar structure and similar biochemical properties. Based on the sequence of a protein, its structure and catalytic activity can therefore be predicted if these properties have already been determined experimentally for evolutionarily related (homologous) proteins. Within such a homologous protein family, positions that determine structure, folding pathway, or function are conserved and can be identified by a conservation analysis within a homologous family. The first step of protein modeling is therefore the search for homologous protein domains with experimentally determined structure (templates).

3.1.1   Domain Structure of Proteins

Proteins have a modular structure. The smallest building blocks are the amino acids, which form the primary structure (amino acid sequence) and the local secondary structure (α-helices, β-strands) (Sect. 2.1.2). Several of these secondary structure elements assemble to form supersecondary structure elements (β-α-β, β-hairpin). The assembly of secondary and supersecondary structures and the loops connecting them forms the tertiary structure of a protein. Quaternary structure is the term used when several proteins form a protein complex. A protein can consist of several domains, defined as functionally and structurally definable areas. Domains therefore form a bridge between tertiary and quaternary structure: if two domains are connected by a linker, they are part of the tertiary structure; otherwise, they form the quaternary structure. The only difference between a monomer consisting of two domains and a dimer formed by association of two domains is the covalent linker.

Often, the individual domains of an enzyme are involved in specific functions. For example, thiamine diphosphate-dependent decarboxylases (DCs) consist of three domains: the N-terminal PYR domain, which binds the pyrimidine ring of the cofactor; the C-terminal PP domain, which binds the phosphate group of the cofactor; and the TH3 domain, which connects the two domains (Fig. 3.2). The active DC protein consists of a dimer, which in turn is formed by two antiparallel PYR-TH3-PP monomers. A functionally similar family, the thiamine diphosphate-dependent transketolases (TKs), also consists of

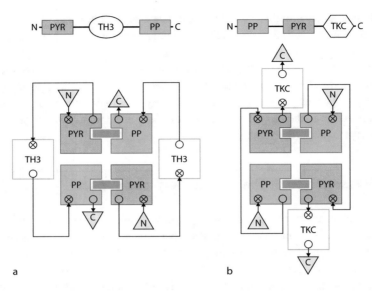

Fig. 3.2 Schematic representation of the quaternary structure of two thiamine disphosphate-dependent enzymes: homodimers of decarboxylases (**a**) and transketolases (**b**). The catalytically active PYR (red) and PP (blue) domains bind the cofactor (gray) and are spatially arranged in the same way in both proteins, although the order of the two domains on the protein sequence differs considerably in both proteins. (Figure adapted from Vogel and Pleiss 2014 and Widmann et al. 2010)

three domains, but the sequence is reversed: an N-terminal PP domain is connected by a loop to a PYR domain, which in turn is connected to a C-terminal TKC domain. Despite the different sequences and tertiary structures of the two protein families, the catalytically active PP and PYR domains are spatially arranged in the same way (Vogel and Pleiss 2014).

3.1.2   Search for Homologous Proteins

Various programs are available for searching for homologous proteins in public sequence databases such as GenBank or UniProt (Table 3.1). BLAST and FASTA start from a single sequence and search for homologous proteins by pairwise sequence comparison. Both are heuristic methods and are characterized by high speed. They are particularly used to search for closely related proteins. BLAST, the tool most commonly used for homology searches, provides the sequence identity to the search sequence, a score, and a significance value (E-value) for each hit. The higher the sequence identity and the score, and the smaller the E-value, the more significant the homology of a hit, which is an important criterion for selecting a suitable homologous protein (*E*-values much smaller than 1 and a sequence identity greater than 20%). Further selection depends on the particular application. For homology modeling, the simulation of protein dynamics or for substrate docking

Table 3.1 Frequently used programs to search for homologous proteins

Name	Web address
BLAST	http://blast.ncbi.nlm.nih.gov/Blast.cgi/
FASTA	http://www.ebi.ac.uk/Tools/sss/fasta/
HMMER	http://hmmer.org/
PSI-BLAST	http://blast.ncbi.nlm.nih.gov/BLAST.cgi/

homologous proteins with experimentally determined structure are required. For this purpose, the BLAST search is restricted to the structure database PDB. Depending on the research question, criteria such as quality, completeness, and the protein conformation have to be considered for the selection. Crystal structures with a resolution below 2 Å are considered as high resolution and are well suited for homology modeling. For substrate docking, the protein should be in a productive conformation that allows binding of a substrate. Therefore, crystal structures with already bound ligands are preferably used for modeling.

If no suitable homologous protein with experimentally determined structure can be found by BLAST searches, programs such as HMMER and PSI-BLAST can be used. These start with a multisequence alignment of several homologous proteins and calculate for each alignment position the probability that a certain amino acid occurs (sequence profile). With this position-specific information, even more distant homologs can be reliably found in a sequence database.

3.1.3 Clustering of Protein Families

When modeling the structure of a protein, it is best to use templates from the same protein family, which can be found using the methods described in Sect. 3.1.2. However, there is a risk that these methods will collect proteins from multiple families with different substrate specificities, selectivities, or domain arrangements. To limit the template search to homologous proteins with similar properties, the homologs are classified into groups with similar sequence by clustering. A systematic comparison of the sequences then allows the identification of positions that are specific for the respective cluster and possibly contribute to the biochemical properties of the respective protein under investigation.

In order to enable a systematic analysis of a large number of protein sequences, they are made available in databases. Protein families are classified on the basis of their sequence or structure (Table 3.2), using hierarchical trees or cluster analyses. When selecting suitable templates, information from these databases is advantageous: if the homologous protein is in the same subfamily, greater similarities in three-dimensional structure and catalytic activity are more likely. For applications such as homology modeling, priority should therefore be given to these proteins.

Table 3.2 Protein family databases

Name	Web address
BioCatNet[a]	https://biocatnet.de/
CATH[b]	http://www.cathdb.info/
InterPro[a]	https://www.ebi.ac.uk/interpro/
Pfam[a]	http://pfam.xfam.org/
SCOP2[b]	http://scop2.mrc-lmb.cam.ac.uk/

[a] Sequence-based classification
[b] Structure-based classification

3.1.4   Outlook: Integration of Sequence Data and Biochemical Data

While publicly accessible databases for DNA and protein sequences as well as protein structures have been available for 40 and 50 years, respectively (amino acid sequences since 1965, protein structures since 1971, DNA sequences since 1980), the first database of the biochemical properties of enzymes (BRENDA; Table 3.3) was established only in 2001. There are three main reasons. While the data models for storing sequence and structural information are relatively simple (one string for sequence information and *xyz coordinates* for structural information, respectively), biochemical properties of proteins are complex and involve a unique description of the biocatalyst, the chemical reaction, reactants and products, and reaction conditions, as well as biophysical properties (thermostability, solubility) and kinetic parameters (k_{cat}, K_m). The second reason is the existence of generally accepted criteria for the quality of sequence or structural data, whereas, despite many attempts of standardization, there are still no generally accepted criteria for the description of enzyme-catalyzed reactions. The third reason is the easy comparison of different sequences by sequence alignments or of different protein structures by superposition, which makes the usefulness of a structured data collection immediately obvious. In contrast, enzyme kinetic parameters such as k_{cat} or K_m values depend not only on the sequence of the enzyme and the reaction conditions but also on the kinetic model used for analyzing the experimental data (usually Michaelis-Menten kinetics), making a direct comparison of kinetic parameters such as k_{cat} or K_m between different enzymes difficult.

In order to make the flood of published biochemical data accessible for systematic analysis, promising approaches are being pursued in various working groups to integrate sequence, structural, and biochemical data into a uniform data model and make it publicly accessible (Table 3.3).

Table 3.3 Databases with biochemical data of enzymes

Name	Web address
BioCatNet	https://biocatnet.de/
BRENDA	http://www.brenda-enzymes.org/
SABIO-RK	http://sabio.villa-bosch.de/
STRENDA DB	https://www.beilstein-strenda-db.org/strenda/

Questions

1. What is the meaning of "homologous proteins"?
2. Which functions might be encoded in the different domains of a multidomain protein? Provide an example of a multidomain protein, and name the functions of the different domains.
3. Why does clustering or tree construction allow to differentiate between protein families, but not BLAST?

3.2   Structural Modeling

While DNA can be sequenced automatically and quickly using next-generation sequencing methods, the process of determining the structure of individual proteins is time-consuming. This is reflected both in the number of experimentally determined structures and in its rate. Currently, the number of protein sequences and structures differs by a factor of 1000, but the number of nucleotides grows by a factor of 100 every decade, while the number of protein structures only increases by a factor of 3.

However, knowledge of the structure of a protein is an essential prerequisite for an in-depth understanding of its biochemical properties and is thus a basis of rational design of mutants or mutant libraries. The study of the three-dimensional structure allows the spatial localization of catalytically active amino acids and the substrate-binding site and thus the understanding of the catalytic mechanism at the molecular level. However, not only the structure and physicochemical properties of the substrate-binding site are responsible for substrate recognition and thus substrate specificity as well as chemo-, regio-, and stereoselectivity but also the access channel of substrate and cosubstrate as well as the possible channels through which the products leave the binding pocket. Knowledge of the quaternary structure is also important, since in many enzymes the substrate-binding site is localized at the interface between two monomers. In addition, the formation of oligomers is a common principle to increase the stability of a protein.

Since proteins with a similar sequence have a similar spatial structure, the tertiary and quaternary structure of a protein can be predicted on the basis of its sequence using the method of homology modeling (template-based modeling) (Orry and Abagyan 2012). Homologous proteins with an experimentally determined structure can therefore be used as templates to model the three-dimensional structure of a target protein.

Homology modeling is divided into four phases (Fig. 3.3):

1. Search for a suitable template
2. Creation of a suitable sequence alignment between target and template
3. Calculation of a structural model
4. Evaluation of quality of the model

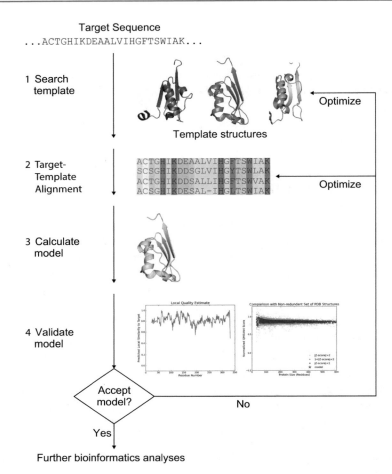

Fig. 3.3 Schematic representation of the steps of homology modeling

If no template is found in phase 1, fragment assembly methods are used to create a spatial target model (Sect. 3.2.5).

3.2.1 Search for a Suitable Template

Protein structures suitable for homology modeling can be found as described in Sects. 3.1.2 and 3.1.3. In general, the higher the sequence identity and the better the resolution of the structure, the better the homology model. However, depending on the particular problem, further criteria have to be considered when selecting a suitable template structure. If the homology model is used for substrate docking, attention should be paid to the binding of relevant cofactors (e.g., copper ions in laccases, NAD(P)(H) in oxidoreductases, or heme in

Table 3.4 Frequently used programs for sequence or structure comparison

Name	Web address
Clustal Omega[a]	http://www.ebi.ac.uk/Tools/msa/clustalo/
COMA[b]	http://www.ibt.lt/en/laboratories/bioinfo_en/software/coma.html/
EMBOSS Needle[c]	http://www.ebi.ac.uk/Tools/psa/emboss_needle/
EXPRESSO[d]	http://tcoffee.crg.cat/apps/tcoffee/do:expresso
MUSCLE[a,b]	http://www.ebi.ac.uk/Tools/msa/muscle/
SALIGN[a,d]	http://salilab.org/salign
T-Coffee[a]	http://tcoffee.crg.cat/apps/tcoffee/

[a]Multisequence alignment
[b]Profile–profile alignment
[c]Pairwise sequence alignment
[d]Structure alignment

cytochrome P450 monooxygenases) and to the presence of a productive conformation of the protein structure to ensure realistic substrate binding. In addition, for models used in protein simulations, the active quaternary structure should be used.

3.2.2   Creation of a Target–Template Alignment

Template-based methods for the structural modeling of a protein or a protein domain require a sufficiently high sequence similarity between target and template in order to evaluate an optimal target–template alignment. Several programs are available for this purpose (Table 3.4). For very similar sequences with a sequence identity of at least 50%, a pairwise alignment of target and template is sufficient to obtain a reliable alignment. However, often the target sequence has low sequence similarity to the most similar protein of known structure. If the sequence identity is between 30% and 50%, a multisequence alignment of target, template, and additional homologous proteins is performed. If the structure of several homologous proteins has been determined experimentally, structure-based sequence alignment methods can also be used, which provide a reliable alignment even at low sequence similarity. The proteins used for multisequence alignments should cover the sequence space between target and template as well as possible. If the sequence similarities are even lower (up to about 15% sequence identity), the methods described in Sect. 3.1.2 can be used to generate a sequence profile for each target and template, which are then compared with each other (profile–profile alignment).

3.2.3   Creation of a Structural Model

Based on the target–template alignment, a structural model of the target protein is created. Various programs and web servers are available for this purpose (Table 3.5). Pairwise,

Table 3.5 Frequently used programs and servers for homology modeling

Name	Web address
CPHmodels Server	http://www.cbs.dtu.dk/services/CPHmodels/
HHpred	http://toolkit.tuebingen.mpg.de/hhpred
I-TASSER	http://zhanglab.ccmb.med.umich.edu/I-TASSER/
M4T	http://manaslu.aecom.yu.edu/M4T/
MODELLER	http://salilab.org/modeller/
ModWeb	https://modbase.compbio.ucsf.edu/modweb/
Pcons	http://pcons.net/
Phyre2	http://www.sbg.bio.ic.ac.uk/phyre2/html/page.cgi?id=index
Protein Model Portal	http://www.proteinmodelportal.org
SWISS-MODEL	http://swissmodel.expasy.org/

multisequence or profile-profile alignment is used to assign positions that correspond to each other on structural level. When a structural model is created, the three-dimensional coordinates of the amino acids are transferred from the template structure to the corresponding amino acids of the target sequence. For identical amino acids, backbone and side chain coordinates are transferred. For different amino acids, only the backbone coordinates are transferred, whereas the conformations of the side chains are determined by an iterative search for optimal packing. Areas such as insertions or deletions in loop regions for which no template information is available are predicted by fragment searches in structural databases or by using template-free methods (Sect. 3.2.5). The obtained model is then energy minimized to obtain an optimized model.

A frequently used web server for homology modeling is the SWISS-MODEL server. In automated mode, only the amino acid sequence of the target protein has to be specified in order to obtain ready-made models with an additional estimation of the model quality (Sect. 3.2.4) in a few minutes. In alignment mode, a target–template alignment can be read in, whereby a target can be reliably modeled even with low sequence similarity to the template (Sect. 3.2.2).

3.2.4 Sources of Error and Estimation of Model Quality

For all homology modeling methods, the better the target–template alignment, the better the resulting model (Marti-Renom et al. 2000). Therefore, the structure in regions with poor or no alignment, such as loops, can only be predicted with low reliability. Also, the orientation of amino acid side chains may differ between target and template. However, with respect to applications such as mutant prediction or docking, the correct orientation of side chains and the correct conformation of loops are essential.

Programs for estimating the model quality are used to determine possible errors in the modeling process (Table 3.6). Statements on the global quality of the model are provided

Table 3.6 Frequently used programs for estimating model quality

Name	Web address
ANOLEA	http://melolab.org/anolea/
DFIRE	http://sparks-lab.org/tools-dfire.html
GROMOS	http://www.gromos.net/
PROCHECK	http://www.ebi.ac.uk/thornton-srv/software/PROCHECK/
QMEAN	http://swissmodel.expasy.org/qmean/
WHATCHECK	http://swift.cmbi.ru.nl/gv/whatcheck/

by QMEAN and DFIRE, while the local quality is estimated with the aid of ANOLEA, ProQres, and GROMOS.

In order to assess the accuracy of homology modeling methods and to advance developments in this field, biennial CASP competitions (Critical Assessment of Protein Structure Prediction, http://predictioncenter.org/) have been organized since 1994. The goal of the participating modeling groups is to develop accurate models of a yet unpublished, experimentally generated structure based on the protein sequence. CASP therefore provides an assessment of the reliability and accuracy of current homology modeling methods.

3.2.5   Fragment Assembly Methods

Fragment assembly methods assemble the protein structure piece by piece from individual fragments between 3 and 20 amino acids in length (Zhang 2007). These fragments, selected from a fragment library, usually have a very low sequence similarity to the corresponding region in the protein to be modeled, which is why several different structures always have to be considered. Furthermore, in contrast to template-based methods, the relative orientation of the individual fragments is unknown, so that there are many possible arrangements during assembly into a complete structure. An optimal combination of fragments and their relative orientation is achieved using Monte Carlo methods, where the quality of each model is scored at each step. This scoring function is a critical step in all template-free methods, since its computation must be fast on the one hand, but on the other hand, it must be able to distinguish correctly folded from incorrectly folded structures. Due to the rapidly increasing number of possible combinations with the number of fragments ("combinatorial explosion"), fragment assembly methods have so far been limited to structure prediction of proteins of up to 250 amino acids.

Template-free methods for protein structure prediction are usually constructed as a pipeline of several methods that build on each other: the search for sequence-like fragments using a sequence profile (PSI-BLAST, HHSearch), the generation of structural models, and the evaluation of the quality of these models using mechanistic or probabilistic methods. The currently most successful method of template-free structure prediction is AlphaFold,

Table 3.7 Frequently used fragment assembly servers

Name	Weblink
AlphaFold	https://alphafold.ebi.ac.uk/
I-TASSER	http://zhanglab.ccmb.med.umich.edu/I-TASSER/
MULTICOM	http://sysbio.rnet.missouri.edu/multicom_toolbox/
QUARK	http://zhanglab.ccmb.med.umich.edu/QUARK
RaptorX	http://raptorx.uchicago.edu/
Robetta	http://robetta.bakerlab.org/
SmotifTF	http://search.cpan.org/dist/SmotifTF/

which applies a pipeline based on machine learning methods to the construction of fragments and their assembly (Table 3.7).

3.2.6 Outlook: Modeling by Molecular Dynamics Simulation

The protein sequence encodes the folding pathway and the protein structure. Thus, both the kinetics and the thermodynamics of folding are determined by the sequence of amino acids and their physicochemical properties. The motion of a protein can be described by mechanistic modeling of the interactions of all atoms (Sect. 3.4). The transition between folded and unfolded states of small, rapidly folding proteins was successfully modeled by long simulations on a millisecond timescale and reproduced experimental data such as folded protein structure, folding kinetics, and thermal stability (Lindorff-Larsen et al. 2011). These successful simulations demonstrate that mechanistic models are in principle capable of describing structure formation of proteins using simple mechanistic principles. With increasing computer power, it is therefore to be expected that the structure of larger proteins and protein complexes can also be predicted in the future.

Questions

4. Some protein entries in the PDB have an identical sequence, but slightly different structures. What might be the difference?
5. What can go wrong in homology modeling?

3.3 Molecular Docking

The computer-aided prediction of the binding of a receptor and a ligand is called molecular docking. The receptor can be a protein or an oligonucleotide; the ligand can be a protein or a small molecule. The most common use case is screening for inhibitors for the development of new medical drugs. Docking methods are also successfully used to study the

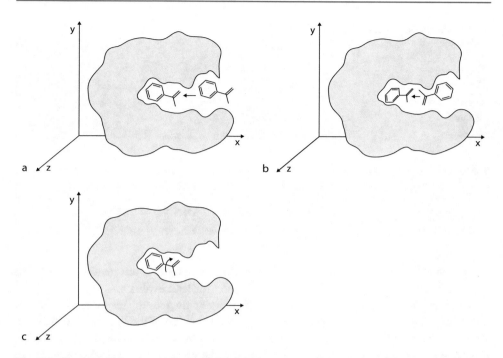

Fig. 3.4 Degrees of freedom to be considered during docking: (**a**) positioning, (**b**) orientation, and (**c**) torsion

interaction of enzymes and their substrates to understand the molecular basis of substrate specificity and regio- and stereoselectivity. The advantage of docking to experimental screening is its higher effectiveness. In addition, a modeled enzyme-substrate complex is prerequisite for successful design of mutants or substrates.

In general, two steps can be distinguished: sampling (predicting the conformation, position, and orientation of the ligand in the binding pocket (Fig. 3.4)) and scoring (estimating the binding affinity (Sousa et al. 2013)).

3.3.1   Sampling

For docking, the structures of the receptor and the ligand are required. If the receptor is a protein, a crystal structure with good resolution or a high-quality homology model is suitable. Also, the protein should be in an active conformation to allow productive binding of the substrate. The spatial structure of the ligand can often be extracted from databases (Table 3.8) or calculated using programs (Table 3.9). In some cases, the crystal structure of the target protein already has a bound substrate or inhibitor that resembles the ligand. In such cases, superposition of the two molecules by hand is possible (e.g., using PyMOL,

Table 3.8 Databases for small molecules

Name	Web address
ChemSpider	http://www.chemspider.com/
PubChem	http://pubchem.ncbi.nlm.nih.gov/
The Cambridge Structural Database	http://www.ccdc.cam.ac.uk/solutions/csd-system/components/csd/
ZINC	http://zinc.docking.org/

Table 3.9 Programs to create the spatial structure of small molecules

Name	Web address
ChemDraw	http://www.cambridgesoft.com/software/overview.aspx
Marvin Space	https://www.chemaxon.com/products/marvin/marvinspace/

https://www.pymol.org/). Even dissimilar molecules provide information about the approximate orientation of the ligand and are useful for later evaluations.

The flexibility of receptor and ligand results in many degrees of freedom to be considered: three degrees of freedom for translation and for rotation of the ligand, as well as one degree of freedom per rotatable bond. Since the number of possible combinations increases rapidly with the number of degrees of freedom to be considered ("combinatorial explosion"), a systematic calculation of all possible degrees of freedom is not feasible. Therefore, different sampling algorithms have been developed for an effective search for optimal receptor–ligand complexes: shape-matching algorithms, systematic search, and stochastic algorithms.

The receptor and the ligand can be assumed to be rigid or (partially) flexible. With increasing flexibility, the model quality increases, but also the complexity and the computational effort.

3.3.2 Scoring

A scoring function is needed for the evaluation of docking results. Depending on the problem, it should calculate the best conformation of a receptor–ligand combination or select high-affinity ligands from a given ligand library. Scoring functions can be divided into three main classes: force field-based, empirical, and knowledge-based. Due to the large number of receptor–ligand combinations to be investigated, the scoring function must be easy to compute; on the other hand, it should be as close to reality as possible and reflect the experimentally determined binding affinities.

The choice of docking programs is very diverse (Table 3.10). They differ in the implemented sampling and scoring methods, performance, accuracy, and user-friendliness.

Table 3.10 Commonly used docking programs for small molecules

Name	Web address
AutoDOCK 4	http://autodock.scripps.edu/
AutoDock Vina	http://vina.scripps.edu/
DOCK	http://dock.compbio.ucsf.edu/
FlexX	https://www.biosolveit.de/FlexX/
FTDock	http://www.sbg.bio.ic.ac.uk/docking/ftdock.html/
GLIDE	http://www.schrodinger.com/Glide/
GOLD	http://www.ccdc.cam.ac.uk/solutions/csd-discovery/components/gold/

3.3.3 Limitations

A good starting point for molecular docking is a high-resolution crystal structure of the receptor, preferably in an active conformation and with known reaction mechanism. If the reaction mechanism and the substrate-binding pocket are unknown, docking results are difficult to interpret. Low-resolution crystal structures or inadequate homology models can also have a negative impact on docking results, as even small changes in receptor structure strongly influence the docking results. Another limiting factor is the flexibility of the molecules involved, such as ligands with a large number of freely rotatable bonds or receptors with mobile substrate-binding loops.

3.3.4 Modeling of Protein Complexes

An essential and often overlooked aspect of modeling is the study of the quaternary structure of the native enzyme. The spatial orientation of protein molecules can have a crucial influence on the prediction of substrate binding and conversion: when the active site is located at the boundary between two protein molecules, when substrate access is formed by the protein–protein interface, or when multiple enzymes cluster together to form an efficient cascade reaction. In addition, the quaternary structure strongly contributes to biophysical properties such as thermostability and solubility.

The modeling of the quaternary structure can be carried out by template-based methods on the basis of experimentally determined quaternary structures of homologous proteins. Protein complexes for which no experimental structural information is available are modeled using docking methods (Table 3.11). Protein–protein docking methods that assume rigid protein structures are fast, but not sufficiently predictive because significant conformational changes can occur when proteins bind. This is confirmed by the CAPRI competition (Critical Assessment of Prediction of Interactions, http://www.ebi.ac.uk/msd-srv/capri/), which has been organized since 2001. All methods that showed high accuracy in predicting protein complexes use flexible docking, often combined with improved

Table 3.11 Frequently used docking programs for protein-protein docking

Name	Web address
ClusPro	https://cluspro.bu.edu/
FRODOCK	http://chaconlab.org/methods/docking/frodock/
GRAMM-X	http://vakser.compbio.ku.edu/resources/gramm/grammx/
Haddock	http://haddock.science.uu.nl/services/HADDOCK2.2/
pyDock	http://life.bsc.es/servlet/pydock/home/
RosettaDock	https://www.rosettacommons.org/
ZDOCK	http://zdock.umassmed.edu/

calculation of the pK value of individual amino acids or statistical analysis of amino acids at the contact surface.

3.3.5 Outlook: Direct Simulation of Substrate Binding

The method of molecular dynamics simulation described in Sect. 3.4 allows energy calculations to be carried out and movements of molecules to be described, provided that no bond is broken or formed. Therefore, this method is in principle suitable to model the non-covalent binding of a ligand to a protein without the need of placing the ligand in an assumed binding pocket (Lawrenz et al. 2015). This method assumes that the simulation time is longer than the binding rate of the ligand. By applying cloud computing, the experimentally determined conformations of bound ligands were reproduced at high accuracy by 100 μs simulations. In addition, critical amino acids affecting the binding kinetics were identified by examining the access pathway of the ligands. The combination of docking methods and extensive molecular dynamics simulations are promising methods for successful drug design. These methods should be directly transferable to the binding of substrates into the substrate-binding pockets of enzymes and therefore represent promising methods for the design of improved enzymes.

Questions

6. Which of two or even more alternative structures of a template protein would you use for modeling substrate specificity?
7. What is the major challenge in docking large, flexible ligands to a protein?

3.4 Mechanistic Models of Protein Structure and Dynamics

Recognition of a substrate by an enzyme is prerequisite for high catalytic activity and high regio- and stereoselectivity. While in docking the substrate-binding pocket of the enzyme is usually assumed to be rigid or only partially flexible, methods have been developed over

Table 3.12 Molecular dynamics programs for protein simulations

Name	Web address
AMBER	http://ambermd.org/
CHARMM	https://www.charmm.org/
Desmond	http://www.deshawresearch.com/resources_desmond.html/
Folding@home	http://folding.stanford.edu/
Gromacs	http://www.gromacs.org/
GROMOS	http://www.gromos.net/
NAMD	http://www.ks.uiuc.edu/Research/namd/
YASARA	http://www.yasara.org/

the past 40 years to take dynamics and flexibility of proteins into account in modeling. This is essential for reliable and quantifiable modeling of substrate binding, since proteins do not have a rigid shape but rather can be seen as mobile nanomachines with a complex dynamics. Interactions with a substrate molecule therefore lead to a slight deformation of the substrate-binding pocket or even major conformational changes due to changes in side chain orientation or displacement of the protein backbone, whereby a lid may open or the orientation of protein domains may change.

3.4.1 Modeling of Conformational Changes

In order to model these conformational changes at a molecular level, the method of molecular dynamics simulation of proteins has been used for over 40 years (Table 3.12). For the simulation of molecules, it is assumed that the motion of the atoms can be described by simple interactions. For example, a chemical bond is modeled as a harmonic spring, whereas for the interactions between atoms that are not covalently bonded, their respective electrostatic partial charges and Van der Waals radius are considered. It is assumed that the properties of the atoms do not change during the simulation and that the motion follows Newton's laws of motion.

The methodological advancements and the available computing capacities now allow the generation of realistic molecular models to simulate enzymes as oligomers in complex with substrates and cosubstrates in arbitrarily complex solvent mixtures over several microseconds at atomic resolution (Dror et al. 2012). This simulation time is usually sufficient to predict and evaluate local conformational changes. For larger, slower conformational changes, methods for accelerated simulation have been developed.

Molecular dynamics simulation methods have been successfully used to experimentally understand and predict specific properties such as substrate specificity and chemo-, regio-, and stereoselectivity at the molecular level and to design enzyme mutants with improved properties. In addition, molecular dynamics simulations are an essential part of the experimental structure elucidation of proteins by X-ray structure analysis or nuclear magnetic resonance spectroscopy.

3.4.2 Modeling of the Biochemical Reaction Mechanism

Enzyme-catalyzed reactions contain at least one step in which a bond is broken or formed. Quantum chemical methods are used to model chemical bonds: fast but highly parameterized semiempirical methods or parameter-free ab initio methods, where, however, the computational effort increases exponentially with the number of electrons.

While very large molecular systems of up to 10^6 atoms can be modeled by molecular dynamics methods, the more elaborated quantum chemical methods are still limited to a few hundred atoms. The advantages of both methods are combined in the so-called QM/MM methods. The motion of the atoms directly involved in the chemical reaction (substrate, catalytic amino acids, cofactors) is calculated by a quantum chemical method, while the remaining atoms (the bulk of the enzyme and the solvent) by molecular dynamics methods. With QM/MM methods, it was possible to predict reaction pathways and calculate activation energies.

3.4.3 Thermodynamic Calculations

While molecular dynamics simulations of protein systems are limited to a time window of microseconds, functionally important processes such as the binding of a substrate, major conformational changes of the protein, or protein folding occur in the time range of milliseconds or seconds. While a direct simulation of such slow processes is not yet possible, methods have been developed to estimate the differences in free energy between different thermodynamic states. Since the free energy represents a potential, the calculation is path independent. This makes it possible to simulate computational paths that are only accessible to simulation. A common application is alchemical changes, where, for example, one amino acid in a protein is gradually converted into another amino acid. The respective change in free energy is then calculated along this pathway. This results, for example, in the difference between the free energies of binding (ΔG) of the wild-type enzyme and a point mutant with and without bound ligand. To compare the calculated free energy differences to an experiment, this simulation method is used in the context of a thermodynamic cycle (Fig. 3.5). In the simulation, the difference between the binding free energies ΔG_B and ΔG_A of the wild-type enzyme and a point mutant with and without bound ligand is calculated; in the experiment, the difference between the binding energies ΔG_1 and ΔG_2 of the wild-type enzyme and the point mutant with the ligand is calculated. Since the calculation of the free energy is pathway independent, the simulated and experimental differences $\Delta\Delta G$ are identical.

Fig. 3.5 Thermodynamic cycle. To study the effect of a mutation on the binding of a ligand L (substrate or inhibitor), four thermodynamic states are defined—the wild-type enzyme E^{WT} and the point mutant E^{mut}— each in complex with the ligand or free. The difference in free energy between the wild-type enzyme and point mutant (ΔG_A and ΔG_B, respectively) is determined by simulation; the difference in binding energies of the ligand to the wild-type enzyme or point mutant (ΔG_1 and ΔG_2, respectively) is determined experimentally. Experiment and simulation can be compared directly, since $\Delta\Delta G = \Delta G_B{-}\Delta G_A = \Delta G_2{-}\Delta G_1$

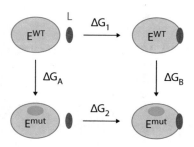

Questions

8. What can we learn from evaluating the $\Delta\Delta G$ between ligands binding to the same enzyme?
9. Can you apply molecular simulation methods to modeling the association rate of a substrate?

3.5   Outlook: Integration of Mechanistic and Kinetic Modeling

The use of mechanistic modeling of enzyme structure, dynamics, and substrate interactions enables deeper insights into the relevant thermodynamic states. The computing power available in the future will make simulations on a millisecond scale possible, so that in addition to thermodynamic quantities, kinetic parameters will also become accessible to molecular simulation, such as the rate constants for the association and dissociation of the enzyme-substrate or the enzyme-product complex. A molecular simulation of kinetic parameters allows the identification of bottlenecks in the enzyme-catalyzed reaction and their elimination by targeted protein, substrate, or solvent engineering. It also allows molecular modeling of enzyme-substrate-solvent systems to be linked to process modeling based on kinetic parameters. The combination of molecular and kinetic modeling represents the core of a scale-bridging modeling strategy in which the complex processes within a reactor are traced back to the atomic structure of the components.

Question

10. How do you imagine that we do enzyme modeling and design in the 2040s?

Take-Home Message
- Sequence data is a valuable source of rapidly growing information, and there are powerful tools available to learn from this data and to use it for the discovery and the design of enzymes.
- The structure of protein domains of small multidomain proteins can be reliably predicted from their sequence.
- The structure of (even large) protein complexes can be predicted by protein–protein docking or by experimental methods such as cryo electron microscopy methods.
- The structure and dynamics of complex molecular systems including one or several proteins, substrate(s), and additives at experimental concentrations as well as solvent can be directly simulated in order to predict conformational transitions, substrate specificity, effect of mutations, or the relative binding affinity of a ligand.

Answers

1. "Homology" originates from the Greek word *homologos* and means an evolutionary relationship of proteins with similar properties (sequence, structure, function).
2. An example for two-domain proteins are cellulases with a catalytically active domain which catalyzes hydrolysis, and a cellulose binding domain, which mediates binding to a cellulosic surface.
3. In clustering and tree construction, all sequences are compared to all others, whereas in BLAST, sequences are searched by their similarity to a single query sequence. In addition, clustering and tree construction are based on the sequence of the whole protein or single domains, whereas BLAST is based on a local sequence alignment.
4. The same protein might have been crystallized under different conditions, or the two entries describe a protein bound to a ligand which slightly changes its structure, as compared to a free, unbound protein.
5. Wrong positioning of side chains, wrong orientation of loops, shift of secondary structure elements, wrong alignment, and template are not homologous.
6. Because binding of a substrate might induce local changes of the structure of the protein's binding site, investigating substrate binding by docking should preferentially start from the structure of a complex of the protein with a substrate-analogous inhibitor.
7. As the number of internal degrees (rotatable bonds) increases, the conformational space that has to be searched by docking increases exponentially. Thus, rigid ligands

with aromatic ring systems can generally be docked more reliably than flexible ligands with alkyl chains.

8. The $\Delta\Delta G$ between two ligands corresponds to the ratio of the binding constants between the two ligands. Thus, we can predict how much better one ligand binds to the enzyme, as compared to the other ligand.

9. This can't be achieved by evaluating free energy differences. But the association of a substrate including the conformational changes of both substrate and enzyme can in principle be modeled by long timescale molecular dynamics simulations. However, the currently available computer resources limit the simulation times to tens of microseconds.

10. Guess now and check your answer in 15–20 years.

References

Dror RO, Dirks RM, Grossman JP, Xu H, Shaw DE. Biomolecular simulation: a computational microscope for molecular biology. Annu Rev Biophys. 2012;41:429–452.

Lawrenz M, Shukla D, Pande VS. Cloud computing approaches for prediction of ligand binding poses and pathways. Sci Rep. 2015;5:7918.

Lindorff-Larsen K, Piana S, Dror RO, Shaw DE. How fast-folding proteins fold. Science. 2011;334: 517–520.

Martí-Renom MA, Stuart AC, Fiser A, Sánchez R, Melo F, Sali A. Comparative protein structure modeling of genes and genomes. Annu Rev Biophys Biomol Struct. 2000;29:291–325.

Orry AJW, Abagyan R. Homology modeling, vol. 857. Humana Press; 2012.

Sousa SF, Ribeiro AJ, Coimbra JT, Neves RP, Martins SA, Moorthy NS, Fernandes PA, Ramos MJ. Protein-ligand docking in the new millennium-a retrospective of 10 years in the field. Curr Med Chem. 2013;20:2296–2314.

Vogel C, Pleiss J. The modular structure of ThDP-dependent enzymes. Proteins. 2014;82:2523–2537.

Widmann M, Radloff R, Pleiss J. The Thiamine diphosphate dependent Enzyme Engineering Database: a tool for the systematic analysis of sequence and structure relations. BMC Biochem. 2010;11:9.

Zhang Y. Template-based modeling and free modeling by I-TASSER in CASP7. Proteins Struct Funct Bioinform. 2007;69:108–117.

Enzyme Kinetics

4

Ana Malvis Romero, Lorenzo Pesci, Selin Kara, and Andreas Liese

What Will You Learn in This Chapter?

In this chapter, you will learn the fundamental concepts of chemical and enzyme catalysis as well as kinetics such as the *lock and key* and the *induced-fit* models, the idea of an enzyme-substrate complex, the Michaelis-Menten equation, or the *steady-state assumption*. Key basic concepts such as catalysts and their classification, reaction rate, enzyme inhibition, and the order of a reaction are also explained. A fundamental point to evaluate enzyme-catalyzed reactions is the determination of kinetic parameters such as the Michaelis-Menten constant and the maximum reaction rate. Their calculation using linearization, numeric, and combined methods is explained in this chapter. Here you will also find the necessary equations and models to determine other relevant parameters such as the turnover frequency and the specificity constant.

Enzyme-catalyzed reactions are nothing but chemical reactions accelerated by a catalyst from nature. Enzyme kinetics describes the rate at which one or more substrates are converted by an enzyme to one or more products. Thermodynamics only specifies the possibility of a reaction, and it is irrelevant whether this reaction takes place uncatalyzed or accelerated with a chemical or biological catalyst. Since enzyme kinetics can be defined as

A. Malvis Romero · L. Pesci · A. Liese (✉)
Institute of Technical Biocalysis, Hambug University of Technology, Hamburg, Germany
e-mail: liese@tuhh.de

S. Kara
Institute of Technical Chemistry, Leibniz University Hannover, Hannover, Germany
e-mail: selin.kara@iftc.uni-hannover.de

© The Author(s), under exclusive license to Springer Nature Switzerland AG 2024
K.-E. Jaeger et al. (eds.), *Introduction to Enzyme Technology*, Learning Materials in Biosciences,
https://doi.org/10.1007/978-3-031-42999-6_4

the chemical kinetics of enzyme-catalyzed reactions, it is helpful to know the basic concepts, properties, and methods of chemical kinetics and catalysis. Briefly, chemical kinetics is the description of the rate at which reactions occur at both the macroscopic and microscopic (reaction mechanism) levels. Its study provides important insights to understand a reaction in detail (and ideally to accelerate the development of a catalyst). Furthermore, chemical kinetics is not only of fundamental scientific interest but is the basis for engineers to design and size a reactor for a particular conversion. In this sense, chemical kinetics is also correlated with economics.

4.1 Basic Concepts of Chemical Catalysis

Kinetics must be clearly distinguished from thermodynamics. Kinetics describes the speed at which a chemical reaction takes place or how quickly the thermodynamically maximum possible conversion, i.e., state of equilibrium, is reached. A catalyst, regardless of whether it is chemically synthetic or a biocatalyst (= enzyme), accelerates a thermodynamically possible reaction without itself being consumed. Thermodynamics, on the other hand, describes the position of equilibrium, i.e., the maximum possible turnover under equilibrium conditions. The principle of Le Chatelier, formulated by Henry Le Chatelier and Ferdinand Braun between 1884 and 1888, states that it can only be changed by shifting the equilibrium, by removing a reactant on the product side from the equilibrium, or alternatively by increasing the concentration of a reactant on the substrate side. Accordingly, it is also called the "principle of least constraint." In addition to the concentration, changes in temperature and pressure can also lead to a shift in the equilibrium.

As we know from chemical thermodynamics, chemical compounds are converted in the reaction direction in which the entropy reaches its maximum. In general, it is not so easy to deal with the entropy function, because the properties of the system and the environment have to be taken into account. It is therefore often easier to work with energy functions, such as the Gibbs energy (or free enthalpy), for which only the properties of the system need to be known to describe a particular process (including chemical processes).

A chemical reaction occurs in the direction in which the Gibbs free energy (G) decreases. In other words, if an object falls because it is subject to gravity (in the direction of lower gravitational energy, also called potential energy), chemical compounds will also "fall" (under certain experimental conditions) in the direction of lower chemical potential (molar Gibbs energy).

Consider the general chemical reaction (Eq. 4.1):

$$a\mathrm{A} + b\mathrm{B} \rightleftharpoons c\mathrm{C} + d\mathrm{D} \tag{4.1}$$

It is possible to show by classical thermodynamics that the Gibbs energy difference for this reaction (Eq. 4.1) can be described by the Van't-Hoff equation (Eq. 4.2):

$$\Delta_r G = \Delta_r G^0 + RT \ln Q \qquad (4.2)$$

$\Delta_r G$ (J mol^{-1}) is the difference in free energy between the initial and final states of the system; $\Delta_r G^0$ (J mol^{-1}) is the difference in free energy between the initial and final states of the system for the substances in their standard states (i.e., 1 bar pressure and unitary activity); R (8.314 J mol^{-1} K^{-1}) is the universal gas constant; T (K) is the absolute temperature; and Q is the reaction quotient (Eq. 4.3), which indicates the reaction progress:

$$Q = \frac{c_{C^c} \times c_{D^d}}{c_{A^a} \times c_{B^b}} \qquad (4.3)$$

This means that under certain conditions, where the substrates are present in relatively high concentrations, the reaction in Eq. 4.1 will progress from left to right, with the free energy decreasing progressively. The function will eventually reach a minimum when $\Delta_r G = 0$ (Eq. 4.2) (chemical equilibrium). In this condition, the quotient Q is equal to the reaction equilibrium constant (K_{eq}):

$$\Delta_r G^0 = -RT \ln K_{eq} \qquad (4.4)$$

Therefore,

- $\Delta_r G^0 > 0$, $K_{eq} < 1$, i.e., the reaction mixture at equilibrium contains more substrates.
- $\Delta_r G^0 < 0$, $K_{eq} > 1$, i.e., the reaction mixture at equilibrium contains more products.

In summary, chemical thermodynamics provides information about which transformations are preferred, but it does not provide any information about how fast they will occur. The information about the reaction rate comes from kinetics. As a simple example, every living thing is composed primarily of carbon, hydrogen, and oxygen. Even if carbon dioxide (CO_2) and water (H_2O) are the thermodynamically preferred compounds, the transformation will not be too fast.

Any chemical reaction can be described by a rate law, which correlates the rate with the concentrations of the species involved. If we consider the chemical equation (Eq. 4.1) as an irreversible reaction, the following formula (Eq. 4.5) results in the rate law:

$$v = k c_A{}^x c_B{}^y \qquad (4.5)$$

In this equation, v (e.g., with the unit mol L^{-1} s^{-1}) is the reaction rate and k is the kinetic constant, which depends on the reaction conditions but not on the concentration. Here, the concentrations of the reagents and their individual reaction orders x and y, which are not related to the stoichiometric coefficients—except for a special case that we will discuss later—are used. The global reaction order is the sum of x and y. It is obvious that the unit of k depends on the global reaction order. The structure of this equation is quite intuitive since at the microscopic level chemical reactions occur that form and break chemical bonds, which can only happen when molecules of a certain energy collide (or vibrate); therefore, higher concentrations ensure a higher probability that these events will occur.

The individual reaction order of a given reaction must be determined experimentally, but it typically takes values of 0, 1/2, 1 or 2. These values depend on the reaction mechanism and more precisely on the number of molecules involved in a given reaction step.

Reactions can proceed through the formation of a variety of intermediates. The elementary reactions take place as described in the following equations. Elementary reactions are characterized by their molecularity. Equation 4.6 shows a monomolecular reaction in which, for example, a single molecule vibrates to break a chemical bond, while Eqs. 4.7 and 4.8 show bimolecular and trimolecular reactions. Higher molecularities are extremely rare.

$$A \rightarrow B \tag{4.6}$$

$$A + B \rightarrow C \tag{4.7}$$

$$A + B + C \rightarrow D \tag{4.8}$$

The reaction order (which has mainly an empirical meaning) should not be confused with the molecularities, which are related to the microscopic mechanism. However, it seems clear that in elementary reactions, the molecularity corresponds to the global reaction order, and the individual orders correspond to the stoichiometric coefficients.

To understand the basic components of a rate law, we first consider the monomolecular reaction (Eq. 4.6). The rate law, together with the differential terms of decreasing substrate concentration and increasing product concentration as a function of time t, is described in Eq. 4.9:

$$v = -\frac{dc_A}{dt} = -\frac{dc_B}{dt} = kc_A \tag{4.9}$$

This differential equation can be easily solved by separating the variables:

$$-\int \frac{dc_A}{c_A} = k \int dt \tag{4.10}$$

$$c_A = c_{A0}e^{-kt} \tag{4.11}$$

The solution (Eq. 4.11) is a negative exponential function with c_{A0} (substrate concentration at time $t = 0$) and strongly dependent on the constant k (in this case with unit time^{-1}).

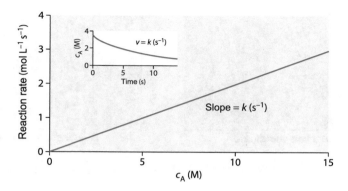

Fig. 4.1 Kinetics of a first-order reaction (Eq. 4.6)

The simplest way to determine the respective reaction orders and the kinetic constants is to perform experiments with simultaneous measurement of the substrate consumption as a function of time and calculation of the reaction rate. For example, Fig. 4.1 shows a typical concentration profile for a first-order reaction and the resulting reaction rate as a function of substrate A concentration. Here, the reaction rate constant k is the slope of the straight line.

By analogous derivations, it is possible to determine linear relations for the other reaction orders.

The minimum average energy required for a reaction to proceed is the activation energy E_a, which also represents the activation barrier of the reaction. Figure 4.2a shows a typical energy-reaction coordinate profile in which it can be seen that substrate and product are separated by an energy pathway and that its maximum corresponds to the activation energy. The species at the top of the "energy mountain" is the transition state, an extremely unstable chemical species with partially broken/formed chemical bonds that exist for only a very short period of time and can evolve into a product. In Fig. 4.2, the *x-axis* represents the reaction coordinate, which corresponds to a geometric parameter (e.g., bond length, angle, etc.) that changes as the reaction proceeds. The lower the E_a, the higher the probability of successful collisions between the molecules and the higher the reaction rate. Accordingly, the kinetic constant can be described by the Arrhenius equation (Eq. 4.12).

$$k = Ae^{-\frac{E_a}{RT}} \tag{4.12}$$

Accordingly, an increase in temperature leads to an increase in the reaction rate. While the exponential factor is a purely energetic expression, the pre-exponential term A (time^{-1}) is a statistical expression indicating the collision frequency between the molecules. The activation energy and the pre-exponential factor can be calculated by determining the kinetic constants at different temperatures and plotting the logarithm of k against $1/T$. This results in a straight line with a negative slope—E_a/R and $\ln A$ as the y-axis intercept (Arrhenius plot).

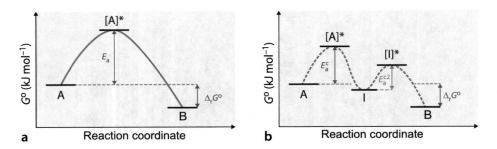

Fig. 4.2 Energy-reaction coordinate profile for an uncatalyzed (**a**) and a catalyzed reaction (**b**)

So far, we have only considered irreversible reactions where the equilibrium position on the product side is at an extreme (theoretically, all chemical reactions should be considered reversible). For reversible reactions, the rate of both the outward and reverse reactions must be considered. Thus, for the reversible reaction (Eq. 4.13), the equation for the reaction rate (Eq. 4.14) becomes

$$A \underset{k_{-1}}{\overset{k_1}{\rightleftharpoons}} B \tag{4.13}$$

$$v = k_1 c_A - k_{-1} c_B \tag{4.14}$$

At equilibrium, the net velocity is zero:

$$v = k_1 c_{Aeq} - k_{-1} c_{Beq} = 0 \tag{4.15}$$

$$K_{eq} = \frac{c_{Beq}}{c_{Aeq}} = \frac{k_1}{k_{-1}} \tag{4.16}$$

This can be easily interpreted from Fig. 4.2a: when $K_{eq} > 1$, there is a higher activation energy for the reverse reaction; thus, it becomes slower ($k_{-1} < k_1$). Consequently, the kinetic equation can also be written as a function of the equilibrium constant:

$$v = k_1 \left(c_A - \frac{c_B}{K_{eq}} \right) \tag{4.17}$$

It is clear from Eq. 4.12 that increasing the temperature accelerates a reaction. Therefore, it seems to be sufficient to heat each reaction vessel to the desired temperature. As a rule of thumb, the reaction rate doubles for every 10 °C temperature increase. Unfortunately, not only is the desired reaction accelerated, but other possible side reactions may also be accelerated. In addition, higher temperatures can accelerate the decomposition of reagents and consequently affect the equilibrium composition. To accelerate a reaction, it is therefore preferable to increase the activation energy by using a catalyst (from the Greek *catalysis, dissolve through*) rather than increasing the population of molecules that have a certain energy.

The term catalyst was defined in 1836 by Jöns Jacob Berzelius defined as a substance that increases the rate of equilibration of a chemical reaction by lowering the activation energy without being consumed in the reaction. At the molecular level, a catalyst is a substance that alters the mechanism of the reaction, often making it more complicated. In this case, the highest activation barrier is lower than the activation barrier of the uncatalyzed reaction.

The reaction profile in Fig. 4.2b schematizes a reaction pathway for a catalyzed reaction. A change in the mechanism can be observed, involving the formation of an intermediate I, which leads to the product after the formation of another transition state [I]*. The key point is the lower activation energy of the catalyzed reaction compared to the uncatalyzed reaction.

Question 1. How would the addition of a catalyst to the reaction A $\rightleftharpoons$ B affect the value of $\Delta_r G^0$?

Question 2. How does a catalyst influence the activation energy of a reaction?

4.2   Homogeneous, Heterogeneous, and Enzymatic Catalysis: Differences and Similarities

The question we would like to address in this section is how the different types of catalysis are defined and in what sense they are similar. This would make it possible in the next sections to classify enzyme catalysis into the general categories of classical chemical catalysis and to apply the same principles of kinetics and reaction engineering.

Catalysts are usually divided into the following three categories from a chemical point of view: i) homogeneous, ii) heterogeneous, and iii) enzymatic. The first category, homogeneous, indicates that the catalyst and reagents are in the same phase. The second category, heterogeneous, indicates that both are in different phases, and the last indicates that the catalytic species is an enzyme. This distinction is mainly for historical but also for professional reasons:

– **Homogeneous catalysis**. Traditionally associated with the profession of chemist(s), this implies that the chemical reaction takes place in a single phase, usually liquid but also gaseous. Homogeneous catalysts include several subcategories such as Brønsted acids and bases, organometallic complexes, and organocatalysts (metal-free organic molecules). One of the most important features of this group of catalysts is their molecular and *single-site catalytic* character, which means that the catalytic effect is exerted by a precise group of atoms.

- **Heterogeneous catalysis**. This category is more associated with the chemical engineering profession and, unlike the previous category, describes processes that take place at the interface between two phases, gaseous-solid or liquid-solid, where the catalyst (e.g., a metal oxide) is in the solid phase. Since these processes take place on a heterogeneous surface, there is not a single active site, but a multitude of them, depending on the surface characteristics of the solid.
- **Enzymatic catalysis**. Traditionally associated with the biology and biochemistry professions (although the boundaries between these and the other chemical professions are blurring at a remarkable rate), it describes the conversion of a substrate into a product, catalyzed by an enzyme. The catalytic events occur in the active site of the enzyme, which includes a precise set of amino acids and inorganic/organic cofactors as needed.

Looking at the three categories above, the distinction for enzymatic catalysis is not so obvious, mainly for three reasons:

- Although enzymes have size differences of one to two orders of magnitude with respect to their substrate molecules, enzymes are soluble in aqueous media. Therefore, they must be considered as homogeneous catalysts.
- On the other hand, the certain similarity with heterogeneous catalysts, based on the molecular and mechanistic aspects, has been demonstrated. The chemical environment in the active site of the enzyme is different from that in the bulk solution (as in heterogeneous catalysts). Moreover, the kinetic behavior of enzymes is strongly characterized by the binding steps, which resemble adsorption processes on solids. In addition, homogeneously dissolved enzymes can be retained using ultrafiltration membranes.
- Enzymes can be used not only in free form but also immobilized or as a component of whole cells, i.e., in a unique heterogeneous form.

Homogeneous catalysts can be immobilized on solid supports to maintain the properties of homo- and heterogeneous catalysts (molecular character but different phases). Therefore, heterogeneous catalysts, although traditionally related to inorganic solids and active surfaces, encompass a wide range of catalysts obtained by the heterogenization of homogeneous catalysts. Accordingly, catalysts should only be categorized in an application-oriented manner. This is because the reaction engineering principles (reaction control and apparatus selection) are the same for chemo- and biocatalysts if they are heterogeneous or homogeneous. In the case of immobilized and thus heterogeneous catalysts, regardless of whether they are chemical or enzymatic, mass transfer limitations must be considered in contrast to their homogeneous use. From this perspective, catalysts should only be divided into the two categories: homogeneous and heterogeneous. This distinction is illustrated in Fig. 4.3.

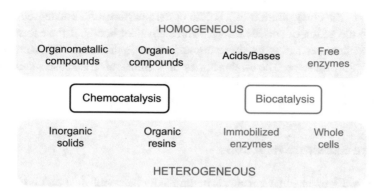

Fig. 4.3 Application-based classification of catalysts (adapted from Yuryev and Liese 2010)

- Question 3. What is an example for heterogenization of a homogeneous catalyst, and what must be considered in comparison with its homogeneous form?
- Question 4. What types of catalysts are a free enzyme and a whole cell?

4.3 Enzyme Kinetic Models

Looking at the three-dimensional structure of a protein = enzyme, three important areas can be identified:

- The outer surface exposing polar groups to the solution
- The inner core, which contains mainly interactions between nonpolar groups (e.g., aliphatic and aromatic side chains of amino acids)
- The active site, which is located at a specific point of the inner core where the catalytic activity take place

Therefore, the substrate must be incorporated into the active site prior to catalysis. This step represents a major concept in enzyme kinetics, and several theories have been proposed to rationalize such an event. Emil Fischer developed the so-called *lock and key* mechanism in 1894, which assumes a perfect structural and functional match between the active site and the substrate molecules, like a lock with its matching key. However, this model cannot explain why, for example, larger substrates that do not fit into the active site can also be converted. As an extended model that also takes this observation into account, the *induced-fit model* was proposed in 1958 by Daniel Koshland Jr. (Koshland 1958). A flexible adjustment of the protein conformation allows the incorporation of reactants with different sizes. Figuratively speaking, an enzyme can be imagined as a "magic" lock that can change its gears to accept a variety of keys.

So far we have used the term "binding." However, more precisely, the binding of the substrate in the active site of the enzyme is more of an adsorption (the IUPAC definition is "an increase in the concentration of a solute at the interface of a condensed and a liquid phase due to the action of surface forces") since covalent bonds, if their formation is not part of the reaction mechanism as, for example, in some hydrolases or transaminases, occur only in the following steps. The establishment of enzyme-substrate adsorption complexes, which are usually characterized by dissociation constants in the micro- or millimolar range (Leskovac 2003), is a common feature of all enzyme kinetic models.

4.3.1   One-Substrate Reactions

The seminal work on enzyme kinetics is undoubtedly that published by Leonor Michaelis and Maud Leonora Menten in 1913, entitled "Kinetics of Invertin Action" (Michaelis and Menten 1913). Their paper reports not only the first coherent rate equation for an enzyme-catalyzed reaction but also the first fit to experimental data using product inhibition (Sect. 4.2). Before we begin our discussion of the equation, it is worth noting that the equations were developed at a time when enzymes were still called "ferments," meaning that their molecular composition was unknown and their concentrations could therefore not be defined. The reaction system studied by Michaelis and Menten was an invertase that catalyzes the cleavage of sucrose to fructose and glucose. Their experiments revealed two key points of enzymology:

- The initial reaction rate (v_0) increases linearly with the amount of enzyme added.
- v_0 shows a hyperbolic behavior with increasing substrate concentration.

The need to measure initial reaction rates ($\leq 10\%$ of conversion) is clearly a consequence of the second point. Furthermore, the influence of possible product inhibition should be minimized.

The validity of the Michaelis-Menten equation is usually evaluated using the equation developed by George Briggs and John Haldane 1925 to prove the *steady-state assumption* (Briggs and Haldane 1925) because it leads to more general results. It is a two-step mechanism in which an enzyme-substrate intermediate is formed. This is stabilized by the enzyme, which leads to a decisive lowering of the activation energy.

$$\underbrace{S + E \underset{k_{-1}}{\overset{k_1}{\rightleftharpoons}} ES}_{\text{Adsorption}} \underbrace{\rightarrow P + E}_{\text{Transformation}} \tag{4.18}$$

According to Briggs and Haldane, since, as a rule $c_E \ll c_S$, the enzyme-substrate complex (*ES*) reaches a quasi-steady state shortly after the reaction starts, a steady state is formed:

$$\frac{dc_{ES}}{dt} = k_1 \, {}_sc_E - (k_{-1} + k_2) \, c_{ES} = 0 \tag{4.19}$$

$$-\frac{dc_S}{dt} = k_1 c_S \, c_E - k_{-1} c_{ES} \tag{4.20}$$

$$-\frac{dc_E}{dt} = k_1 \, c_S \, c_E - k_{-1} c_{ES} - k_2 c_{ES} = -(k_{-1} + k_2) \, c_{ES} + k_1 \, c_S \, c_E \tag{4.21}$$

$$\frac{dc_P}{dt} = k_2 c_{ES} \tag{4.22}$$

The total enzyme concentration c_{E0} consists of the concentration of currently free enzyme c_E and that bound with the substrate in the enzyme-substrate complex (c_{ES}):

$$c_{E0} = c_E + c_{ES} \tag{4.23}$$

The solutions of the differential equations can be easily obtained with a computer and are shown in Fig. 4.4.

Equations 4.19 and 4.23 result in the term for the enzyme-substrate complex *(ES)*:

$$c_{ES} = \frac{k_1 \, c_S \, c_{E0}}{k_1 \, c_S + k_{-1} + k_2} \tag{4.24}$$

By substituting in Eq. 4.22 for the product formation, the Michaelis-Menten equation is obtained:

$$v = -\frac{dc_S}{dt} = \frac{dc_P}{dt} = k_2 c_{ES} = \frac{k_2 \, k_1 \, c_S \, c_{E0}}{(k_1 \, c_S + k_{-1} + k_2)} = \frac{v_{max} c_S}{K_m + c_S} \tag{4.25}$$

where $v_{max} = k_2 \, c_{E0}$ and $K_m = \frac{k_{-1} + k_2}{k_1}$

Figure 4.5 shows the Michaelis-Menten equation of a one-substrate reaction.

The Michaelis-Menten graph can be divided into three parts with respect to the reaction rate:

– Low substrate concentration, $c_S \ll K_m$: first reaction order range, i.e., the reaction rate is linearly dependent on the substrate concentration.

$$v = \frac{v_{max}}{K_m} c_S \tag{4.26}$$

– Average substrate concentration c_S: range of broken/mixed reaction order, i.e., the reaction rate is not linearly dependent on the substrate concentration.

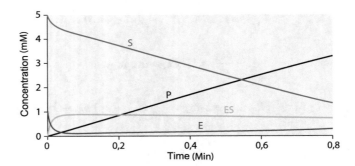

Fig. 4.4 Solutions of the differential equations (Eqs. 4.19–4.23) with: $k_1 = 20$ min^{-1} mM^{-1} L; $k_{-1} =$ min^{-1}; $k_2 = 10$ min^{-1}; and $c_{E0} = 0.5$ μM

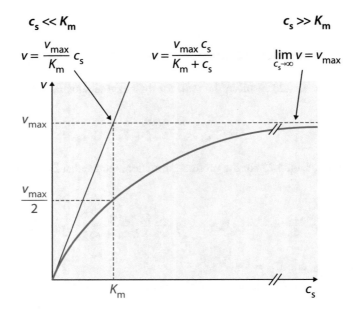

Fig. 4.5 Graphical representation of Michaelis-Menten kinetics (Eq. 4.25). K_m (in mM): Michaelis-Menten constant; v_{max} (in μmol min^{-1}): maximum reaction rate

$$v = \frac{v_{max}\, c_S}{K_m + c_S} \tag{4.27}$$

– High substrate concentration, $c_S \gg K_m$: zero reaction order range, i.e., the reaction rate does not depend on the substrate concentration. All active sites are saturated, and the resulting reaction rate depends only on the enzyme concentration.

Table 4.1 Four kinetic parameters of enzyme kinetics

Parameter	K_m	v_{max}	k_{cat}	$K_a = \frac{v_{max}}{K_m}$
Name	Michaelis-Menten constant	Maximum reaction rate	Turnover frequency	Specificity constant
Unit	mM	$\mu mol \cdot min^{-1} \cdot mg$	min^{-1} or s^{-1}	$L \cdot min^{-1}$

$$\lim_{c_S \to \infty} v = v_{max} \tag{4.28}$$

Four key parameters of enzyme kinetics can be derived from Eq. 4.25 (Table 4.1). These parameters are important for chemists and engineers to evaluate an enzyme-catalyzed chemical process and for biologists to understand in vivo enzyme function and regulation. In general, K_m can be defined as the substrate concentration at which the reaction rate is half of the v_{max} (Fig. 4.5). This relationship can be easily derived by substituting $K_m = c_s$ in Eq. 4.25. K_m (usually in the micro- to millimolar range) is often referred to as the "affinity constant." However, this definition is inaccurate because K_m is only proportional to the enzyme-substrate dissociation constant.

Only if $k_2 << k_{-1}$ (the assumption in the 1913 paper by Michaelis and Menten) will the Michaelis-Menten constant is equal to the dissociation constant. However, it is generally accepted to take K_m as an estimation for the substrate-enzyme affinity, for which higher affinities correspond to a lower K_m value. A second important parameter is v_{max}, which mathematically indicates the asymptote of the rectangular hyperbola and chemically the reaction rate when all enzyme molecules are saturated with substrate molecules in the active site. The volumetric reaction rate ($\mu mol \ min^{-1} \ mL^{-1}$) is distinguished from the mass-specific reaction rate ($\mu mol \ min^{-1} \ mg^{-1}$). The parameter k_{cat} is the turnover number or *turnover frequency*:

$$k_{cat} = \frac{v_{max}}{nc_{E0}} \tag{4.29}$$

The turnover number k_{cat} describes the number of catalytic reactions occurring per time unit per active site of the enzyme (n is the number of active sites per molecule for multimeric enzymes). The k_{cat} values vary widely, e.g., the enzyme carbonic anhydrase has a k_{cat}-value up to $\approx 10^6 \ s^{-1}$, whereas the values for the key photosynthetic enzyme ribulose-1,5-bisphosphate carboxylase are usually in the range of $\approx 10^{-1} \ s^{-1}$ (BRENDA Database 2018). The reciprocal of k_{cat} indicates the time required for a single catalytic reaction. For the simple mechanism given above, $k_{cat} = k_2$.

A fourth important parameter is the specificity constant (K_a), which represents the efficiency of an enzyme with respect to a particular substrate. The specificity constant K_a is equal to $\frac{v_{max}}{K_m}$.

Under conditions where $c_S \ll K_m$, this can be described as the second reaction order rate for conversion of a substrate to product:

$$v = \frac{k_{\text{cat}} \, c_S \, c_{E0}}{K_{\text{m}}} \tag{4.30}$$

When different substrates are to be compared against a particular enzyme, K_a gives an indication of the enzyme's preferences.

4.3.2 Two-Substrate Reactions

Most of the enzymatic reactions involve not only one substrate but two (e.g., in oxidoreductases, transferases, lipases, etc.) that release one or two products.

$$A + B \leftrightharpoons P + Q \tag{4.31}$$

The simplest approach to describe a two-substrate reaction is the two-substrate Michaelis-Menten kinetics (Eq. 4.32) assuming a quasi-irreversible reaction, which is obtained by multiplying two-substrate Michaelis-Menten kinetics. Here, it is assumed that both substrates bind to the same active site and thus only one v_{max} is given. Individual K_{m} values, K_{mA} and K_{mB}, are determined for the two substrates A and B, respectively.

$$v = v_{\max} \frac{c_A}{K_{\text{mA}} + c_A} \frac{c_B}{K_{\text{mB}} + c_B} \tag{4.32}$$

The mathematical treatment of these systems from a mechanistic point of view is naturally more complex. But true to the motto "the simplest model that accurately describes reality is the best," even this simple kinetic approach can be sufficient. If this is no longer the case, one must fall back on detailed, mechanistically based models.

In two-substrate mechanisms, a distinction is usually made between "ordered" and "random" mechanisms, based on the time sequence in which the substrates are bound to the active site and the products are released. The number of species involved is designated by the terms "uni-" (one substrate or product molecule) and "bi-" (two substrate or product molecules). In general, the complex ordered and random mechanisms are quite difficult to deal with. A clear and systematic methodology to describe the multisubstrate enzyme mechanisms comes from a study by William Cleland (Cleland 1963). In this approach, a straight line is drawn separating the reactants in the bulk solution from the enzyme-bound species (Fig. 4.6). The letters A, B, ... describe the order of attachment to the enzyme and P, Q, ... the products.

To derive the various kinetics from the different multisubstrate enzyme mechanisms, the King-Altman method can be used as an alternative to derivation via the differential equations, which is a systematic approach for the treatment of highly complex mechanisms (King and Altman 1956). For this purpose, the network of individual reactions is recorded

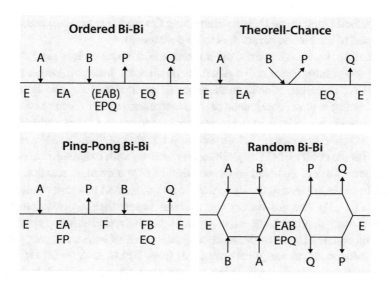

Fig. 4.6 Schematic representation of the basic reaction mechanisms of two-substrate reactions

geometrically, and the respective individual rate constants are combined according to certain rules.

The Theorell-Chance mechanism is typical of alcohol dehydrogenases, where, for example, in the case of oxidation, A would be $NAD(P)^+$, B an alcohol, P an aldehyde, and Q a NAD(P)H (Fig. 4.6). In cases where dissociation of P is required for the binding of the second substrate B, this Bi-Bi mechanism is referred to as ping-pong (Fig. 4.6). Ping-pong bi-bi mechanisms are typical of transferases and lipases, for example. The F means that the enzyme has been chemically modified during the catalytic cycle (in the case of transferases, for example, a functional group is transferred from substrate A to the enzyme).

Question 5. What is meant by enzyme saturation?
Question 6. What would be the simplest way to maintain a zero-order rate?

4.4 Determination of the Kinetic Parameters

Kinetic parameters cannot be measured directly, as they must be determined via the reaction rate equation, which is based on a specific kinetic model. Changes in the concentration of the reactants (at least one reactant) as a function of time are measured. In addition to the reactant or product concentration, other parameters such as pH value, a gas formation, etc. can be used. The only condition is that this key component must be directly linked to the substrate conversion. Care must be taken to ensure that the error of the analytical method is minimized when determining the reactant concentrations. Otherwise, it is transferred to the kinetic parameters to be determined.

Before starting the measurements, other reaction conditions such as pH value, temperature, or buffer concentration must be determined, since all these parameters influence the enzyme activity. Therefore, kinetic measurements should also always be thermostatted.

A typical example of the continuous on-line determination of a reactant concentration is spectrophotometric analysis, where the increase or decrease of the absorbance of a chromophore at a certain wavelength is followed as a function of time (e.g., the absorbance at $\lambda = 340$ nm for the cofactor NAD(P)H in redox reactions with oxidoreductases). However, if chromophores are not involved in the reaction or if two or more reaction components show significant absorbance in the same wavelength range and no other chemical-physical parameter (e.g., pH) changes during the reaction occur, then a discontinuous off-line analysis must be established. In such an analysis, the reaction is stopped at specific times, and the concentrations of the reaction components (substrates and/or products) are determined using standard analytical techniques (e.g., HPLC, GC, NMR, etc.).

An important factor in minimizing the error of discontinuous analysis is the procedure used to stop the reaction. This can be done by the following:

– Cooling (enzyme then works at a significantly lower speed)
– Heating (this leads to thermal denaturation of the enzyme)
– Addition of denaturing agents (e.g. acids, bases, or water-miscible organic solvents)

Basically, there are two methods for determining enzyme kinetics: *initial rate measurement* and *progress curve analysis* of a set reactor experiment.

4.4.1 Measurement of the Initial Reaction Rate

Initial reaction rates can be determined for substrate conversions of ≤10% in the first reaction order range. The exclusion criteria for the use of this method are the presence of severe product inhibition (Sect. 4.5) and/or—in the case of an equilibrium constant $K_{eq} < 1$—that the equilibrium is on the side of the substrates and only a low conversion can be achieved. In the case of significant product inhibition, even low product concentrations have a significant influence on the initial reaction rate. The amount of enzyme used should be relatively low (the exact meaning of "low" depends on the K_m and k_{cat} values of the system) in order to clearly observe the linear range where quasi-stationary conditions prevail. On the other hand, the enzyme amount should be quite high to establish a relatively fast analysis.

Considering a Bi-Bi mechanism (two substrates A and B and two products P and Q; Fig. 4.6), the following measurement plan can be used for the determination of the initial reaction rates (Table 4.2). The constant concentrations are chosen from the range of zero reaction order of the respective reactant in the Michaelis-Menten curve to minimize the influence of pipetting errors. The result is shown in Fig. 4.7 as an activity curve.

Table 4.2 Measurement plan for the determination of the initial reaction rate of a reaction with the Bi-Bi mechanism

c_A	c_B	c_P	c_Q
Varied	Constant	–	–
Constant	Varied	–	–
Constant	Constant	Varied	–
Constant	Constant	–	Varied

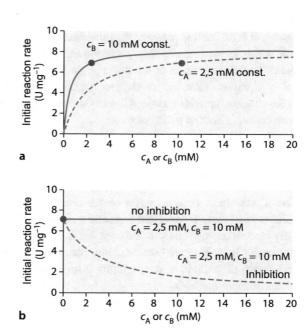

Fig. 4.7 Initial rate measurements for a reaction using the Bi-Bi mechanism; $K_{mA} = 0.5$ mM, $K_{mB} = 2$ mM, $K_{iP} = 0.5$ mM, $v_{max} = 10$ U mg^{-1}. (**a**) Initial reaction rate as a function of the respective substrate concentration A and B at constant concentration of the respective substrate. (**b**) Initial reaction rate as a function of the product P concentration. Points of equal reaction rate ($c_A =$ constant and $c_B =$ constant) are marked with a red dot

Figure 4.7a shows that by establishing a constant concentration of c_B, saturation kinetics can be obtained as a function of c_A (in double substrate kinetics, the term that is a function of c_B becomes a constant). An analogous result is obtained if, when c_B is varied, c_A is constant. Both curves are asymptotic to the same value of v_{max}. Figure 4.7b shows that in the case of product inhibition by P, the reaction rate decreases by increasing c_P. The same principle can be used for the product Q. The reaction rate for the pair of values $c_A =$ constant and $c_B =$ constant with $c_P = c_Q = 0$ must be determined to the same value in all four dependencies within the measurement error. In practice, it is recommended that experiments are performed repeatedly to allow an appropriate statistical treatment of the data. It is also advisable to perform the activity analysis at different enzyme concentrations. The resulting proportionality between reaction rate and c_E confirms the validity of the method.

To perform a Michaelis-Menten curve, usually ≈ 10 measuring points are required. It is recommended not to perform the kinetic measurements over long periods of time if the stability of the enzyme is not known. In the following, we will see how to calculate the kinetic constants from these Michaelis-Menten curves.

4.4.2 Progress Curve Analysis

If the equilibrium constant for the reaction K_{eq} is <1 and/or significant product inhibition occurs, it is difficult to measure the initial reaction rate directly because it is affected even at low substrate conversions. The preferred method in this case is to obtain the data by fitting the differential equation of the appropriate kinetic model to the reaction progress curves, i.e., the progress of substrate and product concentration as a function of time (Sect. 5.3). It is important to take into account the amount of enzyme or enzyme activity used and that the conversion is limited to 5% at most.

4.4.3 Determination of the Catalytic Constants

The direct visual determination of the catalytic constants from the course of a typical Michaelis-Menten curve (initial reaction rate as a function of substrate concentration) is very inaccurate, especially when complex mechanisms with multiple substrates, reversibility, and/or inhibition are present. More precise information can only be obtained from the realization of complementary graphs from additional experiments and with the help of computational methods.

There are different methods to determine enzyme kinetic parameters which can be divided into three groups:

– Linearization methods
– Numerical methods (nonlinear fitting and progress curve analysis)
– Combined procedure

The involvement of computer-aided calculations increase progressively from the first to the third point.

We will now discuss the three groups of determination methods and briefly compare their properties. **Linearization methods** represent the oldest procedure and are usually used when data from initial reaction rate measurements are available. Linearization consists of simple mathematical manipulations of the Michaelis-Menten equation, where two variables plotted against each other give the kinetic constants as the intercepts and slopes of the linear regression.

The three most commonly used linearization methods are as follows:

– Lineweaver-Burk (or double reciprocal) equation (Lineweaver and Burk 1934)

$$\frac{1}{v} = \frac{1}{v_{max}} + \left(\frac{K_m}{v_{max}}\right) + \frac{1}{c_S} \tag{4.33}$$

– Eadie-Hofstee equation (Eadie 1942; Hofstee 1959)

$$v = v_{max} - K_m \frac{v}{c_S} \tag{4.34}$$

– Hanes-Equation (Hanes 1932)

$$\frac{c_S}{v} = \frac{K_m}{v_{max}} + \left(\frac{1}{v_{max}}\right) c_S \tag{4.35}$$

In Fig. 4.8, the Hanes linearization graph is shown.

Although the three methods may seem interchangeable, statistical analyses show strong differences between them. The double reciprocal plot, developed in 1934 (Lineweaver and Burk 1934), is one of the most commonly used and also the one with the largest error; it gives greater weight to experimental points at lower substrate concentrations that have the largest error (Leskovac 2003). The Eadie-Hofstee diagram (Eadie 1942; Hofstee 1959) also has a systematic error, more specifically at lower v and higher v/c_s values. However, it is able to show the presence of unreliable measurements by showing particularly high standard deviations.

The Hanes plot (Hanes 1932) is the best of the three methods because there are no systematic errors. Although linear methods are considered outdated compared to nonlinear methods, careful use of linearization methods can be very useful to observe deviations from the expected saturation kinetics (Leskovac 2003) and can also be used by students at the beginning of their studies who are not yet familiar with more complex computational methods.

Nonlinear methods include the nonlinear fitting of the Michaelis-Menten kinetic parameters to data from the initial reaction rate measurement and, alternatively, to the progress curves of set reactor experiments. The nonlinear regression to the saturation kinetics is a fit of the kinetic model (e.g., two-substrate Michaelis-Menten kinetics) where the kinetic constants are optimized stepwise (e.g. by the least squares method). For this approach, it is necessary to have estimations of the kinetic parameters, which can be obtained, for example, by linearization methods. In addition, assumptions on possible inhibitions and thermodynamic limitations must be considered in the simulation model.

This procedure can be used to fit a kinetic model, i.e., by regression of the kinetic constants to a measured response curve. Libraries of different kinetic models can help to find the best solutions for the problem (Drauz and Waldmann 2002).

In this approach, especially a possible thermal deactivation of the enzyme must be considered, since the course of the reaction can take much longer than the measurement of

Fig. 4.8 Linearization of the Michaelis-Menten equation with the Hanes method

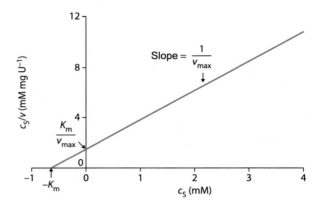

the initial reaction rate. In many cases, there is thermal or reactant triggered enzyme deactivation, which mostly follows first-order kinetics (therefore the enzyme concentration will decrease exponentially). As a result, the enzyme concentration is not constant, but a function of time:

$$c_E = c_{E0}e^{-k_d t} \tag{4.36}$$

c_{E0} is the initial active enzyme concentration, c_E is the current active enzyme concentration, k_d is the deactivation constant at the reaction temperature, and t is the time. The deactivation constant (k_d) can be easily calculated by a logarithmic transformation of Eq. 4.36. Based on this, the half-life $\tau_{1/2}$ can be determined (Eq. 4.37). It indicates the time in which half of the active enzyme concentration is deactivated:

$$\tau_{1/2} = \frac{\ln 2}{k_d} \tag{4.37}$$

The third approach is a combination of initial reaction rate measurements and progress curve analysis. In the combination, a smaller number of experiments are required than in the initial rate approach; it is used for reversible Bi-Bi reactions (e.g., transaminase; Al-Haque et al. 2012). The advantage in combining the two approaches is that performing a single analysis of the progress curve on a complex mechanism is not sufficient to obtain realistic data due to the presence of local minima, leading to erroneous results. Measuring the initial reaction rate would provide a very good estimation of the kinetic constants K_m and v_{max}, which can be used as initial parameters in subsequent regression operations.

Question 7. In what cases must spectrophotometric analysis be avoided for the continuous determination of a reactant concentration?

Question 8. What two approaches can be used to evaluate the reaction kinetics?

4.5   Enzyme Inhibition

An inhibitor is a chemical substance that reduces the rate of an enzyme-catalyzed reaction. *Inhibitors* are associated in vivo with the regulation of enzymatic activities within metabolic networks, but they can generally be any type of natural/synthetic compound that binds with the enzyme resulting in a decrease in enzyme activity. Understanding and quantifying inhibition is central to establish a biotechnological process and overcome the loss of activity. A distinction is made between reversible and irreversible inhibition:

–   **Reversible inhibition**. The inhibitor binds reversibly to the enzyme; depending on the binding partner, different types of inhibition can be defined: (i) competitive, (ii) uncompetitive, (iii) noncompetitive, and (iv) mixed. Substrate excess inhibition and product inhibition are "special cases" of uncompetitive and competitive inhibition.
–   **Irreversible inhibition**. The inhibitor is usually a reactive molecule that irreversibly binds to the enzyme and causes its inactivation. Practically, this type of inhibition can be easily distinguished from reversible inhibition because the full enzyme activity cannot be restored after separation of the inhibitor (e.g., by ultrafiltration or ultracentrifugation) (Kot and Zaborska 2003).

Depending on the inhibition type, one or more kinetic parameters are changed. In the following, reversible inhibitions are described in detail because they are the most common. A general overview of reversible inhibitions by inhibitor I is given in Fig. 4.9. The fundamental inhibition types i–iii are shown in principle.

The case where the kinetic constant k_6 is nonzero in the overall scheme of Fig. 4.9 is per se a rare case. A detailed introduction can be found in Hans Bisswanger (2002). From the various reversible equilibrium, the following constants (with unit mM) can be determined:

$$K_S = \frac{k_{-1}}{k_1} = \frac{c_E\,c_S}{c_{ES}} \qquad (4.38)$$

$$K_{Si} = \frac{k_{-5}}{k_5} = \frac{c_{EI}\,c_S}{c_{ESI}} \qquad (4.39)$$

$$K_{Ik} = \frac{k_{-3}}{k_3} = \frac{c_E\,c_I}{c_{EI}} \qquad (4.40)$$

$$K_{Iu} = \frac{k_{-4}}{k_4} = \frac{c_{ES}\,c_I}{c_{ESI}} \qquad (4.41)$$

Here, "I" stands for "inhibitor," "k" for competitive inhibition, and "u" for uncompetitive inhibition. These equations are related to each other:

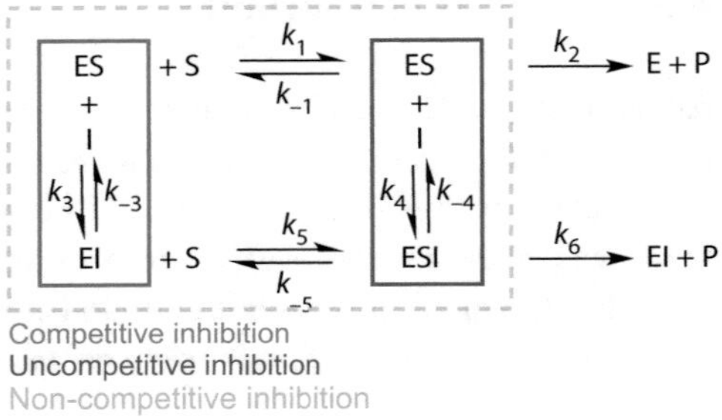

Fig. 4.9 Reversible inhibition by inhibitor I for a Uni-Uni mechanism

$$\frac{K_S}{K_{Si}} = \frac{K_{Ik}}{K_{Iu}} \tag{4.42}$$

The enzyme species are related to the following:

$$E_0 = E + ES + ESI + EI \tag{4.43}$$

Assuming steady state for E and ES and considering the rate of product formation, the following equations are obtained:

$$v = k_2\, c_{ES} + k_6\, c_{ESI} \tag{4.44}$$

$$v = \frac{\left(v_{max,1} + v_{max,2}\, \frac{c_I}{K_{Iu}}\right) c_S}{K_m \left(1 + \frac{c_I}{K_{Ik}}\right) + \left(1 + \frac{c_I}{K_{Iu}}\right) c_S} \tag{4.45}$$

If k_6 is zero, the inhibition is "complete", and it results in the following:

$$v = \frac{v_{max,1}\, c_S}{K_m \left(1 + \frac{c_I}{K_{Ik}}\right) + \left(1 + \frac{c_I}{K_{Iu}}\right) c_S} \tag{4.46}$$

In the following, we discuss the different types of reversible complete inhibition: competitive, uncompetitive, and noncompetitive. The main mathematical terms for each type are shown in Eq. 4.46. The term $\left(1 + \frac{c_I}{K_{Ik}}\right)$ represents the area highlighted in blue in Fig. 4.9, and the term $\left(1 + \frac{c_I}{K_{Iu}}\right)$ represents the area highlighted in red.

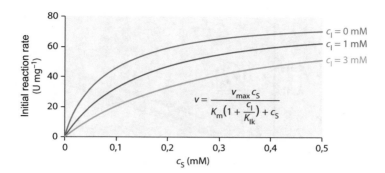

Fig. 4.10 Michaelis-Menten kinetics, graphical representation of competitive inhibition; $K_m = 0.07$ mM, $K_{Ik} = 1$ mM

4.5.1 Competitive Inhibition

The inhibitor I binds reversibly only to the free enzyme and forms an EI complex, which is quantified by the dissociation constant K_{Ik} (Eq. 4.40; blue color in Fig. 4.9). Since the binding does not affect the enzyme-substrate complex, v_{max} is not affected. However, the inhibitor alters the enzyme-substrate equilibrium, resulting in an increased k_m value (Fig. 4.10).

In the presence of a certain inhibitor concentration, the Michaelis-Menten graph shows a typical hyperbolic shape. However, this can only be described with a so-called "apparent" K_m value: $K_m^{app} = K_m \left(1 + \frac{c_I}{K_{Ik}}\right)$

The sum $\alpha = 1 + c_I/K_{Ik}$ is considered as a "correction factor" for competitive inhibition. The inhibition increases for (i) high c_I and (ii) strong EI binding. Linearization procedures can be applied to the kinetic equation from which the K_{Ik} can be calculated with a Hanes plot (Eq. 4.47 and Fig. 4.11).

$$\frac{c_S}{v} = \frac{K_m \left(1 + \frac{c_I}{K_{Ik}}\right)}{v_{max}} + \left(\frac{1}{v_{max}}\right) c_S \tag{4.47}$$

The constants can be determined from the linear plot by analyzing the slope and the x, y-axis intercepts as in the simpler Michaelis-Menten equation. In addition, K_{Ik} can be calculated from a secondary order obtained by plotting K_m^{app} of as a function of c_I. This leads to Fig. 4.12, when K_m^{app} is measured at different concentrations of I, and K_m are known.

Due to the high similarities between substrates and products in enzymatic reactions, a special case is competitive inhibition by the product. The equation can be easily obtained by replacing c_I with c_P and K_k with K_{IP}. To assess the significance of product inhibition, the principle of Lee and Whitesides can be used by looking at the ratio K_{IP}/K_m (Lee and

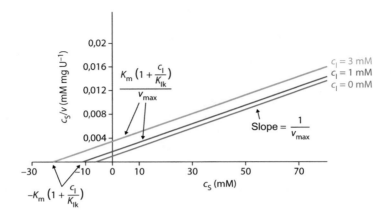

Fig. 4.11 Hanes linearization graph for competitive inhibition

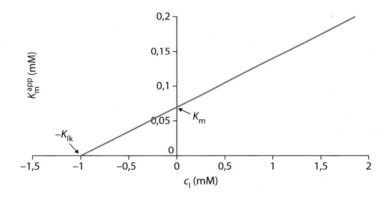

Fig. 4.12 Secondary graph for the calculation of K_{Ik}

Whitesides 1986). As a rule of thumb, a ratio of $K_{IP}/K_m > 1$ guarantees high conversions in a short time, whereas at $K_{IP}/K_m < 1$, the reaction does not proceed efficiently, and molecular biological optimization of the enzyme or, alternatively, a reaction engineering step such as in situ separation of the product must be considered.

For reactions containing two or more substrates/products, product inhibition is often an uncompetitive or noncompetitive one.

4.5.2 Uncompetitive Inhibition

In contrast to competitive inhibition, an uncompetitive inhibitor binds only to the enzyme-substrate complex (red color in Fig. 4.9). In this case, both K_m and v_{max} are altered by the inhibitor I (Fig. 4.13). The reduction of the K_m value does not mean that the enzyme has an

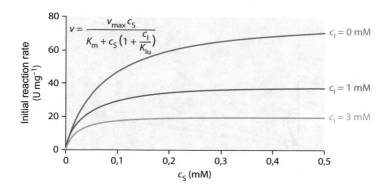

Fig. 4.13 Modification of the Michaelis-Menten graph in the presence of an uncompetitive inhibitor; $K_{\mathrm{m}} = 0.07$ mM, $K_{Iu} = 1$ mM, $v_{\max} = 80$ U mg^{-1}

increased affinity for the substrates but only that the ES equilibrium is shifted to the right by the formation of an additional complex.

The "correction factor" α is multiplied by c_s in this inhibition type. Also in this case, the kinetic constants can be calculated using linearization methods and secondary plots. The Hanes linearization is shown in Fig. 4.14.

$$\frac{c_S}{v} = \frac{K_{\mathrm{m}}}{v_{\max}} + \left(\frac{1 + \frac{c_I}{K_{Iu}}}{v_{\max}}\right) c_S \tag{4.48}$$

A special case of uncompetitive inhibition is substrate excess inhibition. This can be easily described by replacing the inhibitor I with an additional substrate molecule S (formation of a catalytically inactive ES_2 complex). The effect is more pronounced at higher substrate concentrations, with a decrease in the reaction rate being observed (Fig. 4.15).

The double reciprocal order and the Dixon graph are very useful in this context to calculate the uncompetitive substrate inhibition, as described in Eqs. 4.49 and 4.50:

$$\frac{1}{v} = \frac{\left(1 + \frac{c_S}{K_{Su}}\right)}{v_{\max}} + \frac{K_{\mathrm{m}}}{v_{\max}\, c_S} \tag{4.49}$$

$$\frac{1}{v} = \frac{1}{v_{\max}}\left(1 + \frac{K_{\mathrm{m}}}{c_S}\right) + \frac{c_S}{v_{\max}\, K_{Su}} \tag{4.50}$$

Using these two linearization graphs, it is possible to determine both K_{Su} and K_{m} as the x-axis intercept by drawing asymptotes of the curves at low and high substrate concentrations (Fig. 4.16).

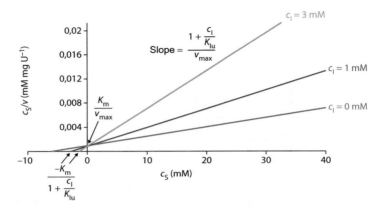

Fig. 4.14 Hanes linearization graph for uncompetitive inhibition

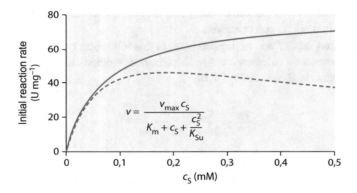

Fig. 4.15 Substrate excess inhibition; $K_{Su} = 0.5$ mM

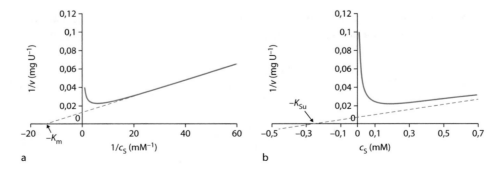

Fig. 4.16 Double reciprocal plot (**a**) and Dixon graph (**b**) for uncompetitive substrate inhibition (substrate excess inhibition)

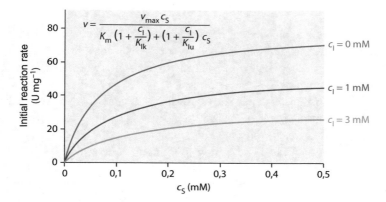

Fig. 4.17 Michaelis-Menten graph for noncompetitive inhibition; $K_{Ik} = 1$ mM, $K_{Iu} = 2$ mM

4.5.3   Noncompetitive Inhibition

The (mixed-)noncompetitive inhibition type (green color in Fig. 4.9) is found in multisubstrate/multiproduct systems and only very rarely in single-substrate reactions. The inhibitor usually increases the K_m value and reduces v_{max} (Fig. 4.17).

The kinetic equation contains two correction factors α and α', which multiply c_s and K_m in the denominator, respectively. For the Hanes linearization graph, this gives the following:

$$\frac{c_S}{v} = \frac{K_m \left(1 + \frac{c_I}{K_{Ik}}\right)}{v_{max}} + \left(\frac{1 + \frac{c_I}{K_{Iu}}}{v_{max}}\right) c_S \tag{4.51}$$

Question 9. If "I" is an enzyme inhibitor that binds reversibly to an enzyme at a site which is not its active site, what kind of inhibitor is *not* "I"?

Question 10. How could irreversible inhibition be distinguished from reversible inhibition in a certain enzymatic reaction?

Take-Home Message
- A catalyst is a substance that increases the rate of a reaction without modifying the standard Gibbs energy change in the reaction.
- Catalytic reactions can be homogeneous (catalyst and reagents are in a single phase) or heterogeneous (multiphase reactions that occur either in one phase or at the interface of the phases).

(continued)

- Depending on the preparation of the enzyme, soluble or immobilized, the resulting catalysis is run homogeneously or heterogeneous, respectively.
- Biocatalysis is the application of catalysts from nature (whole cells or enzymes) for chemical synthesis.
- Four kinetic parameters are key to evaluate fundamental enzyme-catalyzed reactions: Michaelis-Menten constant (K_m), maximum reaction rate (v_{max}), turnover frequency (k_{cat}), and specificity constant (K_a).
- Kinetic parameters can be determined from the Michaelis-Menten curve by linearization, numerical, and combined methods.
- Enzyme inhibitors can reduce or completely inhibit the enzymatic activity either reversibly or irreversibly.

Answer 1. The addition of a catalyst would not affect the free energy between the initial and final states of the system for A and B in their standard states; therefore, $\Delta_r G^0$ would not change.

Answer 2. A catalyst lowers the activation energy of a catalyzed reaction by providing a different mechanism for the reaction. It enhances the reaction rate and applies to the forward and reverse direction of the reaction.

Answer 3. Enzyme immobilization onto a solid matrix is an example of heterogenization of homogeneous catalysts. In this case, mass transfer limitations need to be considered.

Answer 4. Free enzymes are homogeneous catalysts, solubilized in a liquid phase, e.g., aqueous media. In the case of whole cells, these are heterogeneous biocatalysts.

Answer 5. Enzyme saturation means that in all the active sites of the enzyme, the substrate is bound. In this case, the resulting reaction rate just depends on the amount of enzyme in the reaction system. This happens at high substrate concentrations.

Answer 6. The simplest way to maintain a zero-order rate is to add reactant at a rate that exactly matches the depletion rate at high substrate concentrations.

Answer 7. If chromophores are not involved in the reaction or if two or more reaction components show absorbance at the same wavelength, then continuous spectrophotometric analysis must be replaced by discontinuous off-line analytics.

Answer 8. Reaction kinetics are tipically analyzed by the initial rate method, based on the measurement of the initial reaction rate at conversions lower than 5%, or the progress-curve method, which takes into account the concentration-time data obtained over the progression of the reaction.

Answer 9. "I" must not be a competitive inhibitor. Competitive inhibitors compete for the active site of the enzyme, preventing substrate binding. Uncompetitive and noncompetitive inhibitors do not compete with substrate for binding in the active site.

Answer 10. When irreversible inhibition takes places, the enzyme activity cannot be fully restored after removing the inhibitor from the reaction, which is not the case with reversible inhibition. Therefore, both types of inhibition could be distinguished by studying the enzyme activity after removing the inhibitor.

References

Al-Haque N, Santacoloma PA, Neto W, Tufvesson P, Gani R, Woodley JM. A robust methodology for kinetic model parameter estimation for biocatalytic reactions. Biotechnol Prog. 2012;28(5): 1186–96.

Bisswanger H. Enzyme kinetics: principles and methods. Weinheim: WILEY-VCH Verlag GmbH; 2002.

BRENDA Database. Technische Universität Braunschweig. www.brenda-enzymes.org. Accessed 3 May 2018

Briggs GE, Haldane JBS. A note on the kinetics of enzyme action. Biochem J. 1925;19(2):338–9.

Cleland W. The kinetics of enzyme-catalyzed reactions with two or more substrates or products. Biochim Biophys Acta. 1963;67:188–96.

Drauz K, Waldmann H. Enzyme catalysis in organic synthesis. 2nd ed. Weinheim: Wiley-VCH; 2002.

Eadie GS. The inhibition of cholinesterase by physostigmine and prostigmine. J Biol Chem. 1942;146:85–93.

Fischer F. Influence of configuration on the action of enzymes. Ber Dtsch Chem Ges. 1894;27(3): 2985–93.

Hanes CS. Studies on plant amylases. I. The effect of starch condenstation upon the velocity of hydrolysis by the amylase of germinated barley. Biochem J. 1932;26(5):1406–21.

Hofstee BHJ. Non-inverted versus inverted plots in enzyme kinetics. Nature. 1959;184(4695): 1296–8.

King EL, Altman C. A schematic method of deriving the rate laws for enzyme catalyzed reactions. J Phys Chem. 1956;60(10):1375–8.

Koshland DE Jr. Application of a theory of enzyme specificity to protein synthesis. Proc Natl Acad Sci U S A. 1958;44(2):98–104.

Kot M, Zaborska W. Irreversible inhibition of jack bean urease by pyrocatechol. J Enzyme Inhib Med Chem. 2003;18(5):413–7.

Lee LG, Whitesides GM. Preparation of optically active 1,2-diols and a-hydroxy ketones using glycerol dehydrogenase as catalyst: limits to enzyme-catalyzed, synthesis due to non-competitive and mixed inhibition. J Org Chem. 1986;51(1):25–36.

Leskovac V. Comprehensive enzyme kinetics. Springer; 2003.

Lineweaver H, Burk D. The determination of enzyme dissociation constant. J Am Chem Soc. 1934;3: 658–66.

Michaelis L, Menten ML. Kinetic of invertin action. Biochem J. 1913;49:333–69.

Yuryev R, Liese A. Biocatalysis: the outcast. ChemCatChem. 2010;2(1):103–1.

Further Reading

Cornish-Bowden A. Fundamentals of enzyme kinetics. 3rd ed. London: Portland Press; 2004.

Segel I. Enzyme kinetics: behavior and analysis of rapid equilibrium and steady state enzyme systems. New ed. New York: Wiley; 1993.

Enzyme Reactors and Process Control

5

Daniel Ohde, Steffen Kühn, and Andreas Liese

What You Will Learn in This Chapter?

The economic application of enzymes requires selection of the most appropriate reactor that might even lessen the impact of any inhibition or deactivation of the enzyme. When considering the basic reactor types in batch or continuous operation, general rules can already be derived making use of idealized, mathematical descriptions. This will be the first step in enabling production on an industrial scale. An insight into the calculation of different reactor types is given on the basis of ideal conditions, in order to be able to determine the most important parameters for evaluating an enzyme-catalyzed process and to distinguish between the basic reactors. In addition, selection of the most appropriate reactor in the case of inhibition phenomena and the different process control strategies are discussed. Examples are presented at the end of the chapter to show the industrial relevance of enzyme-catalyzed reactions, whereby the entirety of thermodynamic and kinetic parameters must always be taken into account for the economic viability of an industrial process.

Since the discovery of the DNA double helix structure by Watson and Crick in 1953, the application of enzymes in industrial biotechnology for the synthesis of a variety of

D. Ohde · A. Liese (✉)
Institute of Technical Biocatalysis, Hamburg University of Technology, Hamburg, Germany
e-mail: daniel.ohde@tuhh.de; liese@tuhh.de

S. Kühn
Evonik Operations GmbH, Essen, Germany
e-mail: steffen.kuehn@evonik.com

© The Author(s), under exclusive license to Springer Nature Switzerland AG 2024
K.-E. Jaeger et al. (eds.), *Introduction to Enzyme Technology*, Learning Materials in Biosciences,
https://doi.org/10.1007/978-3-031-42999-6_5

chemical and biological products has become indispensable. Over the years, three main ways have emerged to ensure and optimize the process integration of enzymes. On the one hand, the optimization of the enzymes used has been made possible at the structural level. Here, strategies for overexpression of genes and the development as well as design of novel protein properties by controlled switching on and off of different gene sequences play a central role (Woodley et al. 2013). On the other hand, the accessibility of a wide range of substrates by adapting process control and reaction conditions has been considered. Due to the limited capacity of raw materials, the development of environmentally friendly and sustainable process strategies has become the third focus. A central aspect of the industrial applicability of enzymes is a basic understanding of the possible reactor concepts with their respective advantages and disadvantages. Whether the economic viability of the developed process is given must be examined individually for each process. In order to realize the main advantage of the high regio- and enantioselectivity of enzymes compared to inorganic catalysts in the process, the choice of the reactor is essential. A clear distinction is made between reactors considered in an idealized way and the consideration of real phenomena. In the idealized operation, the operating modes can first be distinguished from each other, and parameters for the characterization of process-relevant parameters can be identified. In order to adapt the reactor operation to real conditions, it is decisive, among other things, in which form the biocatalyst is applied in the process. As homogeneously distributed components in the reaction medium, free or isolated biocatalysts have fundamentally different properties with regard to mass transport compared to immobilized biocatalysts (Chap. 11). It must also be taken into account whether the biocatalyst is inhibited by products formed or present substrates. Possible parallel or subsequent reactions also play a role for reactor selection. Furthermore, it is also important to differentiate between thermo-dynamic and kinetic phenomena in the consideration. While the kinetic consideration deals with the reaction rate of the catalyzed reaction, the thermodynamic consideration provides information, for example, about the maximum possible conversion and possibilities for shifting the reaction equilibrium. Building on the structural and kinetic description of enzymes (Chaps. 2 and 3), this chapter describes the basic process engineering aspects mathematically.

5.1 Parameters for the Description of Ideal Reactors

Regardless of the type of reactor used, the evaluation of an enzyme-catalyzed process is carried out via a large number of parameters. The seven most important parameters include conversion, selectivity, yield, catalytic activity (turnover frequency), productivity, space-time yield, and total turnover number. The parameters are first presented before the individual reactor concepts are discussed in more detail.

For this purpose, a simple irreversible reaction is considered as an example (Eq. 5.1):

$$|\nu_S|\, S \rightarrow |\nu_P|\, P + |\nu_{NP}|\, NP \qquad (5.1)$$

S symbolizes the substrate, P the desired product, and NP the undesired by-product. The associated stoichiometric coefficients are labeled with $|\nu_i|$, where i describes the index of the respective component. Based on Eq. 5.1, the parameters to be discussed can be derived.

The conversion χ describes the amount of S reacted in relation to the amount of S_0 used at the beginning of the reaction. For the conversion, it does not matter whether the desired product P or the undesired by-product NP is synthesized. The mathematical description of the conversion is given in Eq. 5.2:

$$\chi = \frac{n_{S,0} - n_S}{n_{S,0}} \qquad (5.2)$$

$n_{S,0}$: initial amount of substance of component S at the beginning of the reaction at time t_0

n_S: amount of substance of substrate S at time $t > t_0$

The selectivity σ denotes the amount of substance of the desired product P formed per amount of substance of S consumed (Eq. 5.3).

$$\sigma = \frac{n_P - n_{P,0}}{n_{S,0} - n_S} \cdot \frac{|\nu_S|}{|\nu_P|} \qquad (5.3)$$

$n_{P,0}$: initial amount of substance of the product P at the beginning of the reaction at time t_0

n_P: amount of substance of the product P at time $t > t_0$

The yield η of the process can be determined from the product of conversion and selectivity (Eq. 5.4). The yield expresses the amount of P formed in relation to the amount of S used.

$$\eta = \chi \cdot \sigma = \frac{n_P - n_{P,0}}{n_{S,0}} \cdot \frac{|\nu_S|}{|\nu_P|} \qquad (5.4)$$

An additional differentiation is made between the analytical (calculated here) and isolated yield. The analytical yield refers to the yield in the considered reaction step alone. In this case, the work-up or separation of the product from the solvent is neglected. The isolated yield refers to the total yield including the work-up steps. If several sequential processing steps are required in the examined process, the total yield is calculated multi-plicatively using each individual step's yield. In principle, the aim is to achieve the highest possible yield in the process in order to improve the economic efficiency of the process.

There is a linear relationship between the process parameter conversion, selectivity, and yield, which shows increased yields with increasing selectivity. The selectivity acts as a slope, which influences the yield.

In the case of enzyme-catalyzed reactions, the catalytic activity is expressed in units $(1\ U = 1\ \mu mol\ min^{-1})$, which can be determined either by substrate consumption or by product formation under constant and defined conditions. Therefore, in order to give a comparison of catalytic activities for a given catalyst, the defined reaction conditions such as temperature, pH, ion concentration, etc. must be maintained. In addition to the catalytic activity, the *turnover frequency* (TOF) is used as a parameter for evaluation. It indicates how efficiently the desired reaction is catalyzed. For this purpose, the time-dependent substrate turnover is mathematically considered in relation to the amount of substance of the catalyst (Eq. 5.5).

$$TOF = \frac{n_S}{t \cdot n_{Cat}} \tag{5.5}$$

n_S: amount of substance of the substrate converted (mol)
n_{Cat}: amount of substance of the catalyst used (mol)
t: reaction time (s)

With the catalytic activity, a comparison of different catalysts is possible without considering the mass of the catalyst. In enzyme-catalyzed reactions, if Michaelis-Menten kinetics are present, TOF can be described as k_{cat}^{-1}, which is equivalent to the rate-determining step for the formation of the product from the enzyme-substrate complex (see Chap. 4).

The catalytic activity, TOF, is to be distinguished from the catalytic productivity (turnover number; TON). TON is the ratio of the amount of substance of product formed to the amount of catalyst used. Alternatively, the mass of product formed in relation to the mass of catalyst used can be considered to describe the TON (Eq. 5.6). This is true for one batch conversion. If the overall amount of product is addressed, which can be generated until the catalyst is deactivated, e.g., in a continuously operated process or via repetitive or fed-batch operations, TON becomes TTN = *total turnover number*.

$$TON = \frac{n_P}{n_{Cat}} = \frac{m_P}{m_{Cat}} \cdot \frac{M_{Cat}}{M_P} \tag{5.6}$$

n_{Cat}: amount of substance of the catalyst used (mol)
m_p: mass of the product formed (g)
m_{Cat}: mass of the catalyst used (g)
M_p: molecular weight of the product (g mol^{-1})
M_{Cat}: molecular weight of the catalyst (g mol^{-1})

TON does not contain any time dependency and serves as a dimensionless characteristic number, which describes the maximum possible amount of material or mass of product that can be obtained for a given amount of substance or mass of the catalyst. Higher TON for a specific reaction makes the biocatalyst more interesting to be used for an industrial process, since the production costs are reduced by longer operating times of the biocatalyst. In

Table 5.1 Catalytic productivity (*turnover number*; TON) for different products with associated ranges for product costs. (According to Tufvesson et al. 2011)

Product	Product costs [€ kg⁻¹]	Range of catalytic productivity (=TON) [kg kg⁻¹]		
		Cell dry mass	Isolated enzyme	Immobilized enzyme
Pharma	>100	10–35	250	50–100
Fine chemicals	>15	70–230	670–1700	330–670
Speciality chemicals	0.25	140–400	1000–4000	400–2000
Bulk	0.05	700–2000	5000–20,000	2000–10,000

Table 5.1, TONs are given for different product categories from the pharmaceutical sector to bulk chemicals for processes using whole cells, isolated enzymes, and immobilized enzymes (Tufvesson et al. 2011). TON should take values from at least 10 when using whole cells for the synthesis of pharmaceutical products to 20,000 when using isolated enzymes for the production of bulk products. In such a comparison, on the other hand, the different reaction conditions or catalyzed reactions must always be taken into account as they can differ significantly from application to application.

In addition to the catalytic productivity (TON), the space-time yield (STY) is an important parameter for a process (Eq. 5.7). In contrast to the catalytic productivity, it is not the amount of catalyst used that is considered here, but the reaction volume. In addition, the STY is time-dependent as the residence time in continuously operated reactors is considered, whereas the TON is time-independent. The STY describes the amount of product formed in the continuous process, which is present in the reaction volume of the reactor.

$$STY = \frac{m_P}{\tau \cdot V_R} \tag{5.7}$$

τ: residence time or reaction time (h)
V_R: reaction volume used (L)

Since product losses are often to be expected in the subsequent product purification, the STY should always be as high as possible. Depending on the product costs, values for the STY in the range of >100 g L⁻¹ day⁻¹ for high-priced pharmaceutical products to values >500 g L⁻¹ day⁻¹ for bulk chemicals are required (Liese et al. 2006; Yuryev et al. 2011).

For applications in the pharmaceutical sector, industry depends on reliable enzyme-catalyzed synthesis of enantiomerically pure products, since chemical catalysts often cannot be used for enantioselective synthesis compared to enzymes. The increased interest in enantiomerically pure products is related to the different properties of enantiomers that cannot be converted into each other. In order to produce one enantiomer selectively, catalysts are needed that specifically form only the desired enantiomer. The selective formation of a chiral molecule is called asymmetric synthesis. A successful example of asymmetric synthesis using biocatalysts is the use of alcohol dehydrogenases, which synthesize chiral alcohols, diols, or hydroxy esters by reduction of prochiral ketones (Daußmann et al. 2006). Another possibility for the synthesis of chiral alcohols is the addition of hydrocyanic acid to aldehydes or ketones using oxynitrilases or hydroxynitrile lyases as biocatalysts, respectively (Daußmann et al. 2006). In addition, a kinetic racemate resolution can be performed to obtain a desired chiral product. This case will be discussed in more detail below. The difference in a kinetic resolution of a racemate compared to the reaction in Eq. 5.1 is caused by different reaction rates in the parallel reaction of the two substrate enantiomers. This feature results from the composition of the substrate, which consists of an equimolar mixture of two enantiomers. The equimolar substrate mixture is referred to as the racemate. Equation 5.8 shows the irreversible case of a kinetic racemate cleavage in which the *(S)*-enantiomer ((S)-S) on the one hand and the *(R)*-enantiomer ((R)-S) on the other hand can react. It is crucial for successful kinetic racemate cleavage that one enantiomer is reacting faster than the other making it necessary to have a catalyst with high selectivity. In this case, high selectivity with respect to the *(R)*-enantiomer is assumed, resulting in $k_1 \gg k_2$. The target product of the kinetic racemate cleavage is either the slower reacting enantiomer of the racemate ((S)-S) or the synthesis product of the faster reacting enantiomer ((R)-P).

$$(R) - S \xrightarrow{k_1} (R) - P$$
$$(S) - S \xrightarrow{k_2} (S) - P \tag{5.8}$$

k_1: reaction rate constant of the reacting *(R)*-enantiomer ((R)-S) (mol s^{-1})
k_2: reaction rate constant of the reacting *(S)*-enantiomer ((S)-S) (mol s^{-1})

The greatest process technological challenge of a kinetic racemate cleavage is to overcome the maximum possible yield of $\eta = 0.5$, which is the limit given by the racemate used as substrate. Moreover, the maximum possible yield is only achievable if the biocatalyst used is highly selective in converting only one enantiomer of the racemate. In order to increase the maximum possible yield, a dynamic kinetic racemate resolution (DKR) can be used. In this process, the remaining enantiomeric mixture of *(R)*-S and *(S)*-S is racemized in situ by means of a second, chemical or biocatalytic catalyst, whereby yields of $\eta > 0.5$ can be achieved. A prerequisite for carrying out a DKR in the presented example case is the interest in the synthesis product of the preferentially proceeding reaction to *(R)*-P. The use of DKR is generally feasible as soon as, in addition to a high selectivity for the kinetic racemate cleavage, at least a tenfold higher reaction rate for the racemization reaction is present (Martin-Matute and Bäckvall 2007).

To describe the racemate cleavage, specific parameters are introduced in addition to those presented previously. These consist of the enantiomeric excess *(ee)* and the enantioselectivity (E). Enantiomeric excess is the excess of an enantiomer present relative to the total amount of both enantiomers (Eq. 5.9).

$$ee_S = \frac{n_{(S)-S} - n_{(R)-S}}{n_{(S)-S} + n_{(R)-S}}$$

$$ee_P = \frac{n_{(R)-P} - n_{(S)-P}}{n_{(R)-P} + n_{(S)-P}} \tag{5.9}$$

ee_S: enantiomeric excess with respect to the substrate $(-)$
ee_P: enantiomeric excess with respect to the product $(-)$

According to Eq. 5.9, the racemic mixture has an enantiomeric excess of $ee = 0$. If the slower reacting substrate or product enantiomer is pure in a kinetic racemate cleavage, the enantiomeric excess is $ee = 1$. The ee can therefore be used to assess the reaction progress with respect to the desired target enantiomer, which can be the substrate and/or the product. In contrast to the asymmetric synthesis, the conversion χ of the reaction must also be specified for the evaluation in order to take the reaction progress into account at a given ee.

The enantioselectivity E describes the ratio of the reaction rate constants of the two enantiomers to each other but can also be determined for an idealized irreversible reaction with the aid of the conversion and ee_P or ee_S (Eq. 5.10; Faber 2011). A highly selective reaction is characterized by high enantioselectivities such as $E > 100$. At $E = 100$, one molecule of the competing enantiomer is converted for every hundred molecules of the preferred enantiomer. The higher the enantioselectivity of a reaction, the more selective is the conversion of the faster reacting substrate. For industry, reactions that have at least an $E = 35$ are interesting for process development (Liese and Kragl 2013).

$$E = \frac{k_1}{k_2} = \frac{\ln[1 - \chi \cdot (1 + ee_P)]}{\ln[1 - \chi \cdot (1 - ee_P)]} = \frac{\ln[(1 - \chi) \cdot (1 - ee_s)]}{\ln[(1 - \chi) \cdot (1 + ee_s)]} \tag{5.10}$$

The influence of enantioselectivity on the enantiomeric excess and the conversion of the reaction is an important factor in the evaluation of a kinetic and a dynamic kinetic racemate cleavage and is illustrated in Fig. 5.1 for a batch reactor. First, the case of kinetic racemate cleavage is discussed (Fig. 5.1a). The higher the enantioselectivity, the faster a high ee_S value is obtained or the longer a high ee_P is maintained with respect to the conversion. In the ideal case ($E =$ ideal), the optically pure substrate enantiomer ($ee_S = 1$) can be obtained in the kinetic racemate cleavage with a highly selective reaction at the conversion of $\chi = 50\%$. Accordingly, only the unreacted substrate enantiomer is still present. At the same time, under highly selective conditions, if the optically pure product enantiomer is of interest, the maximum yield with $ee_P = 1$ can also be obtained at $\chi = 50\%$. This behavior can also be explained by the high enantioselectivity, since only one synthesis product is obtained from the preferentially proceeding reaction. At reduced enantioselectivities ($E = 35$), a higher conversion (>0.5) is required to achieve optically pure substrate enantiomer. However, the ee_P is also reduced from the competing reaction taking place. If a low enantioselectivity reaction ($E = 10$) is considered, the conversion to reach the optically pure substrate enantiomer shifts further to higher values, and a lower ee_P is possible for the product enantiomer from the beginning. In comparison, Fig. 5.1b shows

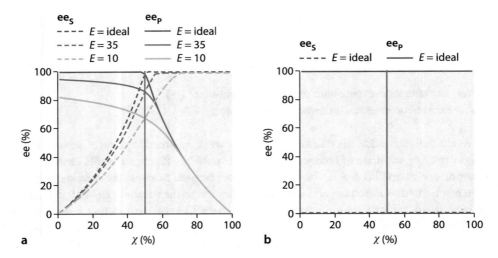

Fig. 5.1 Enantioselectivity (E) for different dependencies of the conversion (χ) and the enantiomeric excess *(ee)* in a batch reactor

the ideal course for a dynamic kinetic racemate cleavage, wherein the advantage of increased conversions due to the racemization reaction can be seen. In the ideal case shown, it is therefore possible to achieve 100% conversion and 100% yield of the optically pure product enantiomer ($ee_P = 1$).

With the parameters explained, the basis for understanding the evaluation of a process is laid. In the following, the basic reactor types are presented and compared on the basis of the explained parameters.

Questions
1. What are the seven most important parameters to evaluate an enzyme-catalyzed process?
2. Does TOF or TON consider the time-dependent substrate conversion?
3. Your product mixture contains 40% of the (S)- and 60% of the (R)-enantiomer. What is the value of the enantiomeric excess for the desired (R)-enantiomer?

5.2 Types of Ideal Enzyme Reactors

Three idealized reactor types are used to carry out enzyme-catalyzed reactions. Idealized conditions include having a complete mixing of the reactor without taking into account real phenomena such as a diffusion limitation or a deactivation of the catalyst. From the idealized basic reactor types, more complex reactor modifications and modified modes of operation can be derived to adapt to more complex reaction and process cases. The basic reactor types differ from one another according to their respective mode of operation:

1. Discontinuous stirred tank reactor (STR) or ideally mixed batch reactor
2. Continuously operated stirred tank reactor (CSTR); ideally mixed
3. Plug flow reactor (PFR) or flow tube reactor with ideal plug flow

In discontinuous batch operation (1), the amount of reaction medium in the reactor is kept constant after the start time $t = 0$, i.e., no further addition or removal of reactants is possible. Therefore, the ideal batch reactor represents a closed system. As soon as the amount of substrate has been reacted to the desired conversion of the reaction, the reaction must be stopped, and the reactor emptied and filled with new reactants before a new reaction can take place. In an ideal operation, there is homogeneous mixing of the reactants, whereby the parameters of temperature, system pressure, and concentrations of the reactants are constant at any given time at any location in the reactor. Time-dependently, a decrease in the substrate concentration is observed from t_0 to $t_2 > t_0$ with a simultaneous increase in the product concentration. The concentration curves of substrate and product as a function of time and location are listed in Fig. 5.2. Due to the discontinuous mode of operation, it is not possible to achieve a steady-state operating point. In industry, a frequently used modification of the batch reactor is the fed-batch reactor, which operates in a semicontinuous mode. The main feature of this mode of operation is the time-dependent addition of reactants after the time $t = 0$ in a feed stream containing substrate (feeding stream). The difference between such a fed-batch reactor and a batch reactor lies in the change of the reaction volume over time, while the reactor volume stays constant. In the fed-batch operation, a semi-open system is created, since an exchange of reactants takes place across the system boundary in one direction. Although semicontinuous operation also requires a termination of the reaction, because no reactants are removed during the reaction, higher product concentrations can be achieved in the reactor by having a constant substrate supply. However, a fed-batch operation is limited by the volume of the reactor, i.e., it is not possible to feed as much substrate as might be desired. Furthermore, the time-dependent volume increase caused by having a feed stream changes the concentrations of all reactants in the reaction volume as well. In both batch and fed-batch operation, depending on the enzyme stability, as many reaction cycles as possible are carried out without enzyme exchanges in order to be able to operate the process economically. This mode of operation is called repetitive batch which makes a repeated use of the enzyme by retention. Between each cycle, the reactor is first emptied, except of the enzyme, and the reactor contents are transferred for product purification. The reactor is then refilled with substrate solution for repeated product synthesis.

The continuous mode of operation in the CSTR (2) is characterized by a constant supply and removal of reactants in the course of the reaction. This distinguishes this operation from the fed-batch operation. Accordingly, in an ideal operation, the amount of reaction medium in the reactor is kept constant, and there is no need to stop the reaction with high enzyme stability over a long period of time. Ideally, due to homogeneous mixing, an abrupt change in the concentrations occurs immediately after the start of continuous dosing of the reactants. From the abrupt change onward, the concentrations everywhere in the reactor are

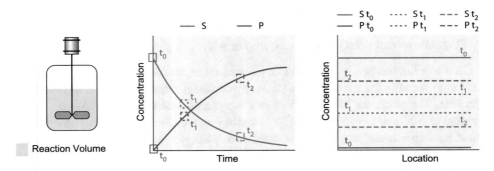

Fig. 5.2 Concentration profiles in batch operation as a function of time *(t)* and location

locally and temporally equal to the concentrations of the reactor outlet (Fig. 5.3). This operating condition is referred to as operation under outflow conditions, i.e., in contrast to batch and fed-batch operation, a stationary operating point or a steady-state equilibrium is reached. An important parameter in CSTR is the time over which a molecule dissolved in the reaction volume remains in the reactor. This time is called residence time and depends not only on the reaction volume but also on the volumetric flow rate of the inlet and outlet (Eq. 5.11).

$$\tau = \frac{V_R}{F} \tag{5.11}$$

τ: residence time (min)
V_R: reaction volume (m^3)
F: flow rate in inlet and outlet (m^3 min^{-1})

The residence time of each added molecule until it leaves the reactor is, under ideal conditions, the same in the CSTR, since the flow velocity of the feed and the effluent are the same.

In the PFR (3), no mixing takes place in the ideal case in contrast to the batch, fed-batch, and CSTR operation. Due to a prevailing plug flow in an ideal operation, i.e., neglecting axial diffusion and complete mixing in the radial direction, each molecule moves through the flow tube at the same velocity. Just as in the CSTR, the flow tube has a feed and discharge (= feed, F), whereby the residence time of all molecules in the reactor and the reactor volume is constant. The concentration curves, however, show the characteristic difference between both reactors. Accordingly, there is no operation under outflow conditions. This difference results from the different configuration of the PFR causing the substrate concentration to decrease with the spatial coordinate, i.e., along the length of the flow tube reactor from z_0 to $z_2 > z_0$ (Fig. 5.4). At the inlet of the PFR, the initial substrate concentration is present, which reduces due to the conversion to the product by the prevailing enzyme-catalyzed reaction until it exits the reactor. When time is considered, constant concentrations are observed in the flow tube reactor at each location. This

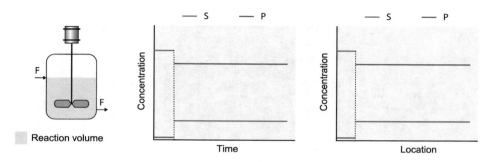

Fig. 5.3 Concentration profiles in the continuous stirred tank reactor (CSTR) as a function of time and location

behavior can be explained by means of differential consideration of the spatial coordinate. If we imagine a batch reactor with differential small volume (dV_R) passing the flow tube from inlet to outlet along the spatial coordinate, the substrate concentration of the differential batch reactor in the PFR decreases. On the other hand, due to the ideal behavior, the prevailing concentration at a given residence time is constant. Accordingly, the temporal and spatial behaviors of the batch reactor and the PFR are reversed. Therefore, from a reaction engineering point of view, the PFR is the continuous version of the batch reactor and not the CSTR. A well-established modification of the PFR is the packed bed reactor (PBR), in which immobilized catalyst preparations are used as a densely packed bed. As long as low flow velocities through the packed bed and negligible mass transfer phenomena are present, the PBR can be considered idealized as a PFR.

Questions
1. What are the fundamental three reactor types?
2. Under ideal conditions, which one of the basic reactor types that operate in a continuous mode shows the same performance as the batch reactor from a reaction engineering viewpoint?

5.3   Mathematical Balancing of Ideal Reactors

Based on the presented process-relevant parameters and the reactor types for the description of enzyme-catalyzed reactions, the balancing of the different reactor types is discussed in the following. The balancing forms the basis for understanding the processes in the respective reactor system and makes it possible to calculate the time courses of all concentrations or quantities of substances. A general balance equation can be described independently of the reactor type as an accumulation term within a defined balance limit. The system boundaries describe the range in which the change of the concentrations or substance quantities of interest takes place. The accumulation term is to be understood as

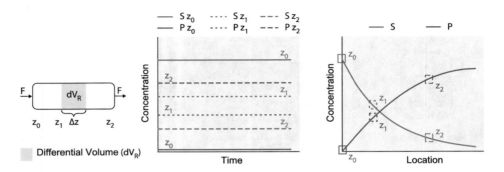

Fig. 5.4 Concentration profiles in the flow tube reactor (PFR) as a function of time and location *(z)*

the time-dependent change of concentrations, substance quantities, and masses in a defined volume element. All occurring factors are considered, which can lead to a change of the accumulation term. These phenomena consist of a convective, a diffusive, and a reactive term (Eq. 5.12).

$$\text{accumulation} = \text{convection} + \text{diffusion} + \text{reaction} \tag{5.12}$$

In Eq. 5.12, the convective component can be described as the sum of the change in reactants in the selected volume element caused by flow. Diffusive effects describe the sum of an exchange of reactants between adjacent volume elements. The reaction initiated by the biocatalyst is considered within the reactive term. The description of the kinetic parameters of an enzyme-catalyzed reaction is explained in more detail in Chap. 4 and is reflected in the reactive term of the balance. In the reactive term, it is therefore possible to take into account the kinetics, existing inhibitions, as well as enzyme deactivations of the reaction under consideration. For the consideration of ideal reactors, no energetic changes such as the heat supply/dissipation or the enthalpy change in exothermic reactions are included here in order to describe the basic phenomena first. However, as soon as the reaction is carried out at higher or lower temperatures, a consideration of the energy balances in the reactor is required. Likewise, the energy input of the stirrer must be considered in the energy balances.

In the following, the changes in mass and concentration of the substrates and products involved in the reaction are considered in more detail. For each component present in the reactor, a mass balance equation (Eq. 5.13) can be formulated on the basis of Eq. 5.12:

$$\frac{dn_j}{dt} = \sum \dot{n}_{j,\text{in}} - \sum \dot{n}_{j,\text{out}} + R_j \cdot V_R \tag{5.13}$$

$\frac{dn_j}{dt}$: change over time of the accumulated total amount of reactant j (mol min^{-1})

$\sum \dot{n}_{j,\text{in}}$: mole flow of component j supplied by convection or diffusion (mol min^{-1})

$\sum \dot{n}_{j,\text{out}}$: mole flow of component j discharged by convection or diffusion (mol min^{-1})
R_j: reaction rate related to the amount of substance (mol min^{-1} m^{-3})
V_R: reaction volume (m^3)

For the balancing of reactions carried out in highly diluted solutions, a constant density in the reactor is assumed for simplicity. In this case, the concentration of the reactants under consideration is much smaller than the concentration of the solvent used (Eq. 5.14).

$$c_S \ll c_{\text{solvent}} \tag{5.14}$$

c_S: substrate concentration (mol m^{-3})
c_{solvent}: concentration of the solvent used in the reactor (mol m^{-3})

This assumption is valid, for example, for synthesis in dilute aqueous or buffered reaction mixtures, but not in solvent-free reactions. By neglecting a change in density, reaction volume is constant making it possible to express the individual components of the mass balance equation (Eq. 5.13) in concentrations (Eq. 5.15).

$$\frac{dc_j}{dt} = \frac{\sum \dot{n}_{j,\text{in}} - \sum \dot{n}_{j,\text{out}}}{V_R} + R_j \tag{5.15}$$

$\frac{dc_j}{dt}$: change in concentration of component j over time (mol m^{-3})
c_j: concentration of component j (mol m^{-3})

Using the example of an irreversible one-substrate reaction with $c_j = c_S$ and no inhibition, the reactive term $R_j = R_S$ of the substrate decrease is defined by the Michaelis-Menten kinetics (Chap. 4) (Eq. 5.16). Due to the decreasing substrate concentration, the reactive term decreases and gets negative.

$$R_S = -v\,(c_E, c_S) = -c_E \cdot \frac{a_E \cdot c_S}{K_m + c_S} \tag{5.16}$$

R_S: substrate related reaction rate (mol min^{-1} m^{-3})
c_E: enzyme concentration (g m^{-3})
a_E: mass-specific enzyme activity (mol min^{-1} g^{-1})
K_m: Michaelis-Menten constant (mol m^{-3})

Considering that the reaction volume stays constant, the dependence on the substrate concentration can be expressed by the conversion χ (Eq. 5.2) and the initial substrate concentration $c_{S,\,0}$ (Eq. 5.17).

$$(c_E, \chi) = -c_E \cdot \frac{a_E \cdot c_{S,0} \cdot (1 - \chi)}{K_m + c_{S,0} \cdot (1 - \chi)} \tag{5.17}$$

$c_{S,\,0}$: initial substrate concentration (mol m^{-3})

Based on the derived reaction term (Eq. 5.17), the conversions within the various reactor concepts can be described mathematically. The presented balancing can be extended by taking into account the inhibition kinetics from Chap. 4, which will be omitted here.

5.3.1 Balancing of an Ideally Mixed Batch Reactor (STR)

For the balancing of the conversion in a batch reactor, only the accumulation and reaction terms need to be considered in the balance equation (Eq. 5.13) when there is a constant balance volume because no convective and diffusive flows occur. No component is added to or removed from the reactor during the reaction, and there is no exchange of reactants with the environment because it is a closed stirred tank. The balance space in the ideal stirred tank is always taken to be the reaction volume, which is the free liquid volume, in contrast to the empty reactor volume, which might include also any carriers like immobilized enzyme to be added, e.g., by the latter point, the reaction volume in that reactor volume is reduced. Since simple Michaelis-Menten kinetics is assumed here (Eq. 5.17), the balance to describe the substrate concentration in the batch reactor can be expressed in the following equation (Eq. 5.18):

$$\frac{dc_S}{dt} = -c_{S,0}\frac{dX}{dt} = -v\left(c_E,\chi\right) \tag{5.18}$$

By separating the variables, the integral expression in Eq. 5.19 is generated.

$$\int_0^\chi \frac{d\chi}{v\left(c_E,\chi\right)} = \int_0^t \frac{dt}{c_{S,0}} \tag{5.19}$$

In the idealized case as considered here, a constant enzyme stability over the entire course of the reaction is assumed for simplicity. There is no deactivation of the enzyme, which is why it is only a function of the conversion and can be integrated directly. Taking into account the Michaelis-Menten kinetics, Eq. 5.20 is the result for the batch reactor.

$$\frac{c_{S,0}}{K_m}\cdot\chi - \ln\cdot\left(1-\chi\right) = \frac{v_{max}}{K_m}\cdot t \tag{5.20}$$

v_{max}: maximum reaction rate (mol min^{-1})

Equation 5.20 shows the time-dependent conversion of an idealized one-substrate reaction as a function of the initial substrate concentration and the enzyme-specific constants in the batch reactor. Within the presented balance equation, it is possible to describe a dimensionless operating time. Often, the applied concentration of the enzyme is a factor that is relevant in the evaluation of the process to achieve a specified conversion in a specified reaction time. Using the concentration of the enzyme and the mass-specific activity, the dimensionless operating time from Eq. 5.21 can be transformed, allowing the catalyst concentration to be varied directly as a parameter.

$$\frac{v_{max}}{K_m} \cdot t = \frac{c_E \cdot a_E}{K_m} \cdot t \tag{5.21}$$

c_E: enzyme concentration (g m^{-3})
a_E: mass-specific enzyme activity (mol min^{-1} g^{-1})

Using Eq. 5.21 in combination with Eq. 5.20, it is possible to identify the factors influencing the conversion in the batch reactor. For the operation of the batch reactor, the ratio of the initial substrate concentration $c_{S,0}$ to the K_m value of the enzyme, the enzyme concentration c_E, and the operating time t are crucial. The three parameters described can be varied differently depending on the application in order to achieve a certain conversion in the batch reactor. On the one hand, low enzyme concentrations require a longer period of time to achieve a defined conversion, and vice versa. On the other hand, high conversions are achieved more rapidly with low ratios of the initial substrate concentration to the K_m value of the enzyme. However, when high $c_{S,0} \cdot K_m^{-1}$-values are used, the catalyst concentration must be increased to ensure comparable reaction times relative to low $c_{S,0} \cdot K_m^{-1}$-values. Accordingly, the operation of the reactor can be adjusted depending on the parameters shown. The influence of the ratio of the initial substrate concentration $c_{S,0}$ to the K_m value of the enzyme on the conversion and the operating time is illustrated in Fig. 5.5.

The derivation of a dimensionless balance equation shown above is performed analogously for the two continuous basic reactor types in the following. The main differences in the balance equations result from the process designs discussed in Sect. 5.2.

5.3.2 Balancing of a Continuously Operated Ideal Flow Tube Reactor (PFR)

Starting from the balance equation (Eq. 5.15), the term shown in Eq. 5.22 can be set up as the balance equation for an ideal flow tube reactor considering steady-state operation. Due to the plug flow, there is no mixing of the reactor contents in the flow tube. Rather, a differential volume element is chosen as the balance space, requiring a local consideration to determine the change in conversion. The indices z and $z + \Delta z$ describe two different locations in the flow tube. The respective differential volume element forms the balance space for the change of all components.

$$-c_{S,0} \frac{d\chi}{dt} = 0 = \frac{F}{V} \cdot c_{S,0}(1-\chi)|_z - \frac{F}{V} \cdot c_{S,0}(1-\chi)|_{z+\Delta z} - \nu \ (c_E, \chi) \tag{5.22}$$

F: ingoing or outgoing volume flow (feed) in the PFR (m^3 min^{-1})

Fig. 5.5 Batch operation as a function of the initial substrate concentration ($c_{S,0}$) and the K_m value

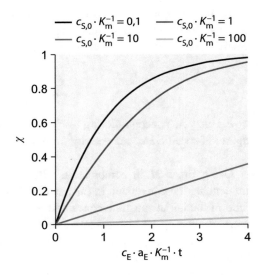

V: infinitesimal, differential volume of the considered balance space (m³)

The balance volume is determined by the cross-sectional area and the location difference in the flow pipe under consideration (Eq. 5.23).

$$V = A \cdot \Delta z \tag{5.23}$$

A: cross-sectional area of the PFR (m²)
Δz: length of the infinitesimal differential volume element (m)

Therefore, for a differential volume element, the limit value of the location-dependent quantities can be determined (Eq. 5.24).

$$\lim_{\Delta z \to 0} \left[\frac{F \cdot c_{S,0}}{A \cdot \Delta z} \cdot \left((1-\chi)|_z - (1-\chi)|_{z+\Delta z} \right) \right] = \nu \ (c_E, \chi) \tag{5.24}$$

A: cross-sectional area of the considered volume element in the flow tube (m²)
Δz: change in location of the considered balance area (m)

The location-dependent change in substrate concentration can thus be described directly in mathematical terms (Eq. 5.25).

$$-\frac{F \cdot c_{S,0}}{A} \frac{d(1-\chi)}{dz} = \nu \ (c_E, \chi) \tag{5.25}$$

After simplification with separation of the variables for the integral description in the flow tube, Eq. 5.26 is obtained.

$$\int\limits_0^X \frac{d\chi}{v\,(c_E,\chi)} = \int\limits_0^z \frac{A}{F \cdot c_{S,0}}\,dz \tag{5.26}$$

Just as in the previously considered case, an expression for the description of the conversion is obtained for the idealized flow tube after integration considering Michaelis-Menten kinetics (Eq. 5.27). The conversion in the PFR depends on the enzyme-specific constants, the initial substrate concentration, and the residence time.

$$\frac{c_{S,0}}{K_m} \cdot \chi - \ln \cdot (1 - \chi) = \frac{v_{max}}{K_m} \cdot \tau \tag{5.27}$$

τ: residence time (min)

The mathematical comparison of the two reactor concepts presented clearly shows the difference between batch and PFR from Sect. 5.2. Instead of the time-dependent change of the substrate concentration in the batch reactor, the local coordinate is responsible for the change of the substrate concentration in the PFR. For the continuously operated PFR, it is also possible to define a dimensionless residence time (Eq. 5.28).

$$\frac{v_{max}}{K_m} \cdot \tau = \frac{c_{Cat} \cdot a_s}{K_m} \cdot \tau = \frac{m_{Cat} \cdot a_s}{F \cdot K_m} \tag{5.28}$$

m_{Cat}: biocatalyst mass used (g)

With the simplification presented in Eq. 5.28, the conversion can be determined by having a defined catalyst mass in the reactor and a constant feed stream with a defined substrate concentration. Analogous to the mathematical description of the batch reactor, the variation of the three mentioned parameters allows the determination of the optimal operating point for the PFR. For different ratios of the initial substrate concentration to the K_m value of the enzyme, the same behavior as in the batch reactor is observed based on the mathematical relationships described (Fig. 5.5). However, the reaction time is substituted by the residence time in the PFR.

5.3.3 Balancing of an Ideally Mixed, Continuously Operated Stirred Tank Reactor (CSTR)

The description of an ideally mixed CSTR is also carried out on the basis of the balance equation (Eq. 5.15). The assumptions made previously for the change from the mass-dependent expression to the concentration-dependent expression apply again. As before in the PFR, the continuous mode of operation results in the accumulation term of the balance equation being zero. A local consideration is not necessary due to the complete mixing in the entire reaction volume analogous to the batch operation. The resulting balance is shown in Eq. 5.29.

$$-c_{S,0}\frac{d\chi}{dt} = 0 = \frac{F}{V} \cdot c_{S,0} - \frac{F}{V} \cdot c_{S,0}(1-\chi) - v(c_E,\chi)$$
(5.29)

By simplification, Eq. 5.30 is obtained.

$$\frac{\chi}{v(c_E,\chi)} = \frac{\tau}{c_{S,0}}$$
(5.30)

For a single-substrate reaction with Michaelis-Menten kinetics, the conversion can be determined as a function of residence time in the CSTR, enzyme kinetic parameters, and initial substrate concentration (Eq. 5.31).

$$\frac{\chi}{1-\chi} + \frac{\chi \cdot c_{S,0}}{K_m} = \frac{v_{max}}{K_m} \cdot \tau = \frac{c_E \cdot a_E}{K_m} \cdot \tau$$
(5.31)

High conversions are only possible in the CSTR at increased catalyst concentrations or longer residence times compared to the batch operation, if Michaelis-Menten kinetics are taken as a basis. This behavior can be observed especially at low $c_{S,0} \cdot K_m^{-1}$ values (Fig. 5.6).

5.3.4 Balancing Under Consideration of Inhibition Phenomena

The mathematical description in batch, PFR, and CSTR can be extended by considering inhibition phenomena (Chap. 4) in the reactive part of the balance equation. The derivation is analogous to the presented one-substrate reaction with Michaelis-Menten kinetics, whereby the reaction rate is influenced by inhibition. Common inhibition types are substrate surplus inhibition and competitive product inhibition. For the mentioned inhibition types, the resulting equations are summarized in Table 5.2.

Despite the integration of inhibition terms within the mathematical description, the behavior in real reactor operation can only be estimated. On the one hand, this is due to superimposed nonideal or real phenomena, which have not yet been taken into account. On the other hand, significantly larger reactors are used in industrial scale applications, whereby, for example, an idealized plug flow in the PFR may no longer be present in comparison with small laboratory reactors due to a changed flow behavior. To obtain the most accurate model representation for reactors, nonideal phenomena are also considered. A selection of the most important phenomena is listed below but is not discussed in detail. One major factor is the preparation of the biocatalyst used. Compared to the free enzyme, the catalyst can be fixed, which is referred to as immobilization. Various methods can be

Fig. 5.6 CSTR operation as a function of the initial substrate concentration ($c_{S,0}$) and the K_m value

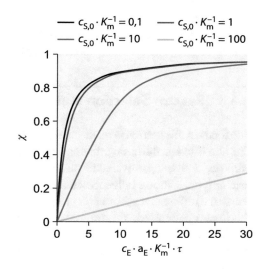

Table 5.2 Mathematical description of basic reactor types in single-substrate reactions. (According to Illanes 2008)

	PFR and batch: $\frac{c_{Cat} \cdot a_{sp}}{K_m} \cdot \tau =$	CSTR: $\frac{c_{Cat} \cdot a_{sp}}{K_m} \cdot \tau =$
Michaelis-Menten	$\frac{\chi \cdot c_{S,0}}{K_m} - \ln(1-\chi)$	$\frac{\chi \cdot c_{S,0}}{K_m} + \frac{\chi}{1-\chi}$
Competitive product inhibition	$\chi \cdot \left[\frac{c_{S,0}}{K_m} - \frac{c_{S,0}}{K_I} \right] - \left[1 + \frac{c_{S,0}}{K_I} \right] \cdot \ln(1-\chi)$	$\frac{\chi \cdot c_{S,0}}{K_m} + \frac{\chi}{1-\chi} + \frac{c_{S,0}}{K_I} \cdot \frac{\chi^2}{1-\chi}$
Substrate surplus inhibition	$\frac{\chi \cdot c_{S,0}}{K_m} - \ln(1-\chi) + \frac{c_{S,0}}{K_m} \cdot \frac{c_{S,0}}{K_I} \cdot \chi \cdot (1 - 0,5 \cdot \chi)$	$\frac{\chi \cdot c_{S,0}}{K_m} + \frac{\chi}{1-\chi} + \frac{c_{S,0}}{K_m} \cdot \frac{c_{S,0}}{K_I} \cdot \chi \cdot (1-\chi)$

used for immobilization, such as adsorption onto a support material. Due to immobilization, the catalyst usually exhibits increased stability to elevated operating temperatures. On the other hand, the challenges are to prevent mass transfer limitation or deactivation of the catalyst due to entrapment in or binding to the support. Another real phenomenon is the deactivation of the biocatalyst by the solvent used or the reactants in the process (Liese and Hilterhaus 2013; Jesionowski et al. 2014). The ideal reactor concepts already presented can be used in modified or combined form, depending on the application. For example, when using immobilized enzymes, the catalyst can be placed in a fixed-bed reactor (PBR). In this case, the PBR exhibits the characteristics of the PFR under ideal conditions, although mass transfer limitations often play a role in the real case and lead to deviations from the ideal behavior (Liese and Hilterhaus 2013). Membrane reactors with ultrafiltration membranes having a suitable exclusion limit are often used to retain free, i.e., homogeneously dissolved, enzymes in order to make the catalyst available for production for longer. In this process, the substrate solution enters the reactor and the membrane is only located at the outlet, where the product is obtained as filtrate. Unlike the product, the catalyst cannot pass through the membrane, which decouples their residence times (Gallucci et al. 2011).

5.4   Reactor Selection and Process Control

The focus in this section is on the application of the basic reactor types to various situations. The aim is to use the basic principles described to enable the selection of a reactor. For this purpose, a simple, irreversible reaction as in Eq. 5.1 is considered. Depending on the reactor type, a strong influence on the time-dependent behavior in the reactor is observed due to the different concentration profiles and the integrated balance equation, respectively. Two main parameters are discussed in more detail in the following: the residence time associated with the enzyme concentration used and the conversion achieved in the respective reactor. Based on the three basic reactor types presented, the behavior in batch and PFR is mathematically the same. The difference in continuous operation lies in the required setup time for the batch reactor between successive batch reactions (repetitive batch). These setup times are not necessary in PFR as long as ideal behavior is assumed, i.e., if the enzyme is not deactivated over time. The shorter the setup time between batch reactions, the closer the reactor types converge in terms of productivity. Under this assumption, a comparison between the operating modes of the batch/PFR and the basic type of the CSTR, which differs from it, is made in the following. The residence time, which is directly linked to the catalyst concentration used (Sect. 5.3), is used as a comparative parameter. The ratio of residence times between PFR and CSTR changes at different $c_{S,0} \cdot K_m^{-1}$ values, which allows the preferred reactor type to be selected depending on the initial substrate concentration.

5.4.1   Reactor Selection Without Consideration of Inhibition Phenomena

For a one-substrate reaction with Michaelis-Menten kinetics without inhibition, a higher residence time τ or a higher enzyme concentration c_E in the CSTR compared to the PFR/batch is necessary for low $c_{S,0} \cdot K_m^{-1}$ values in the range of 0.1–10 to achieve the corresponding conversion (Fig. 5.7). This is because the CSTR operates in the steady state at high conversion at low starting material concencentrations due to the outflow conditions described above, in contrast to the PFR, in which high starting material concentrations are present in the steady state for a long stretch of the reactor. At high $c_{S,0} \cdot K_m^{-1}$ values (here: 100), the courses only diverge at high conversions. As long as no inhibition is present, the PFR is therefore preferable to the CSTR, since in production the highest possible conversions ($>80\%$) are to be achieved in a short time or with the use of a comparable low catalyst concentration. However, the PFR requires additional immobilization of the enzyme.

Fig. 5.7 Reactor selection without consideration of inhibition phenomena

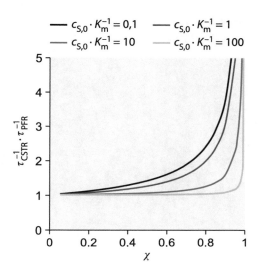

5.4.2 Reactor Selection for Substrate Surplus Inhibition

For a substrate surplus inhibition, the CSTR can be selected as a suitable reactor type based on the local and time-dependent concentration profiles, since the substrate concentration in the reactor is lowest due to working under outflow conditions. Analogous to the previous consideration without inhibition, the evaluation is performed, but by means of the $c_{S,0} \cdot K_I^{-1}$ value, and for clarity, the $c_{S,0} \cdot K_m^{-1}$ value is set to be constant at 10 in the following example. Considering the conversion as a function of the $c_{S,0} \cdot K_m^{-1}$ value, high $c_{S,0} \cdot K_I^{-1}$ values also distinguish the CSTR as the recommended reactor type (Fig. 5.8). If low initial substrate concentrations are used, the effect of substrate surplus inhibition is reduced, and the PFR/batch represents still an effective reactor type. This can be observed from increasing residence times of the CSTR compared to the PFR/batch operation at $c_{S,0} \cdot K_I^{-1}$ values between 0.1 and 1. It is also possible to use a fed-batch reactor (as a modification of the batch reactor) by feeding a substrate feed to enable low substrate concentrations in the reactor.

5.4.3 Reactor Selection for Product Inhibition

If a product inhibition prevails, reactors in batch or PFR operation are preferred in the complete range of $c_{S,0} \cdot K_I^{-1}$ values for a constant $c_{S,0} \cdot K_m^{-1}$ value of 10 (Fig. 5.9). This can be explained by the comparatively low product concentration for a long range in these reactor types in contrast to the CSTR, operating under outflow conditions. From a process engineering point of view, modifications of the individual reactors in which the product can be selectively removed during operation (in situ product removal; ISPR) are also useful

Fig. 5.8 Reactor selection for substrate surplus inhibition

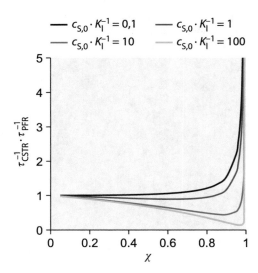

here. Typical ISPR processes are adsorption, distillation, extraction, or stripping using gases.

5.4.4 Reactor Selection for Parallel Reactions of the Substrate and Subsequent Reactions of the Product

If there is in addition to biocatalytic reaction a parallel reaction of the substrate, which causes the formation of an undesired by-product, the operation mode of the CSTR is superior to the operation mode in the batch reactor or PFR. This is based on the low substrate concentration in the concentration profile of the reaction in the CSTR compared to the other two basic reactor types. If, on the other hand, a subsequent reaction of the product is expected, the batch reactor and the PFR are again preferred, since, in contrast to the CSTR, the product concentration increases only slowly.

On the basis of the parameters shown for the selection of a suitable basic reactor type for enzymatic reactions, the influence of the reaction at hand and the behavior of the reactants with regard to the enzyme becomes clear. Accordingly, the effectiveness of the synthesis depends to a large extent on the type of reactor selected, with them having the greatest differences at high conversions as are necessary in industry.

Questions
  3. Which basic reactor should you use, if there is a substrate surplus inhibition present? Please rank the reactors from best to worse.
  4. Which basic reactor should you use if there is a product inhibition present? Please rank the reactors from best to worse.

Fig. 5.9 Reactor selection for product inhibition

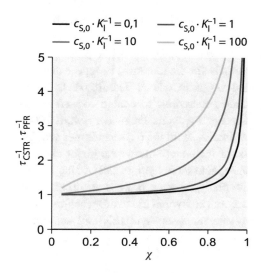

5. How can you easily modify a batch process to better deal with a substrate surplus inhibition? Please name the variation of the basic reactor type.
6. What are the two major advantages of using immobilized enzymes as biocatalyst compared to free enzyme?
7. Please name four typical ISPR processes.

5.5 Examples of Industrial Enzymatic Processes

The applicability of developed processes in industry depends to a large extent on their economic viability. In principle, a distinction is made on a case-by-case basis as to whether chemical catalysts or enzymes are more suitable for synthesis and what type of process control should be used. In this section, the use of enzyme-catalyzed reactions is discussed with examples from industry. Mainly, batch reactors are used for enzyme-catalyzed syntheses. Here, the synthesis of the active ingredient sitagliptin is discussed in more detail as an example, focusing on the trade-off between chemically catalyzed and enzyme-catalyzed reactions. As a continuously operated process, the enzyme-catalyzed production of high fructose corn syrup (HFCS), the largest enzyme biotransformation by volume, is discussed. Finally, the synthesis of fatty acid esters is used to demonstrate the need to consider modified variants with integration of ISPR technologies in the process design in addition to the basic reactor types.

5.5.1 Synthesis of Sitagliptin

The production of the active pharmaceutical ingredient sitagliptin used in drugs such as Januvia for the treatment of type 2 diabetes was developed by Merck in cooperation with Solvias and Codexis (Desai 2011). The process has been progressively optimized over three generations to enable cost-effective and "green" synthesis (Fig. 5.10; Table 5.3; Greener Synthetic Pathways Award 2006). Sitagliptin is a chiral-centered β-amino acid derivative, with the *(R)*-enantiomer acting as the active ingredient. Stereoselective synthesis of the enantiomer is the major challenge of the process. Industry is interested in the phosphate monohydrate salt of sitagliptin for pharmaceutical application (Fig. 5.10d). In the first-generation process, incorporation of the chiral center into the achiral β-keto ester occurs via asymmetric hydrogenation using a ruthenium chemical catalyst. This is followed by the multistep synthesis and subsequent hydrolysis of N-benzyloxy-β-lactam, from which sitagliptin (f) is synthesized by coupling the resulting β-amino acid intermediate (b) with triazole (c) (Hansen et al. 2005). In total, starting from trifluorophenylacetic acid (a), eight steps are required for synthesis, which allowed 100 kg of sitagliptin for the first clinical trials with an overall yield of 45–52% (Hansen et al. 2005; Dunn 2012). The disadvantage of the described synthesis route is on the one hand the use of high molecular weight reagents for the integration of the chiral center and on the other hand the poor atomic efficiency of EDC (N-(3-dimethylaminopropyl)-N-ethylcarbodiimide hydrochloride), which is responsible for the coupling but also for an increased waste production ($250 \ kg \ kg_{product}^{-1}$) (Desai 2011).

In the second-generation process developed by Merck, the integration of the chiral center is carried out on the enamine amide dehydrositagliptin, thus improving the complex introduction of chirality of the first process. For this purpose, trifluorophenylacetic acid (a) is converted and isolated in a reactor via its intermediate, the β-ketoamide, to dehydrositagliptin (e) with a yield of 82% and a purity of 99.6% by weight (Hansen et al. 2005). Subsequent rhodium-catalyzed stereoselective hydrogenation of dehydrositagliptin achieved a yield of 98% with an *ee* of 95%. Activated carbon is used for the separation and recycling of the rhodium catalyst. The total isolated yield increased to values up to 65%, while at the same time the number of reaction steps performed was reduced to five (Hansen et al. 2005). In addition, the resulting waste production was reduced to $50 \ kg \ kg_{product}^{-1}$, a fifth of that of the first-generation process (Hansen et al. 2005).

Further optimization potential was seen in the stereoselective hydrogenation based on the process of the second generation. The use of the chemical catalyst with coordinated transition metal requires the complete separation from the resulting product to perform the next reaction step as well as working under high pressures, which results in high process costs (Desai 2011). In the third generation process, an increase in efficiency was achieved by substituting the rhodium catalyst with the enzyme-catalyzed amination using a transaminase (Savile et al. 2010). Here, the chirality is stereoselectively introduced by the enzyme into the β-ketoamide intermediate, thus saving an additional process step. A total product

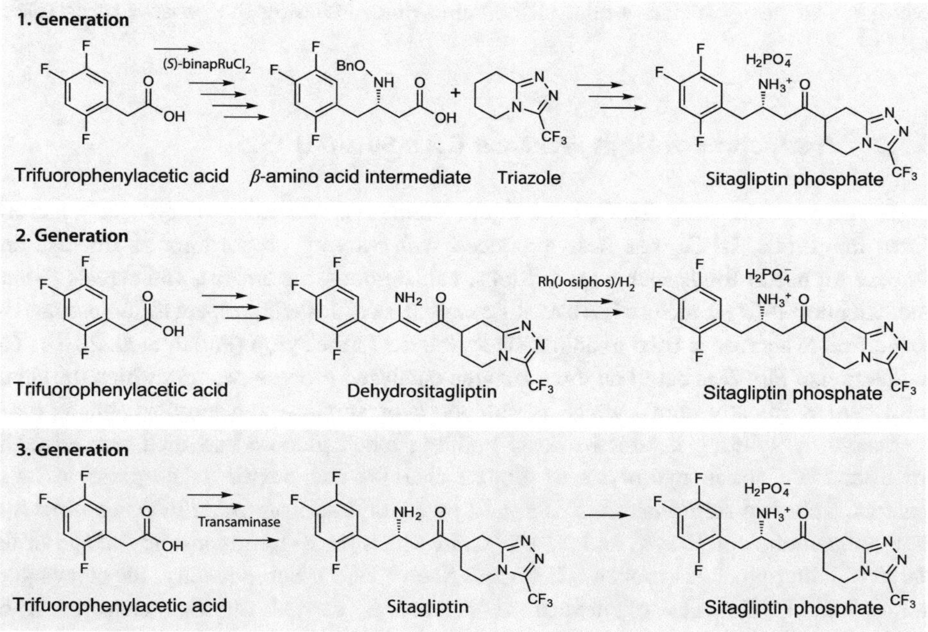

1. Generation

Trifuorophenylacetic acid   β-amino acid intermediate   Triazole   Sitagliptin phosphate

2. Generation

Trifuorophenylacetic acid   Dehydrositagliptin   Sitagliptin phosphate

3. Generation

Trifuorophenylacetic acid   Sitagliptin   Sitagliptin phosphate

Fig. 5.10 The three generations of processes for the synthesis of the active pharmaceutical ingredient sitagliptin

Table 5.3 Comparison of the three generations of processes for the synthesis of sitagliptin. (After Desai 2011; Dunn 2012; Hansen et al. 2009; Savile et al. 2010)

	1. Generation	2. Generation	3. Generation
Catalyst[a]	(S)-binapRuCl$_2$	Rh(COD)Cl$_2$ and t-Bu Josiphos	Transaminase/PLP
Process steps[b]	8	5 (3 are performed in a reactor)	4
Waste [kg·kg$_{Product}^{-1}$]	250	50	40.5
Yield [%]	45–52	65	75
ee [%]	–	95 (>99.9 after additional crystallization)	>99.95

[a]For the insertion of the chiral center
[b]Starting from trifluorophenylacetic acid

concentration of up to 200 g L^{-1} sitagliptin with an *ee* > 99.95% and an overall yield of 75% can thus be achieved (Savile et al. 2010). The main advantages of the enzyme-catalyzed synthesis route over the chemically catalyzed process can be summarized to an increase in productivity by more than 50%, a further waste reduction by 19%, and the avoidance of heavy metals while, at the same time, lowering the process costs (Desai 2011).

5.5.2 Production of High Fructose Corn Syrup (HFCS)

High fructose corn syrup (HFCS) is a liquid alternative sweetener to crystalline sucrose. Since the 1960s, HFCS has been produced with varying proportions of fructose and glucose for use in foods such as soft drinks, baked goods, ice creams, and cereals (Visuri and Klibanov 1987). Fructose has a relative sweetness of 1.3 with respect to sucrose, and its proportion to sucrose is used to adjust the sweetness of the syrup (Parker et al. 2010). The production of HFCS is based on three enzyme-catalyzed process steps, in which the starch from cereals (mostly corn), which is composed of amylose and amylopectin, is made accessible by splitting it into the basic building block glucose and then isomerized to fructose. First, starch hydrolysis to oligosaccharides and dextrin takes place in batch reactors. This two-step process is catalyzed by α-amylase from *Bacillus licheniformis* at high temperatures of 105 °C and 90–95 °C, in which the α-1,4-glycosidic linkage of the starch building blocks is broken (Crabb and Shetty 1999). Subsequently, the cleavage of α-1,4- and 1,6-linkages of dextrin to glucose is carried out in batch mode by amyloglucosidase from *Aspergillus niger* at 60 °C (Parker et al. 2010). Both process steps are referred to as starch liquefaction because liquid glucose syrup is obtained as an intermediate product. Starch liquefaction is operated with free enzymes, which are used directly in the process and are not recycled due to their low procurement costs. In the third step, the isomerization of glucose to fructose takes place in fixed-bed reactors, which are in a parallel or series configuration, catalyzed by immobilized glucose isomerase from *Streptomyces murinus* at 60 °C (Parker et al. 2010). Prior to entering the fixed-bed reactors, the produced glucose is purified via filtration, adsorption, and ion exchange to avoid clogging of the packed bed reactors. In order to counteract the formation of by-products during the isomerization process, high enzyme concentrations of 1800 kg per fixed-bed reactor with correspondingly low residence times are used, which can be operated for up to 687 days (Liese et al. 2006). These operating times still to this day represent a unique feature in biocatalyzed processes. Furthermore, the isomerization reaction of glucose to fructose symbolizes the largest enzyme-catalyzed manufacturing process worldwide with productivities of 8000–10,000 kg per kg enzyme (Liese et al. 2006). The equilibrium conversion of the isomerization is just under 55%, but the proportion of fructose can subsequently be increased up to 90% using chromatographic separation processes (Visuri and Klibanov 1987). The operating parameters of the process are shown in Table 5.4.

Table 5.4 Operating parameters of the industrial synthesis of high fructose corn syrup. (HFCS; after Crabb and Shetty 1999; Liese et al. 2006; Parker et al. 2010; Visuri and Klibanov 1987)

Product	HFCS with up to 90% fructose[a]
T_{max} [°C]	105 °C [b]
Residence time in the packed bed reactor [h]	0.17–0.33
Half-life time of the glucose isomerase [d]	>100 [c]
TON [kg·kg^{-1}]	8000–10,000 [c]
Production [kg·d^{-1}]	>1,000,000

[a]After chromatographic separation
[b]In starch liquefaction
[c]In isomerization

5.5.3   Synthesis of Fatty Acid Esters

Fatty acid esters are widely used in personal care products, pharmaceuticals, foods, and cleaning agents, among others (Ansorge-Schumacher and Thum 2013). The variety of products using fatty acid esters is made possible by the variable chain length. The synthesis of the desired fatty acid esters is possible both chemocatalytically and enzyme-catalyzed. However, the chemocatalytic production route results in the formation of by-products due to process conditions at temperatures above 180 °C. Here, a subsequent product purification is required for the fatty acid esters to be used as the end products, making it necessary to perform several additional downstream processing steps, which increases the process costs (Thum 2004). Alternatively, the fatty acid esters can be synthesized enzyme-catalyzed with the aid of lipases. The mild reaction conditions <75 °C using lipases significantly reduce by-product formation and allow working without the use of solvents (Anderson et al. 1998). Compared to the chemocatalytic process, no work-up steps are necessary, which allows a single-stage process in a batch reactor (Sect. 5.2) or in a fixed-bed reactor (if there is no mass transfer limitation, similar to PFR Sect. 5.2). In addition, the use of only natural raw materials in products is particularly advantageous in the cosmetics industry. Nevertheless, the mild reaction conditions in the enzyme-catalyzed synthesis pathway also pose challenges, since the reactions for the synthesis of fatty acid esters from alcohols and acids are equilibrium-limited. To achieve high conversions, the resulting water of reaction must be removed from the reaction as completely as possible by a selective in situ separation (ISPR). According to the principle of Le Chatelier, the equilibrium can be shifted to the product side, whereby theoretically high conversions of the fatty acid esters of >99% can be achieved. In this case, operation under vacuum conditions for evaporation of the faster boiling water is most suitable for the batch reactor. Nevertheless, there are limits to the use of conventional batch and fixed-bed reactors when considering the present viscosities of the substrates and products in fatty acid ester synthesis. In order to work solvent-free in a fixed-bed reactor, only the use of low melting raw materials is possible, as high melting and high viscous substances generate a high pressure drop (Ansorge-Schumacher and Thum 2013). Similarly, conventional batch reactors cannot be used due to the high mechanical stress for the enzyme (Ansorge-Schumacher and Thum 2013). In industry, therefore, either a circulating reactor, consisting of a fixed-bed reactor with a receiver vessel where a vacuum is drawn, or a bubble column reactor is used (Hilterhaus et al. 2008). In contrast to the fixed-bed circulating reactor, the reaction and separation of the produced water are not spatially separated in the bubble column reactor. The advantage of the bubble column reactor over the conventional batch reactor or fixed-bed approach is that mixing automatically occurs by the gassing for water stripping. This reduces the mechanical load on the enzyme, the reaction water formed is separated by stripping in situ, and there is a macroscopic reduction in viscosity in the system (Hilterhaus et al. 2008). Figure 5.11a schematically shows the operation in the bubble column reactor. It belongs to the class of multiphase reactors, which is illustrated by the example of the synthesis of highly viscous polyglycerol-3-laurate from polyglycerol-3 and lauric acid (Fig. 5.11b). In this case, the reaction takes place in a four-phase system. In addition to two immiscible liquid phases (polyglycerol-3 and lauric acid), the immobilized lipase in the form of Novozyme 435 forms a solid phase, and the introduction of air or nitrogen represents the fourth phase. Over the time course of the process, there is a viscosity increase by a factor of 20 (from 30 mPa s to 700 mPa s; Hilterhaus et al. 2008). In the bubble column reactor, in addition to the synthesis of polyglycerol-3-laurate, the synthesis of myristyl myristate could be demonstrated (Fig. 5.11c). Both reactions are only possible using immobilized enzymes as this ensured the stability of the catalyst over nine cycles for polyglycerol-3-laurate and six cycles for myristyl myristate with half-live times of 20 h and 157 h, respectively (Hilterhaus et al. 2008, 2016). The shorter half-life time for the higher viscosity product polyglycerol-3-laurate can be explained by the faster desorption of lipase from the carrier material. The stronger surfactant properties of polyglycerol-3-laurate compared to myristyl myristate are responsible for this. Reduction of desorption from the carrier is possible by modification of the enzyme preparation, for example, by silicone coating or cross-linking via glutaraldehyde (Wiemann et al. 2009). The operating parameters of the fatty acid ester syntheses shown in the bubble column reactor are summarized in Table 5.5.

Take-Home Messages
- An enzyme-catalyzed process is evaluated using kinetic and thermodynamic parameters that need to be addressed independently.
- Choose the most appropriate reactor based on the standard reaction engineering principles. Different reactor types exist to lessen the impact of any inhibition, side reaction, or deactivation of the enzyme.
- Based on balancing, conversions within the various reactor concepts can be described mathematically giving insights into the production on an industrial scale.

(continued)

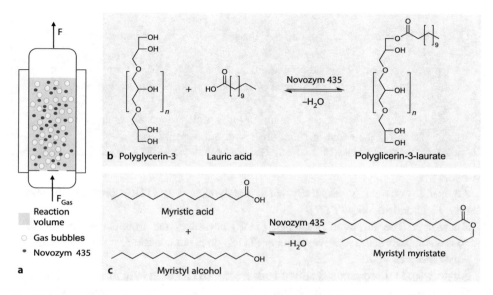

Fig. 5.11 Enzyme-catalyzed synthesis of fatty acid esters. (**a**) Schematic representation of the four-phase reaction in the bubble column reactor. (**b**) Synthesis of polyglycerol-3-laurate. (**c**) Synthesis of myristyl myristate

Table 5.5 Operating parameters of the fatty acid ester syntheses of polyglycerol-3-laurate and myristyl myristate in the bubble column reactor. (According to Hilterhaus et al. 2008)

Product	Polyglycerol-3-laurate	Myristyl myristate
T_{max} [°C]	60	60
Ratio of substrates	1:1[a]	1:1[b]
Half-life time of Novozyme 435 [h]	20 (9 batch cycles)	157 (6 batch cycles)
TON [kg kg^{-1}]	-[c]	2500
STY [kg L^{-1} day^{-1}]	3	7

[a]Mass-related (w/w)
[b]Mole-related (x/x)
[c]Not known

- The CSTR operates under outflow conditions, meaning the concentrations in the outlet are identical to any point in the reactor volume under ideal conditions.
- The steady state of a CSTR is described at high conversion with an overall minimum concentration of the starting material and a maximum product concentration.

(continued)

- The PFR is from a reaction engineering point of view the continuous version of a CSTR.
- A process can be evaluated on a small scale under near ideal conditions, but real phenomena need to be considered when moving to production on an industrial scale.
- Enzyme immobilization or retention via ultrafiltration membranes should be considered for the economic viability of an industrial process.

Answers

- Answer 1: conversion, selectivity, yield, catalytic activity (TOF), productivity, space-time yield and stability (TTN).
- Answer 2: The turnover frequency (TOF) considers the time-dependent substrate conversion, while the turnover number (TON) does not contain any time dependency.
- Answer 3: $ee_P = 0.2$.
- Answer 4: (1) discontinuous stirred tank reactor (STR) or ideally mixed batch reactor; (2) continuously operated stirred tank reactor (CSTR); (3) plug flow reactor (PFR).
- Answer 5: PFR, the same behavior as in the batch reactor is observed, since the time and in place performance in the concentration profiles are just exchanged.
- Answer 6: CSTR > PFR = STR. Under steady-state operation is the residual substrate concentration at a minimum.
- Answer 7: PFR = STR > CSTR. CSTR always operates in the steady state at the maximum product concentration.
- Answer 8: fed-batch reactor.
- Answer 9: Immobilized enzymes are easier to recycle/reuse, and the catalyst usually exhibits increased stability to the process conditions. However, very often diffusion limitations are observed.
- Answer 10: adsorption, distillation, extraction, and stripping using gases.

References

Anderson EM, Larsson KM, Kirk O. One biocatalyst-many applications: the use of Candida Antarctica B-lipase in organic synthesis. Biocatal Biotransformation. 1998;16(3):181–204.

Ansorge-Schumacher MB, Thum O. Immobilised lipases in the cosmetics industry. Chem Soc Rev. 2013;42(15):6475–90.

Crabb WD, Shetty JK. Commodity scale production of sugars from starches. Curr Opin Microbiol. 1999;2(3):252–6.

Daußmann T, Hennemann H-G, Rosen TC, Dünkelmann P. Enzymatic technologies for the synthesis of chiral alcohol derivatives. Chem Ing Tech. 2006;78(3):249–55.

Desai AA. Sitagliptin manufacture: a compelling tale of green chemistry, process intensification, and industrial asymmetric catalysis. Angew Chem Int Ed. 2011;50(9):1974–6.

Dunn PJ. The importance of green chemistry in process research and development. Chem Soc Rev. 2012;41(4):1452–61.

Faber K. Biotransformations in organic chemistry: a textbook. Springer Science & Business Media; 2011.

Gallucci F, Basile A, Hai FI. Introduction—a review of membrane reactors, membranes for membrane reactors: preparation. Optim Select. 2011:1–61.

Greener Synthetic Pathways Award (2006) U.S. Environmental Protection Agency. Accessed 14 Feb 2017

Hansen KB, Balsells J, Dreher S, Hsiao Y, Kubryk M, Palucki M, Rivera N, Steinhuebel D, Armstrong JD, Askin D, Grabowski EJJ. First generation process for the preparation of the DPP-IV inhibitor sitagliptin. Org Process Res Dev. 2005;9(5):634–9.

Hansen KB, Hsiao Y, Xu F, Rivera N, Clausen A, Kubryk M, Krska S, Rosner T, Simmons B, Balsells J, Ikemoto N, Sun Y, Spindler F, Malan C, Grabowski EJJ, Armstrong JD. Highly efficient asymmetric synthesis of sitagliptin. J Am Chem Soc. 2009;131(25):8798–804.

Hilterhaus L, Thum O, Liese A. Reactor concept for lipase-catalyzed solvent-free conversion of highly viscous reactants forming two-phase systems. Org Process Res Dev. 2008;12(4):618–25.

Hilterhaus L, Liese A, Kettling U, Antranikian G. Applied biocatalysis: from fundamental science to industrial applications. Weinheim: Wiley; 2016.

Illanes A. Enzyme biocatalysis: principles and applications. Springer Science & Business Media; 2008.

Jesionowski T, Zdarta J, Krajewska B. Enzyme immobilization by adsorption: a review. Adsorption. 2014;20:801–21.

Liese A, Hilterhaus L. Evaluation of immobilized enzymes for industrial applications. Chem Soc Rev. 2013;42(15):6236–49.

Liese A, Seelbach K, Wandrey C. Industrial biotransformations. Weinheim: Wiley; 2006.

Liese A, Kragl U. Influence of the reactor configuration on the enantioselectivity of a kinetic racemate cleavage. Chem Ing Tech. 2013;85(6):826–32.

Martin-Matute B, Bäckvall J-E. Dynamic kinetic resolution catalyzed by enzymes and metals. Curr Opin Chem Biol. 2007;11:226–32.

Parker K, Salas M, Nwosu VC. High fructose corn syrup: production, uses and public health concerns. Biotechnol Mol Biol Rev. 2010;5(5):71–8.

Savile, C. K.; Janey, J. M; Mundorff, E. C.; Moore, J. C.; Tam, S.; Jarvis, W. R.; Colbeck, J. C.; Krebber, A.; Fleitz, F. J.; Brands, J.; Devine, P. N.; Huisman, G. W.; Hughes, G. J. Biocatalytic asymmetric synthesis of chiral amines from ketones applied to sitagliptin manufacture. Science 2010, 329 (5989), 305–309.

Thum O. Enzymatic production of care specialties based on fatty acid esters. Tenside Surfactant Deterg. 2004;41(6):287–90.

Tufvesson P, Lima-Ramos J, Nordblad M, Woodley JM. Guidelines and cost analysis for catalyst production in biocatalytic processes. Org Process Res Dev. 2011;15:266–74.

Visuri K, Klibanov AM. Enzymatic production of high fructose corn syrup (HFCS) containing 55% fructose in aqueous ethanol. Biotechnol Bioeng. 1987;30(7):917–20.

Watson JD, Crick FH. The structure of DNA. Cold Spring Harb Symp Quant Biol. 1953;18:123–31.

Wiemann LO, Nieguth R, Eckstein M, Naumann M, Thum O, Ansorge-Schumacher MB. Composite particles of novozyme 435 and silicone: advancing technical applicability of macroporous enzyme carriers. ChemCatChem. 2009;1(4):455–62.

Woodley JM, Breuer M, Mink D. A future perspective on the role of industrial biotechnology for chemicals production. Chem Eng Res Des. 2013;91:2029–36.

Yuryev R, Strompen S, Liese A. Coupled chemo(enzymatic) reactions in continuous flow. Beilstein J Org Chem. 2011;7:1449–67.

Part II
Methods

Enzyme Identification and Screening: Activity-Based Methods

6

Jessica Rehdorf, Alexander Pelzer, and Jürgen Eck

What You Will Learn in This Chapter

Methods for the identification and characterization of "ideal" enzymes for industrial use in the field of biotechnology have developed rapidly in recent years. The use of activity-based methods for enzyme identification from natural resources such as the metagenome allows the discovery of biocatalysts. The main advantage of activity-based screening is the direct identification of desired enzyme activities completely independent of any knowledge about the sequence or structure of an enzyme. In this way, activity-based screening also enables the identification of new, previously unknown enzymes. This chapter describes current methods for identifying enzyme activities from natural resources. In which resources can new enzymes be searched for? What is the "ideal" enzyme? Why is gene expression so important? Which screening methods are established today? All this is explained in this chapter. The focus is on high- to low-throughput assays such as selective agar plate assays or fluorescence-activated cell sorting, as well as on liquid assays in 96-well plate format or chromatographic methods. By describing the advantages and disadvantages of each method, you will learn when which screening method is appropriate and what requirements must be met to perform an efficient screening.

J. Rehdorf · A. Pelzer
BRAIN Biotech AG, Zwingenberg, Germany
e-mail: jr@brain-biotech.com; ap@brain-biotech.com

J. Eck (✉)
bio.IMPACT, Bensheim, Germany
e-mail: je@bioimpact.de

© The Author(s), under exclusive license to Springer Nature Switzerland AG 2024
K.-E. Jaeger et al. (eds.), *Introduction to Enzyme Technology*, Learning Materials in Biosciences,
https://doi.org/10.1007/978-3-031-42999-6_6

6.1   Introduction

The use of microbial cells or isolated enzymes for the production of fine and specialty chemicals has steadily increased in recent years and is now considered an indispensable tool in modern synthetic chemistry. Using highly active and selective enzymes, biocatalysis also enables synthesis strategies that are often not possible using traditional organic synthesis. The remarkable development in the field of "industrial" or "white" biotechnology is due not least to the increased attention and appreciation by the (chemical) industry. This is also reflected in the growing number of implemented biotechnological synthesis processes. The basis for this was created by a wide variety of technologies that have developed rapidly. These deal with access to genetic resources, the identification of enzymes, the optimization of enzyme properties (enzyme engineering), and the screening of created enzyme libraries.

6.2   The "Ideal" Enzyme

Scientists and process engineers today assume that only a very small proportion of the microbial cells in an environmental sample (habitat) can be cultivated under laboratory conditions. Therefore, for a long time, industry was forced to use enzymes from the comparatively small resource of culturable microorganisms. Due to the low availability of new enzymes, process control and enzyme properties were often not optimally matched. Industrial processes are sometimes subject to harsh conditions (extreme temperatures, pressures, pH values, use of solvents) in which nonnatural substrates are to be transformed in often very high concentrations. However, enzymes are adapted to substrate concentrations and environmental conditions of their natural habitat. Therefore, the enzymes used could not operate at maximum activity. Consequently, enzymes were usually the limiting factor in a production process. To keep the well-known advantages of chemo-, regio-, and enantioselectivity of an enzymatic reaction, enzymes were used, but these processes often ran suboptimally and were hardly economically competitive (Aehle and Eck 2012). Nowadays, with the growing number of enzyme activities available, it is far easier to identify the "ideal" enzyme for a desired biocatalysis (Burton et al. 2002). For an enzyme to be implemented in an industrial process, it must meet several criteria. In addition to activity and selectivity, these include specificity for a particular substrate spectrum, stability under process conditions, efficiency in terms of space-time yield, and manufacturability on a ton scale. If an enzyme meets these criteria, an industrial process can be designed ecologically and economically. Figure 6.1 uses two selected examples to show how different the requirements for an industrial enzyme can be. A detergent enzyme, for example, a protease or an α-amylase, often requires a broad substrate spectrum. It must also tolerate high temperatures, extreme pH values, as well as the presence of surfactants and bleaching agents. A high-performance biocatalyst, e.g., an alcohol dehydrogenase that

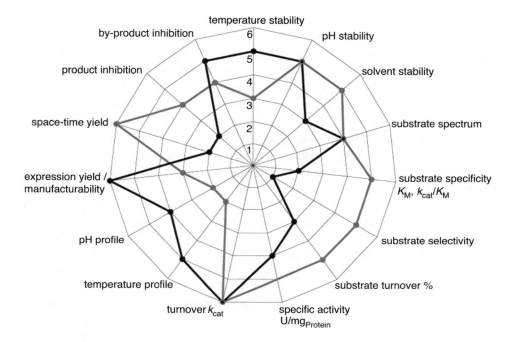

Fig. 6.1 Multiparameter analysis of two different industrial enzymes according to Lorenz and Eck (2005). The required industrial parameters of enzymes and biocatalysts depending on the area of application are indicated. Gray line: biocatalyst for the production of an enantiomerically pure fine chemical, an alcohol dehydrogenase. Black line: technical enzyme for detergent applications, a protease. The classification of the parameters is carried out uniformly in the gradation from 1 to 6, whereby 1 is classified as of subordinate relevance and 6 as extremely relevant for the respective parameter in the field of application

catalyzes the reduction of a prochiral ketone into an enantiomerically pure alcohol, on the other hand, must have high enantio- and regioselectivity, show a good turnover (k_{cat}), and enable a high space-time yield.

According to the principle "you get what you screen for" (Frances H. Arnold), an intelligent multiparameter screening strategy to identify the ideal enzyme is a prerequisite for success. At the end of such an endeavor, the reaction catalyzed by an ideal enzyme should be limited solely by the diffusion of the substrate or product to or from the active site.

6.3   Resources for Enzyme Identification

The three domains of life include bacteria, archaea, and eukarya. With the hydroxynitrillyase from *Manihot esculenta* or *Arabidopsis thaliana* (EC 4.1.2.47) and an esterase from pig liver (EC 3.1.1.1), plants and animals represent a comparably small resource for the identification and development of industrial enzymes. The majority of relevant biocatalysts today originate from microorganisms. In addition to eukaryotes and filamentous fungi, one of the best-known enzymes comes from the yeast *Candida antarctica*, the lipase CalB (EC 3.1.1.3). This serine hydrolase is now produced annually on a multiton scale by major enzyme producers and is used in stereoselective catalysis for the synthesis of chiral alcohols and acyl esters. However, prokaryotes are of greatest importance for the discovery of new biocatalysts. Bacteria and archaea are considered to be the earliest forms of life on Earth (Whitman et al. 1998). Over millions of years, they have adapted to a wide variety of habitats and ecological niches in the course of evolution. They ensure their survival in sometimes extreme locations by means of an impressively diverse repertoire of proteins and specialized enzymes.

Screening of microorganisms from different habitats is one of the frequently used techniques to identify new proteins and enzymes. However, the accessibility of this resource is highly dependent on the cultivability of these microorganisms under laboratory conditions. Naturally, microorganisms often live together in consortia. They are often coordinated in their way of life as well as in the uptake and conversion of nutrients. Therefore, cultivation is only possible if the natural conditions of the habitat such as pH, temperature, carbon source, nitrogen source, essential ions, trace elements, and vitamins are known. Today, scientists assume that only about 1% of microorganisms from a soil sample can be cultivated in the laboratory. A large part of the genetic information and the microbial biodiversity encoded in it is therefore not accessible. In practice, this information is lost. How limited our knowledge of natural biodiversity still is shows the study of Hug et al. (2016). The classical tree of life has dramatically expanded by the candidate phyla radiation (CPR) due to new genomic sampling of previously unknown microbial lineages. This large fraction of diversity is currently only accessible via cultivation-independent genome-resolved approaches.

However, modern technologies now make it possible to extract from various habitats the entire DNA of the microorganisms contained therein (bacteria, fungi, algae, protists, archaea), the so-called metagenome. This principle is shown in Fig. 6.2.

The diversity within these habitats differs depending on the composition of the sample: a sample from nutrient-rich forest soil or pastureland contains on average 6000–8000 genome equivalents per cubic centimeter, whereas the genetic diversity in selected ecological niches is much lower due to high selection pressure and often has only <20 genome equivalents (Table 6.1).

The creation of a metagenome has revolutionized the field of enzyme identification (Handelsmann et al. 1998). Today's metagenome technology provides access to nature's toolbox: microbial biodiversity and its enzymatic repertoire. Cells from a habitat sample are lysed using mechanical and/or chemical methods. Subsequently, the entire DNA of the sample is isolated and purified. The resulting metagenome DNA is integrated into vector systems using a variety of cloning methods, transferred to a suitable expression host that can be easily cultured, and brought to expression (Chap. 9). Currently used expression systems for metagenomic DNA include *Escherichia coli* (*E. coli*) as well as various

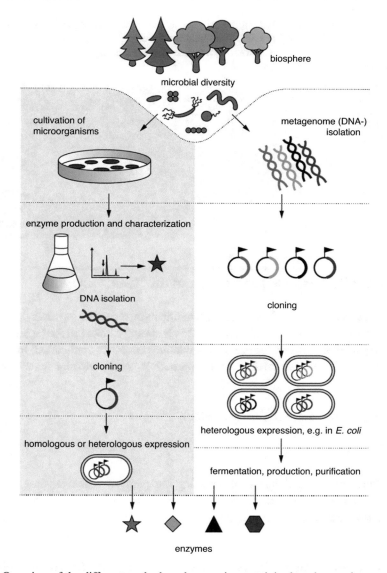

Fig. 6.2 Overview of the different methods and assays in an activity-based screening

Pseudomonas, Bacillus, and *Saccharomyces* strains. In addition to the exploration of natural microbial biodiversity through metagenome technology, it should also be mentioned that methods such as random mutagenesis of an entire genome or defined subregions, as well as directed evolution, represent alternative ways of generating novel enzymes and require suitable screening technologies. Current methods for enzyme engineering and approaches for enzyme design are discussed in detail in Chap. 7.

Table 6.1 Overview of microbial biodiversity in different habitats according to Torsvik et al. (2002). The number of cells per cubic centimeter was determined by fluorescence microscopy. Genome complexity is expressed in total base pairs. The genome equivalents are calculated from the mean relative genome size of *Escherichia coli* with 4.1×10^6 bp

Natural resource (source of DNA)	Number of cells per cm³	Genome complexity in bp	Genome equivalents
Forest soil	4.8×10^9	2.5×10^{10}	6000
Forest soil, cultivable microorganisms	1.4×10^7	1.4×10^8	35
Pasture land	1.8×10^{10}	1.5×10^{10}–3.5×10^{10}	3500–8800
Crop field	2.1×10^{10}	5.7×10^8–1.4×10^9	140–350
Marine sediment	3.1×10^9	4.8×10^{10}	11,400
Salt brine (22% salt content)	6.0×10^7	2.9×10^7	7

6.4 Activity-Based Identification of an "Ideal" Enzyme

The identification of an "ideal" enzyme in a multiparameter screening always requires the determination of a suitable screening setup. This is strongly linked to the target parameters that the enzyme must achieve for use in an industrial process or application. In principle, a distinction is made between three possible approaches, which can also be combined with each other:

– Screening of databases (*in silico screening*)
– Sequence-based screening
– Activity-based screening

In the following, the activity-based screening is explained. A detailed description of the procedure in a sequence-based screening and the use of nucleotide and protein databases for an *in silico* screening is given in Chap. 7.

6.4.1 Functional Gene Expression as a Prerequisite for Activity-Based Screening

The prerequisite for activity-based methods is the successful homologous or heterologous expression of the relevant genes in a suitable screening host. In order to express as many gene sequences of a metagenomic library as possible, the design of a universal expression system is probably the greatest challenge in an activity-based screening. The isolated genes or DNA fragments are cloned downstream of a suitable promoter. Plasmids, cosmids, fosmids, and in selected cases bacterial artificial chromosomes (BACs) can be used for the expression of metagenome DNA. These vectors differ in their copy number and stability, but also in the receptivity of the DNA fragment to be cloned (insert size).

The active production of enzymes is a complex molecular biological process that consists of transcription, translation, and protein folding and may also involve posttranslational modifications, enzyme activation, and protein secretion.

Due to the ease of handling, genetic accessibility, and high transformation efficiency of *E. coli*, this host organism is often the first choice for the expression of genes from a metagenomic library (Chap. 9). The ability to process mRNAs with different translation signals is a major advantage compared to other organisms such as *Bacillus subtilis* and *Streptomyces lividans* (Gabor et al. 2007). For those mRNAs whose translation signals are not recognized by *E. coli* or are subject to a different gene regulation, alternative organisms must be used. In addition to *Bacillus* and *Streptomyces*, these often include *Pseudomonas*, *Mycobacterium*, and *Saccharomyces*. The use of shuttle vectors for the cloning of metagenome DNA also allows expression to be carried out in parallel in different organisms, thus increasing the chance of successful enzyme production. Using so-called *Escherichia coli yeast shuttle vectors*, it is possible to carry out posttranslational modifications such as glycosylations and to produce active enzymes of eukaryotic origin. *Escherichia coli Bacillus shuttle vectors* allow proliferation in Gram-negative and Gram-positive organisms. This allows, for example, the secretion of an enzyme into the culture supernatant by *Bacillus subtilis*. A direct comparison of both organisms in an activity-based screen for new β-galactosidases from a metagenome resource at BRAIN Biotech showed that the number of hit clones can vary widely depending on the expression capacity (BRAIN Biotech AG, unpublished data). In an agar plate screen using X-Gal as a model substrate, 52 lactolytic enzymes could be visualized from 270,000 primary clones (corresponding to >170 bacterial genomes) in *E. coli*, but only 8 in *B. subtilis*. A comparable result was shown by a Japanese research group. Here, a new naphthalene dioxygenase could be actively produced in a screen only in *Pseudomonas putida*. This enzyme was not found in *E. coli* as the screening organism (Ono et al. 2007).

These results demonstrate the different expression capacity of different strains and highlight the importance of choosing a suitable organism for gene expression. In order to further increase the chances of a functional production of biocatalysts, intensive work is now being carried out to establish new expression strains and optimize existing systems. The focus is on research into transcription (promoter selection, regulability of the promoter, influence of various termination signals), translation (initiation, elongation of the polypeptide chain, *codon usage*, internal Shine-Dalgarno sequences, formation of secondary structures, termination), protein folding kinetics, and posttranslational processing of proteins (interaction with chaperones, binding to factors for secretion, protein processing after secretion, formation of disulfide bridges).

6.4.2 Methods for Activity-Based Screening

Activity-based screening focuses exclusively on the measurable activity of an enzyme. Protein sequence and structure do not play a role. This also makes it possible to find enzymes whose activity for a particular reaction was previously unknown. Further, enzymes have the ability to catalyze side reactions in addition to their main reaction. This is often referred to as enzyme promiscuity. This property of enzymes is not surprising as evolution is not the pursuit of perfect enzymes.

A few years ago, a group of scientists used activity-based screening of a bovine stomach metagenomic library to discover a completely new family of multicopper oxidases that are structurally distinct from the previously known enzymes in this group (Beloqui et al. 2006). Other methods, such as sequence-based screening, would probably not have been able to identify these enzymes.

From an industrial point of view, enzymes of the classes oxidoreductases (EC 1), transferases (EC 2), hydrolases (EC 3), lyases (EC 4), and isomerases (EC 5) are of particular importance. Robust and selective assays are required for the identification and characterization of these highly specialized enzymes. The assays must be reliable, enzyme-specific, sensitive, reproducible, and easy and fast to perform. The use of such activity-based assays allows an extremely efficient and targeted search for the desired enzyme property in a metagenome library from an environmental sample.

In principle, an activity-based screening distinguishes between two approaches. On the one hand, a method can be selected that allows a high screening throughput (i.e., the screening of many clones per time) but in return provides less information with regard to activity and specificity of the hit clones (candidates that show the desired activity). On the other hand, there is the possibility to choose an assay that allows to obtain much information, but only a comparatively low throughput can be achieved. A detailed comparison of the methods is shown in Fig. 6.3. The method of choice is largely dependent on the enzyme being sought. If, for example, a new lipase or alcohol dehydrogenase is being searched for, the number of metagenome clones to be screened is comparatively small. Looking at the genome of an average prokaryote, a simple E. coli has about 20–50 gene sequences coding for such enzymes. Therefore, for rarer enzymes such as nitrile hydratases or P450 monooxygenases, which have on average only about 1–10 gene sequences per genome, a much larger resource must be screened to increase the probability of identification. Therefore, a high-throughput method is required, in which between 10^5 and 10^9 clones can be screened in a manageable time window.

In practice, both approaches are often combined in the course of a multiparameter screening. In a primary screening for the identification of an enzyme that catalyzes the hydrolysis of an esterified glycerol, all hydrolase-active clones are identified from a metagenome library with 10^5–10^9 clones by means of selective agar plate assays or fluorescence-activated cell sorting (FACS). At this stage, it is not possible to make a precise statement about the activity or specificity of the individual hit clones. Following the primary screening, the hit clones are validated in a secondary screening with lower

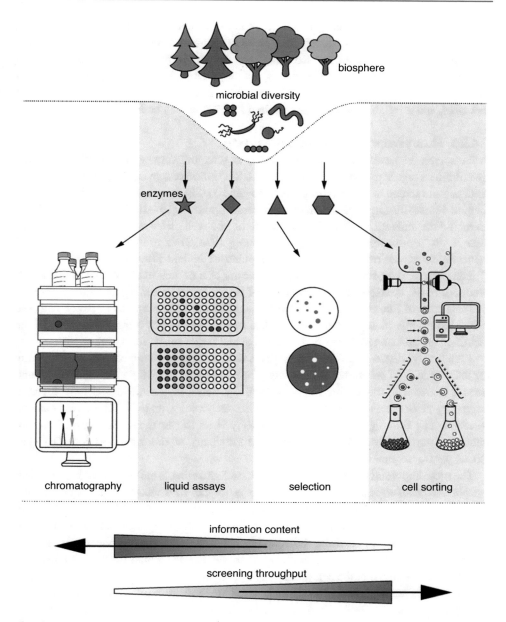

Fig. 6.3 Comparison of the classical cultivation approach and the metagenome technique according to Lorenz (2006). Only a fraction (<1%) of all microorganisms have been cultivated and thus directly used for biotechnological purposes (left). The biosynthetic potential of the entire biodiversity can be made accessible by directly isolating the DNA of all microorganisms from an environmental sample (metagenome, right)

throughput, and the associated gene sequence is elucidated by sequencing. Subsequently, these sequences can be expressed in a suitable expression organism under controlled conditions. Finally, in a tertiary screening, the enzyme properties of the hit candidates (activity, stability, efficiency, specificity) can be intensively characterized using liquid assays or chromatographic methods. In the course of this approach, the information content increases, and a suitable enzyme for a desired application can be filtered out.

6.4.2.1 Fluorescence-Activated Cell Sorting

In the search for new biocatalysts, the choice of screening strategy often depends on the size of the metagenome library to be screened. Since the average number of hit clones in a metagenome screen is about 1–10 candidates per 100,000 clones, an ultrahigh-throughput method is usually required to screen very large libraries (10^8–10^{13} clones). Traditional screening methods such as liquid assays in 96- or 384-well plates or agar plate assays reach their limits. *Fluorescence-activated cell sorting* (FACS) is used today in the field of metagenome screening and also in the screening of mutant libraries in enzyme design. The enzyme activity sought is detected by means of a fluorescent molecule (substrate or product). In the course of the development of FACS-based screening methods, a descriptive selection of fluorescently labeled substrates for different enzyme classes is available today, but usually these are model substrates. In a secondary screening, the activity of the identified enzymes must be verified with the actual target substrate in each case.

Technically, FACS-based screening uses a flow cytometer equipped with special lasers for the detection of substrates or products. Today, there are FACS devices that detect and sort hit clones with the desired enzymatic activity. Thus, they achieve enrichment from a large enzyme or metagenomic library. Using this technology, up to 10^7 clones/hour can be screened. The limiting factor of this technology is no longer the screening method, but rather the comparatively low transformation efficiency of the screening host (10^6–10^8 clones per transformation).

The most important prerequisite in FACS-based screening is the confirmed link between the phenotype of a positive reaction and the genotype. In a metagenome or an enzyme variant screen, this means that the fluorescent molecule must always remain physically linked to the nucleotide sequence coding for the respective enzyme. In practice, there are various possibilities for ensuring the coupling between genotype and phenotype: surface presentation of active enzymes (*cell surface display*), use of reporter proteins, inclusion of the fluorescence signal in the cell (entrapment), and the *in vivo* or *in vitro compartmentalization*. The latter is the most commonly used method in activity-based screening of metagenomic libraries by FACS. Compartmentalization involves packing the entire reaction space into small, usually mono- or polydisperse droplets using a water-oil-water (W/O/W) double emulsion. For this purpose, cells expressing a metagenomic library are enclosed in W/O/W vesicles in the laboratory together with the fluorogenic model substrate and all necessary cosubstrates for an activity assay. The resulting compartmentalization allows for both *in vivo* screening using whole cells producing the metagenomic enzymes and *in vitro* screening using a cell-free expression system to express the metagenome libraries. These

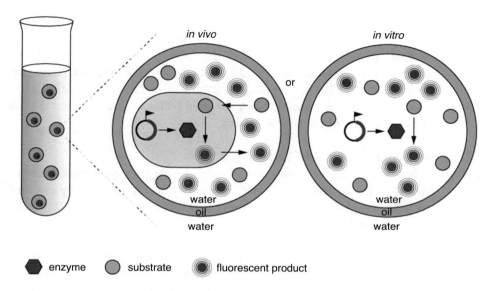

Fig. 6.4 Schematic representation of the procedure for finding new enzymes from a metagenome library using fluorescence-activated cell sorting (FACS) in microcompartments (Rehdorf et al. 2015). In the first step, the library is entrapped in a water-oil primary emulsion and then packaged in water-oil-water double emulsions. This can be done *in vitro* (cell-free, right) or *in vivo* (with whole cells, left). By preparing the double emulsions, the model substrate as well as the product remains in the microcompartment. The detected phenotype can be clearly traced back to the corresponding genotype

microcompartments have an average size of 3–300 μm and allow transcription, translation, as well as biocatalysis within the compartment. The oils used in the W/O/W double emulsion later provide the barrier for the model substrate and the resulting product. This ensures the coupling of the phenotype to the associated genotype (Fig. 6.4). These so-called microdroplets are extremely stable and can even be stored and incubated for several days. Fluorescent microdroplets are subsequently detected, sorted, and enriched in FACS. After several rounds of screening, the DNA of the enriched hit clones can be isolated and sequenced by breaking the double emulsions.

The pioneers of these W/O/W microcompartments for FACS are Andrew D. Griffiths and Dan S. Tawfik (2000). As early as 1998, both described this technology as a tool for the molecular evolution of proteins using the example of a DNA methyltransferase. In addition to being used to screen mutant libraries, microcompartment-based FACS has also been successfully used at BRAIN Biotech to screen for β-galactosidases that selectively and rapidly cleave lactose for the production of lactose-free products. A metagenomic library served as the genetic resource. Cells expressing this library were included in double emulsions for screening. Fluorescein digalactoside served as the model substrate. After a primary screen with three rounds of enrichment, sorted hit candidates were singulated on selective agar with X-Gal (5-bromo-4-chloro-3-indoxyl-β-d-galactopyranoside). All clones were active and showed clear blue staining. Schwaneberg and colleagues optimized the method of *in vitro* compartments and synthesized fluorogenic model substrates for screening different variant libraries of, e.g., monooxygenases (P450 monooxygenases) (Ruff et al. 2012). To completely prevent the diffusion of the fluorescent product from the microcompartment into the environment, researchers have published the further development of the compartments based on hydrogels (Pitzler et al. 2014). In addition, other polymers such as alginates or agarose are now also used for the *in vivo* compartmentation of cells in microbeads for FACS-based screening.

6.4.2.2 Complementation and Selection

Using the selection method, the survival of cells or clones depends on a specific enzyme activity. Due to its robustness, ease of implementation (no expensive equipment and analytical methods required), and high screening throughput (10^8–10^{12} clones), the selection is popularly used as a primary screen. The selection method is particularly used in the metabolization of carbon sources (e.g., rare sugars) and aromatic and aliphatic hydrocarbons (e.g., polyphenols and alkanes) or in the release of nitrogen, sulfur, and phosphates from nonnatural substrates. Here, clones expressing a metagenomic library are plated on selection medium that either contains no additional carbon source or is completely free of nitrogen, sulfur, and phosphate sources. Without the supplementation of the sought enzyme activity that allows the use of the C, N, S, or P source, the clones cannot survive. After several days of incubation, the plates are evaluated. The plasmids of the grown clones are isolated, and the sequence of the associated gene can be elucidated by sequencing. In an activity-based screen for new alkane hydroxylating enzymes, such as P450 monooxygenases or hydroxylases, new enzymes were identified at BRAIN Biotech in a metagenome library (BRAIN Biotech AG, unpublished data). Here, a population of *E. coli* expressing a metagenome library of an environmental sample was offered octane as a sole carbon source. Since *E. coli* cannot metabolize alkanes such as octane, only those clones that terminally hydroxylated the substrate to 1-octanol by a metagenome-derived hydroxylase or P450 monooxygenase were able to survive. This primary alcohol can be used by *E. coli* as a carbon source.

A special case of selection is complementation. Here, an essential gene, which is necessary for cell growth under certain conditions, is deleted in a screening strain (deletion mutant). This strain is then only viable if it receives a metagenomic gene that codes for a complementation-capable and thus functionally identical enzyme. This compensates for the gene loss and the cells are once again viable. This technology is excellently suited for the search for new enzymes for the *de novo* synthesis of essential amino acids or vitamins. Complementation can also be used to identify enzymes for the metabolization of rare sugars or polymers.

6.4.2.3 Determination of Activity on Agar Plates

Plate-based activity assays are also suitable for the identification of enzyme activities in high throughput. Activity determinations on agar plates are qualitative or semiquantitative analyses. Liquid assays, which include the use of chromogenic or fluorescent substrates, are much more suitable for quantitative analyses. On one agar plate (9 cm diameter), 2000–3000 clearly separated single colonies can be cultivated. This easily allows the screening of 10^5–10^6 clones in a manageable period of time. Thus, this method is also suitable for large gene banks. However, it is a prerequisite that the enzyme group sought is secreted. For intracellular enzymes, the cells would have to be lysed or at least destabilized to ensure contact with the substrate in the agar plate. In addition to approach of heterologous expression, it is also possible to cultivate strains isolated from habitats of interest directly on substrate agar plates and characterize the secretion of enzymes.

The characterization of hydrolytic activity is often carried out using agar plates containing an enzyme-specific substrate that can be degraded by the enzymes. This causes a clarification of the agar (halo formation). This clearing in a zone around the growing microorganism is clearly distinguishable from the turbid background of the substrate plate. The size of the halo may be an indication of the amount and/or activity of the secreted enzyme. These substrate agar plates are particularly used for the identification of proteolytic and lipolytic enzymes. Waschkowitz et al. (2009) identified and characterized two metagenome-derived metalloproteases using skim-milk agar plates, among other methods. The casein from the milk powder serves as a substrate for proteases. Degradation of casein leads to clarification in a region around a hit clone, indicating protease activity is present.

For the identification of lipolytic enzymes such as lipases and esterases, agar plates containing a glycerol substituted with three butyl esters (tributyrin) are predominantly used. Again, halo formation occurs as a result of enzyme activity. Tributyrin is used as a substrate by both lipases and esterases and thus does not allow clear differentiation. If differentiation of lipases and esterases is to occur, enzyme-specific substrates must be used. Liquid assays, for example, would lend themselves to such differentiation.

In addition to substrate plates that detect enzyme activity via a halo formation, agar plates can also be used that induce a color change when enzymatic activity is present. The diammonium salt of 2,2′-azino-di-(3-ethylbenzthiazoline-6-sulfonic acid) (ABTS) is a redox dye suitable for the direct as well as indirect detection of oxidoreductases such as oxidases, peroxidases, or laccases. Laccases are copper-dependent enzymes that use molecular oxygen for the oxidation of substrates. From an industrial point of view, laccases are interesting because they are used in the textile and paper industries, among others.

6.4.2.4 Determination of Activity in Liquid Assays

For the analysis of many enzymes in high throughput, color changes and fluorescence are very suitable. Labeled substrates are able to directly show the relationship between enzyme activity and measurable signal. Furthermore, good labeled substrates allow the detection of enzyme activity below the protein detection limit. In addition to the pure identification of enzymes, labeled substrates can also be used to determine enzyme properties such as pH- and temperature-dependent activity, enantioselectivity, tolerance to substances, or enzyme stability.

Suitable substrates have been labeled with a coloring or fluorescent component. Enzymatic reactions do not cause color changes or fluorescence independently. Enzyme reactions that use nicotinamide adenine dinucleotide (NAD) as a coenzyme are an exception, since NAD(P)$^+$/NAD(P)H can be determined photometrically.

Labeled substrates are usually model substrates that allow direct activity detection. Substrates whose color or fluorescence changes significantly in the course of an enzymatic reaction are particularly suitable. An obvious disadvantage of using model substrates is that enzymes are not characterized on the basis of their target substrates from a certain application. Therefore, the transferability of activity to the "real" target substrate is low. Consequently, it is not guaranteed that identified enzymes will ultimately react on the desired target substrates. In this case, coupled assays provide a remedy. In coupled assays, enzymes are provided with the "real" target substrates. In a subsequent reaction, the unreacted substrates or the resulting products are converted into a measurable signal that can be easily detected.

Hydrolases are by far the most widely used enzymes in industry. Proteases catalyze the hydrolysis of peptide bonds and are therefore able to degrade peptides as well as proteins. Peptides linked to a p-nitroaniline via an amide are classical protease substrates. They produce a color change from colorless to yellow at basic conditions, which can be detected photometrically at a wavelength of 410 nm. Characterization of subtilisin-like serine proteases is usually performed using N-succinyl-l-Ala-l-Pro-l-Phe-p-nitroanilide, which releases p-nitroaniline after proteolytic cleavage and can be detected photometrically. Detection of other hydrolases is often by other p-nitrophenyl derivatives. Cleavage of glycosides by glycosidases and cleavage of esters by lipases, esterases, or phospholipases can be identified by the use of p-nitrophenol labeled substrates. For example, a p-nitrophenol labeled ester was used to characterize the optimization of a lipase from *Pseudomonas aeruginosa* with respect to its enantioselectivity (Liebeton et al. 2000; Reetz et al. 1997). The lipase was provided with either an *(S)*-ester or an *(R)*-ester, each of which was labeled using p-nitrophenol. A screen was performed to examine the activities of the lipase variants for both substrates and to analyze the preference of the lipase for one of the two enantiomers. Ultimately, it was possible to generate highly enantioselective variants of the lipase by means of enzyme engineering.

Resorufin is another dye used for labeling substrates. Due to the large extinction coefficient, the sensitivity of the assay is high and thus allows the detection of the lowest enzyme activities. After excitation of the resorufin, it emits light at 584 nm and can thus be used for both spectrophotometric and fluorometric analysis. Substrates for proteases, lipases, and esterases, β-galactosidases, or glucosidases are often labeled with resorufin.

Oxidoreductases are a versatile group of enzymes that catalyze redox reactions. For the identification of new oxidoreductases, numerous detection methods have been established and optimized in recent years which allow a rapid and reliable finding of these enzymes. The activity of dehydrogenases can be quantified directly or indirectly via NAD(P)$^+$/NAD (P)H. As mentioned above, no labeled substrate is required for the detection of this reaction. NAD(P)$^+$ acts as an acceptor of reduction equivalents and is therefore involved in redox reactions. Dehydrogenases use these properties of NAD(P)$^+$ to transfer protons between molecules. Both oxidized NAD(P)$^+$ and reduced NAD(P)H absorb light at specific wavelengths and can thus be determined photometrically. The oxidized $NAD(P)^+$ absorbs light of a wavelength of 260 nm, whereas the reduced $NAD(P)H$ absorbs light of a wavelength of 340 nm. Due to these properties, enzymatic reactions can be followed via $NAD(P)^+/NAD(P)H$. However, if dehydrogenases are to be identified or analyzed in a high-throughput manner, this method is usually not suitable. For this reason, detection methods have been developed that label $NAD(P)H$ via a redox dye to facilitate photometric determination. Such a reaction can be mediated by nitro blue tetrazolium chloride (NBT), which consists of a ditetrazolium salt and a redox dye. Reduction of the ditetrazolium salt in the presence of phenazine methosulfate produces an insoluble blue diformazan dye with an absorption maximum at 560 nm.

Oxidases, in turn, catalyze the transfer of a hydrogen atom from a substrate to molecular oxygen which acts as an electron acceptor in this reaction (oxidation). Water (H_2O) or hydrogen peroxide (H_2O_2) is released in the process. Substrates include sugars such as galactose (galactose oxidase) or glucose (glucose oxidase) but also D/L-amino acids (amino acid oxidases). Oxidase activity is usually detected by the H_2O_2 formed in a coupled enzyme assay. The formation of H_2O_2 can be quantified using a horseradish peroxidase (HRP). This coupled method utilizes the peroxidase activity of HRP which catalyzes the reduction of H_2O_2 and can oxidize substrates. These substrates include chromogenic substrates such as 2,2'-azino-di-(3-ethylbenzthiazoline-6-sulfonic acid) (ABTS) or 3,3'--dimethoxybenzidine (o-dianisidine). The HRP-mediated reaction with the redox indicator ABTS produces a blue-green ABTS radical cation (ABTS-$^+$) whose absorbance is photometrically measurable at 414 nm and proportional to the amount of H_2O_2 formed. ABTS is well suited for high-throughput identification and characterization of oxidases. It is soluble in both aqueous and organic solvents; it can be used in a wide range of pH and temperature, and the blue-green coloration of ABTS-$^+$ is stable. In addition to the HRP/ABTS method, H_2O_2 can also be detected via HRP/o-dianisidine. This gives a brown-red product that can be determined at 436 nm. ABTS or o-dianisidine can also be used for the identification of new peroxidases. In this case, the peroxidases directly catalyze the oxidation of the chromogenic substrates. Other chromogenic substrates oxidized by the activity of peroxidases are 3,3',5,5'-tetramethylbenzidine (color change from blue to yellow) and guaiacol (color change from colorless to brown).

Within the oxygenases, a distinction is made between mono- and dioxygenases. Using molecular oxygen and coenzymes such as FAD and $NAD(P)H$, one or two oxygen atoms are reductively transferred to a substrate with the release of water. The most important representatives of this enzyme group are the cytochrome P450 monooxygenases (e.g., P450-BM3 from *Bacillus megaterium*), the Baeyer-Villiger monooxygenases (e.g., cyclohexanone monooxygenase from *Acinetobacter calcoaceticus*), and the catechol or tryptophan dioxygenases. For the detection of oxygenases in high-throughput methods, labeled model substrates such as p-nitrophenyl and umbelliferone derivatives can be used, which are regioselectively oxygenated. The released p-nitrophenol can be detected spectroscopically at 410 nm, and the blue-fluorescent umbelliferone can be quantified at 455 nm after excitation at 325 nm.

Fluorogenic substrates are an alternative to chromogenic substrates. The conversion of the substrates works analogously to the chromogenic substrates. The main advantage of fluorogenic substrates is a higher sensitivity. This allows to work in smaller volumes with lower enzyme and substrate concentrations as well as with enzymes whose concentration or activity is too low for detection by chromogenic detection. Fluorescence intensity can be quantified using a fluorescence photometer. If a 96-well plate fluorescence photometer is used, fluorogenic substrates can also be used in a high-throughput manner. Ideally, a fluorogenic substrate allows the release of highly fluorescent and water-soluble products whose optical properties are significantly different from those of the substrate. Under these conditions, it is possible to read out the product formation without a further separation step, e.g., by means of HPLC.

Numerous fluorescent dyes are based on coumarin. One of the most common is 7-amino-4-methylcoumarin (AMC). Peptides such as Z-Gly-Gly-l-Arg-AMC, which serve as substrates for proteases, are often linked to an AMC dye at the C-terminal carboxyl group via an amide. After cleavage of the peptide by a protease, the fluorescent dye is released, and excitation at 360 nm allows the emission to be determined at 460 nm. This is possible because the fluorescence of the dye after cleavage from the peptide is significantly different from the fluorescence when bound to the peptide. More sensitive fluorescent dyes are based on fluorescein, rhodamine, and resorufin.

Another class of fluorescent dyes are 4,4-difluoro-4-bora-3a,4a-diaza-s-indacene derivatives (BODIPY). Based on the BODIPY basic structure, a series of different BODIPY dyes with different optical properties were generated (green, yellow, orange, and red fluorescent). The green fluorescent BODIPY FL was used in a study at BRAIN Biotech to identify novel proteases (Niehaus et al. 2011). In this study, the authors used the metagenome of a soil sample to identify three novel proteases with suitable properties for use in detergents. In particular, serine proteases from the subtilase group are well suited for this application due to their properties (Gupta et al. 2002). Prioritized soil samples that had a basic pH were used to create a metagenomic library. Sequence-based and activity-based screening of DNA libraries using skim-milk substrate agar plates resulted in the identification of 18 protease sequences. The identified protease sequences were cloned into expression vectors and subsequently expressed in *E. coli*. BODIPY-FL labeled casein was used to confirm the activity of these new proteases and to characterize them. The labeled substrate was excited at a wavelength of 485 nm, and the emission was determined at 520 nm. In addition to the collection of biochemical properties, the washing performance of the new proteases was analyzed in application tests. Ultimately, three new proteases were identified from the metagenome that showed great potential for application in the detergent industry.

6.4.2.5 Chromatographic Methods

In order to be able to identify the desired catalytic activity after screening a metagenome or an enzyme library, the hit candidates must be intensively investigated and evaluated. The characterization of the enzymes often includes the elucidation of the regio-, stereo-, and enantioselectivity. For the determination of these parameters, analytical procedures are used, for which various methods and measuring devices are employed. The choice of these methods and instruments depends largely on the analytes to be investigated (substrates, products, by-products) and their properties. Very important is the solubility of the compounds in various polar and nonpolar organic solvents, the melting and boiling points, the vapor pressure, and the number and properties of the functional groups in the molecules (e.g., hydroxyl, amino, carboxyl, epoxy, keto, and halogen groups). The most important analytical methods include gas chromatography (GC) and high-performance liquid chromatography (HPLC).

The choice of method depends on the volatility and hydrophilicity of the analytes (Table 6.2). Volatile and mostly hydrophobic compounds are mainly analyzed by GC. The most important prerequisite for GC analysis is that all analytes and solvents used in the injector pass into the gas phase without decomposing. The analytes are therefore initially vaporized in the injector (at approx. 250 °C) and transported in gaseous form by means of a suitable carrier gas (usually helium, nitrogen, or hydrogen) via a solid phase separation column in the temperature controlled GC oven. Various detectors can be used in GC for the detection of the analytes. These detectors include the flame ionization detector (FID) and the mass spectrometer (MS). The FID is suitable for many organic compounds. It is characterized by high sensitivity, robustness, and the possibility of direct quantification. An MS also allows a mass to be assigned to each peak in a chromatogram. In addition to the retention time, the molecular peak is an important parameter in the identification of unknown analytes in mixtures of substances. The sensitivity of the MS can be increased by using quadrupole mass spectrometers. In addition to the FID and the MS, there are also special detectors for the detection of functional groups such as nitrogen and phosphorus (NPD, nitrogen phosphorus detector) or sulfur-containing, nitrated and halogenated compounds (ECD, electron capture detector).

Nonvolatile and predominantly hydrophilic analytes are analyzed by means of HPLC. In contrast to GC, the analytes are transported in a liquid stream (running medium) over a solid phase separation column. The complete solubility and stability of the analytes in the running medium is a prerequisite for LC analysis. For the detection of the separated analytes, LC very often uses a UV detector (diode array detector, DAD). The prerequisite for this is that the analytes have a chromophoric group that can be detected at a specific wavelength. A universally applicable detector for LC analysis is the evaporative light scattering detector (ELSD). It can be used to detect all nonvolatile organic compounds. However, the sensitivity is limited, so that the mass spectrometer is often used in analogy to GC analysis. Another detector for LC analysis which is mainly used in sugar analysis is the refractive index detector (RID). Due to the low sensitivity of the RID, low amounts of sugar are often detected by means of high-performance anion exchange chromatography (HPAEC) coupled to the amperometric PAD detector.

In some cases, it is necessary to derivatize the functional groups of the analytes for better detection or stability. For quantification by GC analysis, this may be necessary for unstable samples that would already decompose in the injector without derivatization. A standard for OH-groups or acid functions represents the silylation, methylation, or the acetylation.

Table 6.2 Overview of current analytical methods for the determination of enzyme activity and regio-, chemo-, and enantioselectivity

Analytes	Method	Options for detection
Volatile, hydrophobic analytes (melting point <250 °C), i.e., fatty acids, terpenes	GC	FID, MS
Nonvolatile, hydrophilic analytes, i.e., amino acids, sugars, or oligosaccharides	HPLC	ELSD, DAD, MS, RID

For primary amino groups, OPA reagents (ortho-phthaldialdehyde) are used, whereas for secondary amino groups, so-called FMOC (fluoromethoxycarbonyl), BOC (*tert-butyl*), or Cbz (benzyloxycarbonyl) derivatives are employed. For the use of the UV detector in LC analysis, non-UV-active substances must also be derivatized with UV-capable chromophores prior to analysis.

Questions
1. What are relevant criteria that an enzyme must fulfill in order to be used in an industrial process?
2. From which group originate the majority of enzymes used today?
3. What is the name for the total genomic DNA that can be isolated from a habitat sample?
4. What is meant by a multiparameter screening?
5. Which expression systems are normally used to create a metagenome library?
6. What are shuttle vectors and what is their advantage?
7. What is the major benefit of an activity-based screening approach with regard to the necessary prerequisites?
8. What is the difference between a high-throughput and a low-throughput screening?
9. What is the disadvantage of a FACS-based screening?
10. How can the use of model substrates be circumvented?
11. What is the difference between selection and complementation?
12. Which two analytical devices are often used to analyze substances, and what is the main difference between both?

6.5   Conclusion and Challenges

The methods and possibilities to identify "ideal" enzymes for industrial use have developed greatly in recent years. Activity-based enzyme screening is still a very powerful approach to discover previously unknown functionalities. In addition, bioinformatics methods, which have made tremendous progress in recent years, allow the prediction of sought enzyme activities from digital sequence databases. The combination of both approaches enables biotechnologists to identify exciting enzymes and thus make biocatalytic industrial processes possible.

As nature's toolbox, the metagenome provides access to an almost unlimited biodiversity and thus represents an important resource for the identification of new biocatalysts. Today, numerous screening options have been established, which can be modified depending on the requirements and the scientific question. The general workflow of an activity-based enzyme screening starts with a high-throughput method providing limited information about the enzymes and ends with a low-throughput method providing much information about the enzymes: one could think of screening throughput as a funnel. FACS-based screening and selection technology is often used for primary screenings due to its high throughput. In a secondary screening using chromogenic or fluorogenic substrates and/or products, the hit candidates are further investigated and biochemically characterized with respect to target parameters for subsequent process implementation. Finally, the exact determination of enzyme activity, selectivity, and specificity is carried out using chromatographic methods like GC and LC.

The greatest challenges of activity-based screenings are the development of a robust, high-throughput screening system and the functional production of new biocatalysts in an expression host. Reliable predictions that allow an accurate prediction of a suitable expression host for a given gene sequence have not yet been developed. Further, an expression host must have a high transformation efficiency and grows under conditions that are neither ideal nor correspond to the later industrial enzyme production using the production host. Strong production hosts are essential for the large-scale and economic production of an enzyme in an industrial setting. In order to ensure the producibility of an enzyme at an early stage, it is advantageous if the activity-based screening can be carried out in the subsequent enzyme production host. However, this is not always possible. Consequently, one of the most important questions, namely, the economic production of a biocatalyst, often remains open during the actual screening campaign.

Another limitation is the investigation of multistep enzyme reactions (reaction cascades). Especially in the biotechnological production of natural substances for the food and cosmetics industry (amino acids, antimicrobial compounds, etc.), designer microorganisms are often used that produce entire enzyme cascades and metabolic pathways. In such a system, the properties of the individual biocatalysts must be coordinated. So far, these can only be implemented to a limited extent in a multiparameter screening (e.g., cofactor dependencies, toxicity of substrates, intermediates and products, unstable intermediates, metabolic flux) and must be investigated separately in most cases. In the end, all enzymes involved in the cascade reaction must be assembled in a production strain and analyzed for the target parameters.

Prediction of the properties (activity, specificity, selectivity) and representability of new enzymes early in the process of enzyme identification would greatly advance the development of high-performance biocatalysts. Great hope is pinned on the field of bioinformatics. *De novo* enzyme structure prediction, artificial intelligence, and machine learning will dramatically accelerate screening for new enzymes.

Take-Home Message
Activity-based screening means identifying new enzymes solely on the basis of their activity. This requires powerful expression strains and reliable detection methods. The search in nature's toolbox, e.g., in metagenomes, requires high-throughput methods at the beginning to identify interesting enzyme activities. In the course of the process, the methods become more specialized in order to characterize multiple parameters of a few hit candidate enzymes in detail. The result of the screening process is a suitable enzyme that must be produced in a production strain in large quantities to make the step to industrial application.

Answers
1. Enzymes often do not fulfill process criteria because industrial processes are artificial and reaction conditions for an enzyme are very different from those in nature. Limitations are often caused by process temperatures, pressure, resistance to solvents, and high concentration of substrates and products. Therefore, relevant criteria for an industrial enzyme include activity and inhibition, stability (e.g., pH, temperature), specificity and substrate spectrum, high production yields, and manufacturability in industrial relevant scales.
2. The majority of relevant biocatalysts today originate from microorganisms; beyond them, prokaryotes play the most important role.
3. The total genomic DNA isloated from an environmental sample (habitat) is called metagenome.
4. Simultaneous screening and/or characterization of more than one relevant enzyme property. This could include, e.g., screening for an enzyme's specific activity and production yield in a production host under defined process conditions (temperature, pH, substrate concentration).
5. Plasmids, cosmids, fosmids, or bacterial artificial chromosomes (BACs).
6. Shuttle vectors allow gene expression in parallel in different expression hosts. This saves time as the construction of several, specific expression vectors is not required and increases the likelihood to express genes from unknown organisms. This is especially beneficial for genes encoding for enzymes that require posttranslational modifications or secretion.
7. Activity-based screening focuses exclusively on the measurable activity of an enzyme. Protein sequence and structure do not play a role. This also makes it possible to find enzymes whose activity for a particular reaction was previously unknown. Enzymes have the ability to catalyze side reactions in addition to their main reaction (enzyme promiscuity).
8. A high-throughput screening allows the investigation of many data points per time, e.g., up to 10^9 clones per day for a certain activity. However, no detailed information on the screened enzymes can be determined (e.g., specificity, by-product formation, product inhibition). In a low-throughput screening using chromatographic methods,

detailed information on different enzyme parameters can be determined. This allows only a few enzymes per day.

9. A FACS-based screening and sorting often makes use of fluorescent model substrates. Model substrates allow the utilization of the FACS method but are normally not the relevant substrate for a target application. Thus, the activity of enzyme on these substrates must be reinvestigated and verified using the target substrate.

10. Coupled assays can be used to avoid model substrates but still be able to detect enzyme activity via a measurable readout. In coupled assays, enzymes are provided with the "real" target substrates. In a subsequent reaction, the unreacted substrates or the resulting products are converted into a measurable signal that can be easily detected.

11. Selection refers to a growth substrate. The screening host can only grow on selection medium if the clone expresses a gene that facilitates generation of the growth substrate. Complementation is a special case of selection, where the screening host lacks an essential gene to survive (deletion mutant). Only if the mutant possesses a gene from the metagenomic library that compensates the deletion will the strain is able to survive and grow.

12. GC and HPLC. GC is used for volatile and hydrophobic compounds, whereas HPLC is often the better choice for nonvolatile and rather hydrophilic compounds.

References

Aehle W, Eck J. Discovery of enzymes. In: Enzyme catalysis in organic synthesis. Wiley-VCH Verlag GmbH & Co. KGaA; 2012. p. 67–87.

Beloqui A, Pita M, et al. Novel polyphenol oxidase mined from a metagenome expression library of bovine rumen: biochemical properties, structural analysis, and phylogenetic relationships. J Biol Chem. 2006;281:22933–42.

Burton SG, Cowan DA, et al. The search for the ideal biocatalyst. Nat Biotechnol. 2002;20:37–45.

Gabor E, Liebeton K, et al. Updating the metagenomics toolbox. Biotechnol J. 2007;2:201–6.

Griffiths AD, Twafik DS. Man-made enzymes—from design to in vitro compartmentalisation. Curr Opin Biotechnol. 2000;11:338–53.

Gupta R, Beg QK, et al. Bacterial alkaline proteases: molecular approaches and industrial applications. Appl Microbiol Biotechnol. 2002;59:15–32.

Handelsmann J, Rondon MR, et al. Molecular biological access to the chemistry of unknown soil microbes: a new frontier for natural products. Chem Biol. 1998;5:245–9.

Hug L, Baker B, Anantharaman K, et al. A new view of the tree of life. Nat Microbiol. 2016;1., Article No. 16048 https://doi.org/10.1038/NMICROBIOL.2016.48.

Liebeton K, Zonta A, et al. Directed evolution of an enantioselective lipase. Chem Biol. 2000;7:709–18.

Lorenz P. Metagenomics for white biotechnology. Chem Ing Tech. 2006;78:461–8.

Lorenz P, Eck J. Metagenomics and industrial applications. Nat Rev Microbiol. 2005;3:510–6.

Niehaus F, Gabor E, et al. Enzymes for the laundry industries: tapping the vast metagenomic pool of alkaline proteases. Microb Biotechnol. 2011;4:767–76.

Ono A, Miyazaki R, et al. Isolation and characterization of naphthalene-catabolic genes and plasmids from oil-contaminated soil by using two cultivation-independent approaches. Appl Microbiol Biotechnol. 2007;74:501–10.

Pitzler C, Wirtz G, et al. A fluorescent hydrogel-based flow cytometry high-throughput screening platform for hydrolytic enzymes. Chem Biol. 2014;21:1733–42.

Reetz MT, Zonta A, et al. Generation of enantioselective biocatalysts for organic chemistry by *in vitro evolution*. Appl Chem. 1997;109:2961–3.

Rehdorf J, Meinhardt S, et al. Regarding the potential of fluorescence-based double emulsion cytometry for screening complex metagenome libraries. In: Blickwinkel evolution. Zwingenberg: BRAIN Aktiengesellschaft; 2015. p. 7–12.

Ruff AJ, Dennig A, et al. Flow cytometer-based high throughput screening system for accelerated directed evolution of S. 450 monooxygenases. ACS Catal. 2012;2:2724–8.

Torsvik V, Ovreas L, et al. Prokaryotic diversity—magnitude, dynamics, and controlling factors. Environ Microbiol. 2002;296:1064–6.

Waschkowitz T, Rockstroh S, et al. Isolation and characterization of metalloproteases with a novel domain structure by construction and screening of metagenomic libraries. Appl Environ Microbiol. 2009;75:2506–16.

Whitman WB, Coleman DC, et al. Prokaryotes: the unseen majority. Proc Natl Acad Sci U S A. 1998;95:6578–83.

Bioinformatic Methods for Enzyme Identification

7

Anett Schallmey

What You Will Learn in This Chapter

Bioinformatic methods for sequence analysis of homologous DNA or protein sequences are nowadays basic tools in biological sciences, whether for determining the sequence identity of two protein or DNA molecules, finding homologous sequences of a known enzyme, or analyzing complex relationships of whole organisms. This chapter describes different bioinformatic methods and respective tools that will enable you to search for new protein sequences, which are homologous to a known search sequence, in public sequence databases, to align those homologous protein sequences in multiple sequence alignments and to infer their phylogenetic relationships in the form of a phylogenetic tree.

7.1 From Gene Sequence to Enzyme Function

The existing sequence information in public sequence databases represents an inexhaustible source for new enzymes and other proteins. Due to many genome and metagenome sequencing projects, this sequence information has increased by a factor of 100 per decade over the last 30 years. However, in order to be able to evaluate the available sequence data appropriately and

---

A. Schallmey (✉)

Institute for Biochemistry, Biotechnology and Bioinformatics, Technische Universität Braunschweig, Braunschweig, Germany

e-mail: a.schallmey@tu-braunschweig.de

© The Author(s), under exclusive license to Springer Nature Switzerland AG 2024

K.-E. Jaeger et al. (eds.), *Introduction to Enzyme Technology*, Learning Materials in Biosciences,

https://doi.org/10.1007/978-3-031-42999-6_7

thus make it usable, a number of different bioinformatic methods are necessary. This starts with the correct annotation of DNA sequences from genome and metagenome sequencing projects to deduce primary protein sequences encoded therein. Based on sequence homologies, possible functions can then be assigned to the protein sequences. However, for a large number of probable protein sequences, no functions have yet been experimentally demonstrated. These sequences are accordingly labeled as hypothetical proteins in public databases. In contrast to activity-based methods of enzyme identification, bioinformatic methods can only be used to find enzymes with given similarity (homology) to already known enzymes. In addition, the prediction of biochemical properties such as substrate spectrum or selectivity of an enzyme based only on sequence homologies has been difficult up to now. Regarding the prediction of enzyme structure, the recent development of AlphaFold2 as a protein structure prediction tool revolutionized the work of protein biochemists as well as structural and computational biologists as it allows prediction of protein structures with high accuracy even for proteins displaying only low or even no homology to proteins with already known structure (Jumper et al. 2021).

7.1.1 Sequence Analysis and Gene Identification

The specific amino acid sequence of each protein determines its structure and thus, in the case of enzymes, also its activity. The basis of this protein sequence is the genetic information, which is given via the nucleotide sequence of the DNA. The identification of new enzymes is therefore usually always carried out indirectly via their gene sequences or via the protein sequences derived from them (Sect. 7.2). This means that the correct identification of genes within known DNA sequences is a prerequisite for the derivation of correct protein sequences.

In order to identify new enzymes on a sequence basis, one can make use of the sequence information already available in public databases such as GenBank (https://www.ncbi.nlm. nih.gov/genbank/). However, sequence information can also come from in-house genome or metagenome sequencing projects—supported by the development of modern sequencing technologies and the associated dramatic reduction in sequencing costs. In this case, genes contained in the DNA sequence must first be correctly identified (annotated). A gene consists of a sequence of triplet codons, which are framed by a start and a stop codon. In addition, the DNA sequence contains further "signals," such as ribosome binding sites or promoter regions, which also indicate the presence of a gene. For eukaryotic DNA sequences, gene identification is further complicated by the presence of noncoding regions (introns). Gene identification in a given DNA sequence can be performed by a database search and the generation of sequence alignments, e.g., using the Basic Local Alignment Search Tool (BLAST) algorithm (Sect. 7.2.1) or ab initio (from scratch, i.e., without sequence comparison) with special computer programs.

BLAST offers different search options: blastn, blastx, and tblastx (Table 7.1). **Blastn** can be used to search a DNA sequence database for homologous sequences to the given DNA sequence. This is useful, for example, if the genome sequences of closely related organisms are already known and annotated. In addition to genes, DNA regions coding for functional RNAs, for example, can also be identified. In contrast, **blastx** can be used to

Table 7.1 BLAST programs in NCBI for searching homologous sequences

Name	Description	Comment
Blastn	Searches for homologous sequences in a DNA sequence database starting from a DNA sequence	Considering both DNA strands
Blastp	Searches for homologous sequences in a protein sequence database starting from a protein sequence	Standard program for the identification of homologous enzymes
Blastx	Searches for homologous sequences in a protein sequence database starting from a translated DNA sequence	Taking into account all possible reading frames during translation; useful for gene identification (Sect. 7.1)
Tblastn	Searches for homologous sequences in a translated DNA sequence database starting from a protein sequence	Taking into account all possible reading frames during translation
Tblastx	Searches for homologous sequences in a translated DNA sequence database starting from a translated DNA sequence	Considering all possible reading frames during translation; useful for gene identification (Sect. 7.1)
PSI-BLAST	Searches for homologous sequences in a protein sequence database based on a multisequence alignment of several homologous proteins	For identification of even remotely homologous enzymes
PHI-BLAST	Searches for homologous sequences in a protein sequence database based on a protein sequence and specific sequence motifs	For identification of even remotely homologous enzymes that share a specific sequence motif
DELTA-BLAST	Searches for homologous sequences in a protein sequence database based on an initial alignment of the search sequence against conserved protein domains	For identification of even remotely homologous enzymes

search directly in a protein database, whereby the given DNA sequence is first translated taking into account the six possible reading frames (three per DNA strand). As a result of a blastx search, all protein-coding regions (genes) of the DNA search sequence as well as the corresponding homologous protein sequences available in the database are obtained. In a **tblastx search,** the given DNA sequence is first also translated but then searched in a likewise translated DNA sequence database (taking into account the six possible reading frames). As with blastx, this enables a search at the protein level and thus simplifies the identification of (even remotely homologous) genes.

With the help of the program **ORF Finder** (Open Reading Frame Finder) of the National Center for Biotechnology Information (NCBI), possible open reading frames (ORFs) and thus possible protein-coding regions can be predicted for a given DNA sequence on the basis of sequence signals (link: https://www.ncbi.nlm.nih.gov/orffinder/). Modern ab initio gene prediction programs, e.g., for annotation of genome or metagenome sequences, additionally combine the search for sequence signals with modern statistical methods (hidden Markov models, HMM). However, even these programs do not provide 100% accurate predictions. This is particularly true when eukaryotic DNA sequences are involved.

The sequence information available in public databases is already annotated. However, since the majority of the sequence information contained has been annotated automatically and the computer programs used for this purpose do not (yet) work without errors, it is recommended that the sequences found during a homology search for new enzymes be checked again manually for correct annotation. In particular, it is important to check the correct identification of the start and stop codon of a gene and thus its length (Sect. 7.5.3). Furthermore, protein sequences from metagenome data, such as those contained in the env_nr database, may have incomplete N or C termini. This is not initially apparent from the amino acid sequence but becomes clear when the underlying gene is checked, e.g., if the start and stop codons are missing.

Questions
1. What does the acronym BLAST stand for?
2. Why is it important to reevaluate correct gene annotation of protein sequences, especially when working with sequences derived from metagenomic sequence databases?

7.2   Homology Search

7.2.1   BLAST

Probably the most commonly used program for searching for homologous sequences is BLAST, which is available via NCBI (https://blast.ncbi.nlm.nih.gov/Blast.cgi) (Altschul et al. 1990). BLAST enables both DNA and protein sequences to be searched in public sequence databases by comparing sequences in pairs on the basis of a search sequence. In addition, versions exist that start not only from one but from several search sequences. BLAST thus represents a whole family of different search programs (Table 7.1).

For the identification of homologous enzymes—starting from an already known enzyme—the search in a protein sequence database is the method of choice (compared to the homology search via the corresponding DNA sequence). Homology searches based on protein sequences are much more sensitive due to the redundancy of the genetic code and the fact that the similarity of protein sequences is much more conserved compared to DNA sequences.

For the protein search by BLAST, various sequence databases are available from which to choose (Table 7.2). Depending on the goal pursued in the homology search, the search can thus be restricted to a specific database. For the identification of new enzymes, however, the nr and env_nr databases from NCBI are the most interesting. The nr database includes sequence data from other protein databases in addition to GenBank and is thus very comprehensive. The env_nr database, on the other hand, offers the possibility of

Table 7.2 Important protein sequence databases for protein BLAST search in NCBI

Name	Description
Nonredundant protein sequences (nr)	Includes all translated gene sequences from GenBank as well as protein sequences from the Protein Data Bank (PDB), Swiss-Prot, Protein Information Resource (PIR), and Protein Research Foundation (PRF), excluding metagenomic data from world-global-sampling (WGS) projects
Reference proteins (refseq_protein)	Includes all NCBI protein reference sequences
UniProtKB/Swiss-Prot	Includes all protein sequences from UniProtKB/Swiss-Prot
Patented protein sequences	Includes only patented protein sequences from GenBank
Protein Data Bank	Includes all protein sequences with known structures from PDB
Metagenomic proteins (env_nr)	Includes only protein sequences from WGS metagenome projects

searching specifically for new enzymes in metagenome sequence data. In this search, however, care must be taken to ensure that potentially interesting protein sequences are subsequently checked for completeness using the underlying DNA sequence, as the env_nr database has a higher proportion of incomplete protein sequences. In addition, the BLAST search can also be performed specifically in translated genome sequence data of individual organisms.

BLAST is a heuristic method that provides local alignments between the search sequence and a corresponding database entry as the result of the homology search. In addition, for each hit found, the actual significance of the homology between the database entry and the search sequence is indicated in the form of the sequence identity, a score, and an E-value. In general, the higher the sequence identity or score and the lower the E-value, the more significant the homology. It is particularly important to assess this significance correctly if remotely homologous sequences are also to be considered, e.g., those found via a *position-specific iterated* (PSI)-BLAST search.

PSI-BLAST enables an iterative BLAST search, whereby a simple BLAST search is first performed in a protein sequence database based on a search sequence (Altschul et al. 1997). Subsequently, a multiple sequence alignment is generated from the homologous sequences found, using the search sequence as a template, and a sequence profile is generated from this in turn. Here, for each amino acid position, the probability that a particular amino acid occurs at that position is calculated. In the next round of BLAST, this sequence profile serves as a search criterion for identifying further homologous sequences in the database and generating a new sequence profile. These steps proceed iteratively, either over a predetermined number of rounds or until no new sequences are found in the database. It is important to set a threshold for the E-value, which delimits what minimum homology a sequence must have to be included in the multisequence alignment and thus the generation of the sequence profile. If this value is set too high (equivalent to a low significance of the homology present), there is a risk that false-positive hits will be included in the alignment, amplifying in subsequent rounds and thus giving an overall false homology result (Jones and Swindells 2002). The advantage of PSI-BLAST, however, is that this method does not only search for homologous enzymes starting from a single search sequence, but practically a sequence profile based on a multisequence alignment serves as the search sequence and the search is carried out iteratively, whereby homologous enzymes can be found even far away. Similarly, *domain enhanced lookup time accelerated* (DELTA)-BLAST also enables iterative BLAST searches using a sequence profile, which however is obtained from an initial alignment of the search sequence against conserved protein domains in the Conserved Domain Database (CDD) of NCBI (Boratyn et al. 2012).

Pattern-hit initiated (**PHI-)BLAST** also allows the unambiguous identification of homologous enzymes in a sequence database, even if their homology to the search sequence is quite low. In this method, in addition to the search sequence, one or more sequence motifs conserved for the enzyme family being searched are included in the search (Zhang et al. 1998). In practical terms, the sequence database to be searched is first reduced to the sequences that have the specified sequence motif, and then a simple BLAST search is performed only in this restricted and thus significantly smaller sequence database. In this way, it is also possible to find distantly homologous enzymes that might be missed in a standard BLAST search due to insufficient homology. However, a prerequisite for the successful performance of a PHI-BLAST search is the knowledge of appropriately conserved sequence motifs for the enzyme family in focus, which can be determined, e.g., via a multiple sequence alignment (Sect. 7.3). The more specifically the sequence motifs can be defined, the more meaningful and reliable the result of the homology search will be.

Questions
3. Why should a homology search in a sequence database for the identification of novel enzymes be better performed on protein sequence level instead of gene sequence level?
4. What is the difference between PSI-BLAST and PHI-BLAST?

7.3   Multiple Protein Sequence Alignment

Multisequence alignments (MSAs) of three or more protein sequences are nowadays required in the biological sciences for a whole range of different analyses. In addition to the search for homologous protein sequences (Sect. 7.2), they are mainly used as a basis for phylogenetic analyses (Sect. 7.4) or to identify conserved regions (e.g., individual amino acids, sequence motifs, or domain structures) in enzyme families. In addition, MSAs are used in structural modeling of proteins (Chap. 3). Various programs are available for the creation of multiple sequence alignments, some of which use different algorithms (Edgar and Batzoglou 2006). Since the creation of an MSA requires significantly more computing power than a pairwise alignment of only two sequences, these algorithms are also largely based on heuristic methods and thus only provide an approximation to the optimal alignment of a given number of sequences.

The most commonly used method for calculating MSAs is progressive alignment. It is based on the combination of pairwise sequence alignments, starting with the most similar sequence pair up to the most distantly related sequence pair. In the first step, the relationship of the sequences to each other is determined and defined in a guide tree, comparable to a phylogenetic tree. Subsequently, the sequences are added to the growing MSA according to the guide tree. If, however, an error is introduced at one point in the pairwise alignment, this will propagate to the final MSA and thus deliver an inaccurate alignment. The correct alignment of sequences with low homology is particularly difficult. Therefore, today's programs use additional strategies, such as iterative or consistency-based methods, to minimize errors and thus generate more reliable MSAs. Nevertheless, the respective result will always be only a more or less accurate approximation. As a user, one must therefore be aware that different MSA programs may also provide different sequence alignments due to different underlying algorithms for the same set of protein sequences. Depending on the further use of the obtained MSA, it is up to the user to check the accuracy or reliability of the alignment.

Iterative and Consistency-Based Methods for the Improvement of Multisequence Alignments

In programs such as MAFFT, MUSCLE, or Clustal Omega, the accuracy of a multisequence alignment (MSA) initially obtained via progressive alignment is improved by iterative refinement *of* the alignment. In this process, the original MSA is repeatedly subdivided into sub-alignments, and the arrangement of the sequences in the respective sub-alignment is optimized.

In contrast, programs like T-Coffee or ProbCons use consistency as information when computing an MSA via progressive alignment. This means that each pairwise alignment of two sequences takes into account how it contributes to the global MSA. Thus, a slightly less optimal alignment of two sequences may well be preferred if this results in a better overall alignment of all sequences.

7.3.1   Selection of a Suitable Alignment Program

When selecting a suitable program for the creation of a multiple sequence alignment, three parameters are of particular importance:

- The biological accuracy of the generated MSA
- The required computing time
- The required processing unit power

The accuracy of the MSA is usually the most important. However, it is quite difficult to objectively check or validate the accuracy of the different programs. In practice, this is

Table 7.3 Frequently used programs for the generation of multiple sequence alignments of homol-ogous protein sequences and their special features and areas of application

Name/web address	Description	Application
Clustal Omega http://www.ebi.ac.uk/Tools/msa/clustalo/	Good accuracy of generated MSAs for very large datasets (>20,000 sequences); for small datasets less accurate than MAFFT, ProbCons, or T-Coffee	Alignment of very large datasets
Expresso http://tcoffee.crg.cat/apps/tcoffee/do:expresso	Alignment based on sequence and structure information (a variant of the T-Coffee method)	Generation of structure-based alignments, e.g., for structural modeling (Chap. 3)
MAFFT http://mafft.cbrc.jp/alignment/server/	Offers a good ratio between required computing power and achieved accuracy of the MSA; selection of additional alignment options possible; also suitable for very large datasets (>20,000 sequences), but with somewhat lower accuracy than Clustal Omega	Standard program for the alignment of fewer sequences as well as large datasets
MUSCLE http://www.ebi.ac.uk/Tools/msa/muscle/	Offers a good ratio between required computing power and achieved accuracy of the MSA; also suitable for large datasets	Standard program for the alignment of a few sequences as well as large datasets (up to 3000 sequences)
webPRANK http://www.ebi.ac.uk/goldman-srv/webPRANK/	Generation of MSAs including the evolutionary relationship of protein sequences; not suitable for clearly distantly related sequences; computationally intensive; not suitable for large datasets	Phylogenetic analysis of small datasets (<100 sequences) with clear (>40%) sequence identity
ProbCons http://probcons.stanford.edu/index.html	Very high accuracy of the generated MSA, but very computationally intensive, especially for large datasets	Standard program for alignment of rather small datasets
T-Coffee http://tcoffee.crg.cat/apps/tcoffee/do:regular	High accuracy of the generated MSA, but computationally intensive, especially with larger datasets; inclusion of additional information possible	Standard program for alignment of rather small datasets

done using benchmark tests based on reference alignments. Not surprisingly, programs that perform best in these benchmark tests are often also very computationally intensive. However, not only the accuracy of the MSA plays a role for the required computing power but also the number of sequences to be compared (the more, the more difficult) and their homology to each other (the lower, the more difficult). Therefore, it always depends on the respective problem which program is best suited for the creation of an MSA (Table 7.3).

According to the literature, ClustalW, for example, was used most frequently for many years because the program required comparatively little computing capacity to create an alignment. However, benchmark tests showed that newer programs such as MAFFT, MUSCLE, T-COFFEE, or PROBCONS, which use different or improved algorithms, provide significantly more accurate MSAs. Due to ever new questions and application areas for multiple sequence alignments and the related continuous development of algorithms for the computation of MSAs, a large number of programs, even for very specific applications, are available nowadays (Chatzou et al. 2015). Only some of these can also be used via a web server. Table 7.3 provides an overview of common online programs for comparing homologous protein sequences and their areas of application.

Questions
5. What is the most common method for the calculation of a multiple sequence alignment (MSA)?
6. Which methods are used by multiple sequence alignment programs to improve the accuracy of resulting MSAs?
7. Which multiple sequence alignment programs can be used to align large sequence datasets (>10,000 sequences)?

7.4 Phylogenetic Analyses

To investigate the relationships of homologous protein sequences, a phylogenetic analysis is performed as standard. The result is a phylogenetic tree illustrating the relationships. The branching of individual branches (topology of the phylogenetic tree) and the distances between individual nodes in the phylogenetic tree describe the relative relationships between individual sequences. Each phylogenetic tree is based on a multiple sequence alignment, which is required for the computer-aided calculation of distances and topologies. The computer programs available for this purpose use different methods to infer phylogenetic trees from MSAs (Table 7.4). Basically, these methods can be divided into distance-based and character-based methods (Table 7.5; Yang and Rannala 2012). Distance-based methods such as *neighbor joining* rely on pairwise calculation of genetic distances between all sequences in the alignment and the creation of a corresponding distance matrix as the basis for calculating the phylogenetic tree. Character-based methods such as *maximum parsimony* or *maximum likelihood* instead directly use the sequence information of homologous regions in the alignment to infer a phylogenetic tree. A newer method, *Bayesian inference*, is also likelihood-based like *maximum likelihood* but yields not only one highly probable phylogenetic tree as a result but several similarly probable phylogenetic trees for comparison. This method is therefore particularly interesting when it comes to tracing the exact evolutionary relationship and origin of sequences.

Table 7.4 Frequently used computer programs for phylogenetic analyses

Name	Explanation	Web address
MEGA	The program package for phylogenetic analyses via distance, parsimony, and likelihood methods including alignment program; can import sequence data directly from GenBank; offers a graphical user interface	http://www.megasoftware.net/
PHYLIP	The program package for phylogenetic analyses using distance, parsimony, and likelihood methods	http://evolution.genetics.washington.edu/phylip.html
PhyML	Fast program for maximum likelihood analyses based on nucleotide or protein sequence data	http://www.atgc-montpellier.fr/phyml/
FastME	Fast program for phylogenetic analyses using distance methods; extension of the neighbor-joining algorithm	http://www.atgc-montpellier.fr/fastme/
IQTREE	Fast and efficient program for maximum likelihood analyses based on nucleotide or protein sequence data	http://www.iqtree.org/
BEAST	Program for phylogenetic analyses via Bayesian inference	http://beast.bio.ed.ac.uk/

Table 7.5 Methods for reconstructing phylogenetic trees

Method	Description	Advantages/disadvantages
Neighbor joining[a]	Uses a fast clustering algorithm to calculate a phylogenetic tree from a distance matrix; uses a substitution model to calculate the pairwise sequence distances	Requires comparatively little computing power; particularly suitable for large datasets of clearly homologous sequences, but error-prone for sequences with low homology
Maximum parsimony[b]	Searches for the phylogenetic tree that requires the fewest nucleotide or amino acid exchanges to describe the sequences present in the MSA and their homologies; does not use a substitution model	Very simple method; however, it does not use existing information about actual sequence evolution; error-prone in the case of more distantly related sequences, since possible multiple exchanges at a sequence position are not taken into account
Maximum likelihood[b] (ML)	Searches for the most probable phylogenetic tree for a given MSA based on a substitution model	Requires significantly more computing power; allows the use of different substitution models to describe the phylogeny; also error-prone when using a too simple substitution model
Bayesian inference[b]	Searches for the most probable phylogenetic trees for a given MSA based on a substitution model; in contrast to ML, model parameters are not unknown constants, but random variables with statistical distribution	Requires high computing power; allows the use of different substitution models to describe the phylogeny; the result is easy to interpret by calculating probabilities for phylogenetic trees and branching, but calculated probabilities often appear too high

[a]Distance-based
[b]Character-based

Distance- and Character-Based Methods for the Determination of Phylogenetic Trees
In distance-based methods, a genetic distance is first calculated for each pair of sequences
in the sequence alignment based on a substitution model. The resulting distance matrix for
all sequences in the dataset then serves as the basis for calculating the phylogenetic tree.

In contrast, character-based methods use the sequence information in the multisequence
alignment directly to calculate the phylogenetic tree. Here, all sequences are compared
simultaneously, but each position in the alignment (character) is considered individually
in turn.

With the exception of *maximum parsimony*, both distance-based and character-based
methods use a substitution model to describe the nucleotide or amino acid exchanges
between the sequences in the MSA. Different models are available, some of which make
different assumptions (e.g., regarding the abundance or distribution of nucleotides or amino
acids). Therefore, for the reconstruction of actual evolutionary relationships (e.g., of
organisms), it is important to use a substitution model that best describes the sequence
information in the MSA. In contrast, for the construction of a phylogenetic tree of
homologous protein sequences for the graphical representation of the relationships of the
underlying sequences, the actual evolutionary development of the proteins plays a rather
minor role. Nevertheless, the "appearance" of the underlying MSA as well as the method
used also have a major impact on the subsequent topology of the phylogenetic tree.
Therefore, for a given set of homologous protein sequences, it makes sense to test different
programs for generating the multiple sequence alignment as well as different methods for
calculating the phylogenetic tree and to subsequently compare the phylogenetic trees
obtained (see also Sect. 7.5.4). The various phylogenetic trees obtained may well differ
in detail.

In order to investigate the reliability of a phylogenetic tree, a bootstrap analysis is often
performed. In this statistical method, pseudoreplicates, usually 100 or 1000, of the original
sequence dataset are created, and phylogenetic trees are also constructed from them via the
same method as the original phylogenetic tree. By comparative analysis of all bootstrap
phylogenetic trees, a percentage frequency is calculated for each branching point (node) in
the phylogenetic tree, indicating the proportion of all bootstrap phylogenetic trees that
actually contain that branch. In practice, branches with at least 70% bootstrap support are
considered reliable. Another possibility is to create a consensus tree from all bootstrap
trees, which only shows the branches that are also contained in all individual trees.

In the representation of phylogenetic trees, an additional distinction is made between
two different types. In a *rooted tree*, a precursor sequence common to the input sequences
is calculated during phylogenetic tree construction, from which the phylogenetic tree is
built (Fig. 7.1). In practice, the determination of such a precursor sequence requires the
inclusion of at least one additional but more distantly related sequence (outgroup) in the
sequence dataset. This outgroup sequence, if chosen correctly, should appear in the
phylogenetic tree close to the precursor sequence and be the furthest away from all other

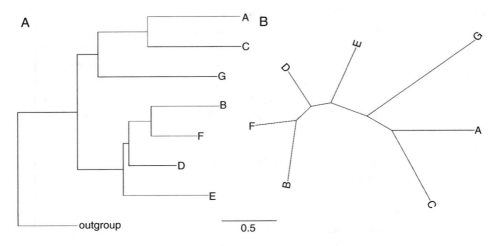

Fig. 7.1 Phylogenetic tree topologies for a *rooted* (**a**) and an *unrooted* (**b**) phylogenetic tree

sequences. An *unrooted tree*, on the other hand, graphs the distances and relationships between input sequences without knowledge or consideration of a common ancestor.

Questions
8. Which two general types of methods are distinguished to infer a phylogenetic tree from a multiple sequence alignment?
9. What is the substitution model required for?
10. What is additionally required for generation of a rooted phylogenetic tree compared to an unrooted one?

7.5   Example: Identification of New Halohydrin Dehalogenases

Halohydrin dehalogenases (HHDHs) are enzymes that have so far only been described in a few bacterial species. They catalyze the reversible dehalogenation of vicinal haloalcohols with formation of the corresponding epoxides and are biotechnologically relevant. For example, HHDHs are used to remove toxic haloalcohols in food applications or serve as enzymes in biocatalysis to produce various β-substituted alcohols. Phylogenetically, HHDHs belong to the short-chain dehydrogenase/reductase (SDR) superfamily, with which they share some structural and mechanistic features, even though the reactions they catalyze are fundamentally different. Because of the significant homology between HHDHs and SDR enzymes, unambiguous discrimination of the two groups of enzymes in a simple BLAST search is nontrivial but essential for the identification of new HHDH sequences in sequence databases. Until recently, only the gene and protein sequences of five different HHDHs were known, which were identified by classical microbial activity-based screening. In order to increase the number of available HHDHs for biotechnological applications, attempts were made to identify new HHDH sequences in public databases via a homology search (Schallmey et al. 2014).

7.5.1 MSA for the Identification of Specific Sequence Motifs

The five previously described HHDHs were known to have a catalytic triad consisting of serine, tyrosine, and arginine. In contrast, the catalytic triad of most SDR enzymes is composed of serine, tyrosine, and lysine. In addition, unlike SDR enzymes, HHDHs do not possess a nucleotide cofactor binding pocket, as HHDHs do not require a corresponding cofactor for their activity. This nucleotide cofactor binding pocket is formed in classical SDR enzymes by the glycine-rich motif $T-G-X_3-(G/A)-X-G$. Instead, HHDHs carry a nucleophile binding pocket at this position, which is flanked by aromatic amino acids in the solute crystal structures of the known HHDHs HheC and HheA2. Thus, first sequence-specific information to distinguish HHDHs from SDRs was known.

In a multiple sequence alignment of the five HHDHs and selected SDR enzymes generated via MAFFT, the amino acid residues of the catalytic triads of the HHDH and SDR enzymes aligned very well. In addition, compared to the SDR enzymes, the HHDH sequences contained an aromatic residue instead of the central glycine or alanine of the glycine-rich motif of the SDRs. This information was used in a first step to distinguish new HHDHs from homologous SDR enzymes in an MSA of homologous sequences obtained via a blastp search starting from an HHDH search sequence. In this way, several new, initially putative halohydrin dehalogenases could be found, whose activity was later confirmed experimentally. However, this search was quite laborious, as each sequence in the alignment still had to be manually checked with respect to its catalytic triad and the presence of a glycine/alanine residue or an aromatic amino acid at the same position in the nucleophile/nucleotide binding site. Therefore, in order to more easily distinguish HHDHs and SDR enzymes in a BLAST search, one or more HHDH-specific sequence motifs should be identified. To this end, a new MSA was generated from the already known and all new HHDHs via MAFFT and examined with regard to conserved regions in the alignment. It was found that for all HHDHs, the three amino acids of the catalytic triad have fixed distances to each other. Thus, on the one hand, the sequence motif $S-X_{12}-Y-X_3-R$ could be established. Furthermore, it turned out that the already mentioned aromatic amino acid in HHDHs is also flanked by a conserved motif $(T-X_4-(F/Y)-X-G)$, which is similar to the glycine-rich motif of SDR enzymes but clearly contributes to the distinction between HHDHs and SDRs. Despite individual differences in the various MSAs, these sequence motifs were also confirmed by sequence alignments generated via webPRANK and Clustal Omega (Fig. 7.2).

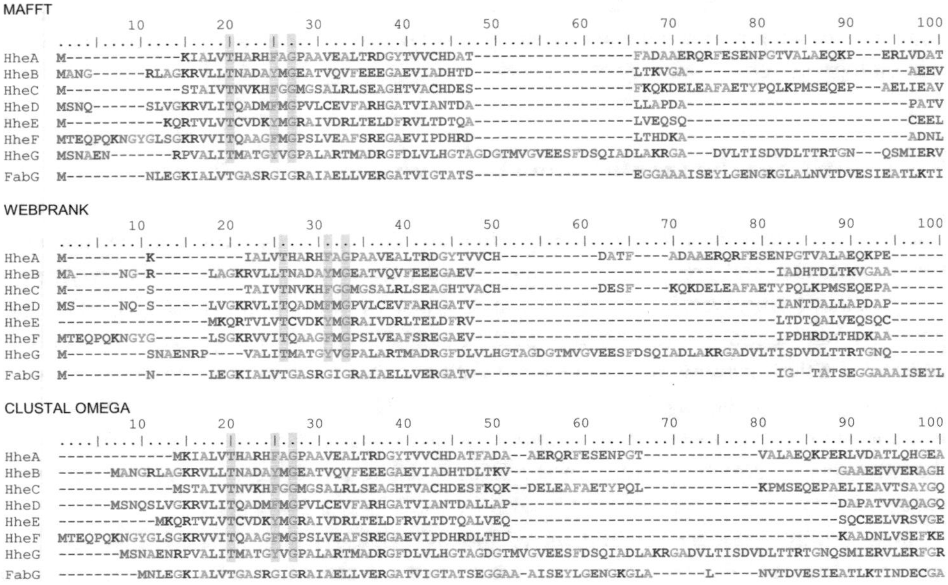

```
MAFFT
            10        20        30        40        50        60        70        80        90       100
     ....|....|....|....|....|....|....|....|....|....|....|....|....|....|....|....|....|....|....|....|
HheA M-------------KIALVTHARHFAGPAAVEALTRDGYTVVCHDAT-----------------FADAAERQRFESENPGTVALAEQKP---ERLVDAT
HheB MANG------RLAGKRVLLTNADAYMGEATVQVFEEEGAEVIADHTD------------------LTKVGA-------------------------AEEV
HheC M------------STAIVTNVKHFGGMGSALRLSEAGHTVACHDES------------------FKQKDELEAFAETYPQLKPMSEQEP---AELIEAV
HheD MSNQ------SLVGKRVLITQADMFMGPVLCEVFARHGATVIANTDA------------------LLAPDA-----------------------PATV
HheE M-----------KQRTVLVTCVDKYMGRAIVDRLTELDFRVLTDTQA------------------LVEQSQ----------------------CEEL
HheF MTEQPQKNGYGLSGKRVVITQAAGFMGPSLVEAFSREGAEVIPDHRD------------------LTHDKA----------------------ADNL
HheG MSNAEN-------RPVALITMATGYVGPALARTMADRGFDLVLHGTAGDGTMVGVEESFDSQIADLAKRGA---DVLTISDVDLTTRTGN---QSMIERV
FabG M---------NLEGKIALVTGASRGIGRAIAELLVERGATVIGTATS-----------------EGGAAAISEYLGENGKGLALNVTDVESIEATLKTI

WEBPRANK
            10        20        30        40        50        60        70        80        90       100
     ....|....|....|....|....|....|....|....|....|....|....|....|....|....|....|....|....|....|....|....|
HheA M---------K----------IALVTHARHFAGPAAVEALTRDGYTVVCH-----------DATF----ADAAERQRFESENPGTVALAEQKPE------
HheB MA------NG-R------LAGKRVLLTNADAYMGEATVQVFEEEGAEV-----------IADHTDLTKVGAA-------
HheC M---------S----------TAIVTNVKHFGGMGSALRLSEAGHTVACH-----------DESF----KQKDELEAFAETYPQLKPMSEQEPA------
HheD MS-----NQ-S------LVGKRVLITQADMFMGPVLCEVFARHGATV-------------------IANTDALLAPDAP------
HheE ------------------MKQRTVLVTCVDKYMGRAIVDRLTELDFRV-------------------LTDTQALVEQSQC------
HheF MTEQPQKNGYG-------LSGKRVVITQAAGFMGPSLVEAFSREGAEV-------------------IPDHRDLTHDKAA------
HheG M---------SNAENRP----VALITMATGYVGPALARTMADRGFDLVLHGTAGDGTMVGVEESFDSQIADLAKRGADVLTISDVDLTTRTGNQ------
FabG M--------N------LEGKIALVTGASRGIGRAIAELLVERGATV--------------------------IG--TATSEGGAAAISEYL

CLUSTAL OMEGA
            10        20        30        40        50        60        70        80        90       100
     ....|....|....|....|....|....|....|....|....|....|....|....|....|....|....|....|....|....|....|....|
HheA -----------MKIALVTHARHFAGPAAVEALTRDGYTVVCHDATFADA--AERQRFESENPGT-------------VALAEQKPERLVDATLQHGEA
HheB ------MANGRLAGKRVLLTNADAYMGEATVQVFEEEGAEVIADHTDLTKV---------------------------GAAEEVVERAGH
HheC -------------MSTAIVTNVKHFGGMGSALRLSEAGHTVACHDESFKQK--DELEAFAETYPQL-------------KPMSEQEPAELIEAVTSAYGQ
HheD ------MSNQSLVGKRVLITQADMFMGPVLCEVFARHGATVIANTDALLAP-------------------DAPATVVAQAGQ
HheE -----------MKQRTVLVTCVDKYMGRAIVDRLTELDFRVLTDTQALVEQ----------------------------SQCEELVRSVGE
HheF MTEQPQKNGYGLSGKRVVITQAAGFMGPSLVEAFSREGAEVIPDHRDLTHD-----------------------KAADNLVSEFKE
HheG -------MSNAENRPVALITMATGYVGPALARTMADRGFDLVLHGTAGDGTMVGVEESFDSQIADLAKRGADVLTISDVDLTTRTGNQSMIERVLERFGR
FabG ---------MNLEGKIALVTGASRGIGRAIAELLVERGATVIGTATSEGGAA-AISEYLGENGKGLA------L-----NVTDVESIEATLKTINDECGA
```

Fig. 7.2 *N-terminal* section of multiple sequence alignments of selected halohydrin dehalogenases (HheA to HheG) for the identification of HHDH-specific sequence motifs. Residues of HHDH-specific sequence motif T-X$_4$-(F/Y)-X-G are highlighted by gray shading. The SDR enzyme FabG was included for comparison. The three MSAs were generated using the programs MAFFT, webPRANK, and CLUSTAL Omega

7.5.2 Homology Search in Public Databases

A simple BLAST search in the NCBI nr database with one of the known HHDHs as search sequence yields, due to the existing homology, a large number of SDR enzymes in addition to new putative HHDH sequences. Without further sequence information to distinguish HHDHs from SDRs, the true HHDHs in the list of homologous sequences could subsequently be identified only by appropriate activity assays. Since this takes a lot of time and resources, only a rather small number of homologous sequences could typically be investigated experimentally. Thus, the probability of finding many new HHDHs this way would be rather low. The homology search, however, can be improved by knowledge of additional HHDH-specific sequence information. As described in Sect. 7.5.1, two HHDH-specific sequence motifs could be identified for halohydrin dehalogenases via a multiple sequence alignment. The first, T-X$_4$-(F/Y)-X-G, comprises amino acids of the nucleophile binding pocket of HHDHs. In contrast, SDR enzymes carry a glycine-rich motif (T-G-X$_3$-(G/A)-X-G) for nucleotide cofactor binding. The second HHDH-specific sequence motif, S-X$_{12}$-Y-X$_3$-R, involves the amino acids of the catalytic triad of halohydrin dehalogenases: serine, tyrosine, and arginine. SDR enzymes instead classically possess a catalytic triad consisting of serine, tyrosine, and lysine, although the distance between serine and tyrosine can vary. Accordingly, HHDHs can be distinguished from SDR enzymes via these two sequence motifs. By using either one of the two sequence motifs together with a HHDH search sequence as the starting point of a PHI-BLAST search, the number of SDRs in the list of homologous sequences can already be significantly reduced. The combination of both sequence motifs in the PHI-BLAST search, on the other hand, yields as a result exclusively (putative) HHDH sequences with homology to the search sequence and at the same time enables a significantly deeper search of the sequence space available in the database than a simple BLAST search. In comparison, a PSI-BLAST search also finds the more distantly homologous HHDHs, i.e., it also searches the existing sequence space in much greater depth. However, the resulting list of homologous sequences still includes SDR enzymes, as these also show clear homology to halohydrin dehalogenases. Thus, in our specific example, i.e., with knowledge of specific HHDH sequence motifs and a clear discrimination between homologous HHDH and SDR enzymes, a PHI-BLAST search provides the better result.

7.5.3   Verification of Correct Gene Annotation

Of the new HHDHs identified by homology search in the nr or env_nr database, some sequences did not contain the typical ATG start codon according to the original annotation. Therefore, the correct annotation was first checked for all sequences found, searching for an alternative ATG start codon within the original gene sequence (Fig. 7.3). Truncated gene versions, however, were only considered if the amino acid sequences they encoded still contained the sequence motif T-X_4-(F/Y)-X-G located near the N-terminus. In addition, the presence of a *Shine-Dalgarno* sequence, a few nucleotides upstream of the new ATG start codon was checked. This A/G-rich sequence serves as a ribosome binding site during translation in prokaryotes. By checking the gene annotation of the sequences found, it was possible to identify an alternative ATG start codon with a corresponding ribosome binding site for all HHDH genes that did not carry the typical ATG start codon in their original annotation.

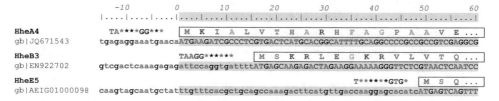

Fig. 7.3 Verification of the correct annotation of translation starts with three HHDH-coding sequences contained in GenBank

7.5.4   Phylogenetic Analysis of New Halohydrin Dehalogenases

To investigate the phylogeny of all newly identified HHDHs including the previously known enzymes, different phylogenetic trees were constructed based on two different MSAs (generated via MAFFT and PRANK + F, respectively) and using two different phylogenetic algorithms (PhyML and FastME, respectively). Moreover, bootstrap analysis was performed afterward in each case. As a result, all the obtained phylogenetic trees were quite similar in their topology, especially regarding the main branches in the phylogenetic tree. Accordingly, HHDHs can be divided into a total of six phylogenetic subtypes (A + C, B, D, E, F, and G) (Fig. 7.4). In contrast, the exact splitting of the individual enzymes within the subtypes differed in the different phylogenetic trees, so that the actual phylogenetic relationship of HHDHs of the same subtype cannot be conclusively clarified. Based on the performed bootstrap analyses, these results were confirmed overall.

> **Take-Home Message**
> - Public sequence databases are a rich source for the identification of novel enzyme catalysts.
> - BLAST is an easy tool to search for homologous sequences in public sequence databases based on a starting protein or nucleotide sequence. BLAST derivatives such as PSI-BLAST or PHI-BLAST help to find also remotely homologous enzymes.
> - A large variety of programs is available for the generation of multisequence alignments of small as well as large datasets of homologous protein sequences. All programs, however, can only provide an approximation to the optimal alignment of a given sequence dataset.
> - The phylogenetic relationships of homologous protein sequences can be illustrated by a phylogenetic tree. A multisequence alignment of those homologous protein sequences forms the basis for tree calculation.

Answers
1. BLAST is an abbreviation for Basic Local Alignment Search Tool.
2. Protein sequences from metagenome data may have incomplete N or C termini.
3. Homology searches based on protein sequences are much more sensitive due to the redundancy of the genetic code and the fact that the similarity of protein sequences is much more conserved compared to DNA sequences.
4. Both methods enable the identification of also distantly homologous enzymes to a certain search sequence. However, while PSI-BLAST uses a sequence profile derived from a multiple sequence alignment as search query and performs iterative rounds of homology searches, PHI-BLAST uses a single search sequence in combination with one or more sequence motifs conserved for the enzyme family being searched.

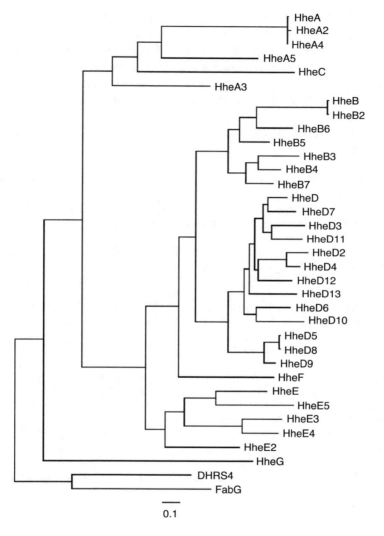

Fig. 7.4 Phylogenetic tree, generated with PhyML based on a PRANK + F MSA, showing the relationships of the different HHDH sequences. FabG and DHRS4 (SDR enzymes) were used as outgroup sequences

5. The most commonly used method for calculating multiple sequence alignments is progressive alignment.
6. Iterative and consistency-based methods are used to improve the accuracy of multiple sequence alignments.
7. The programs Clustal Omega or MAFFT can be used to generate reliable multiple sequence alignments of large datasets including more than 10,000 sequences.

8. Distance- and character-based methods are distinguished for the determination of phylogenetic trees from multiple sequence alignments.
9. A substitution model is required to describe the nucleotide or amino acid exchanges between the sequences in the multiple sequence alignment.
10. For generation of a rooted phylogenetic tree, an outgroup (i.e., a more distantly related sequence) has to be included in the underlying multiple sequence alignment.

References

Altschul SF, Gish W, Miller W, Myers EW, Lipman DJ. Basic local alignment search tool. J Mol Biol. 1990;215:403–410.

Altschul SF, Madden TL, Schäffer AA, Zhang J, Zhang Z, Miller W, Lipman DJ. Gapped BLAST and PSI-BLAST: a new generation of protein database search programs. Nucleic Acids Res. 1997;25(17):3389–3402.

Boratyn GM, Schäffer AA, Agarwala R, Altschul SF, Lipman DJ, Madden TL. Domain enhanced lookup time accelerated BLAST. Biol Direct. 2012;7:12.

Chatzou M, Magis C, Chang J-M, Kemena C, Bussotti G, Erb I, Notredame C. Multiple sequence alignment modeling: methods and applications. Brief Bioinform. 2015:1–15.

Edgar RC, Batzoglou S. Multiple sequence alignment. Curr Opin Struct Biol. 2006;16:368–373.

Jones DT, Swindells MB. Getting the most from PSI-BLAST. Trends Biochem Sci. 2002;27(3): 161–164.

Jumper J, Evans R, Pritzel A, Green T, Figurnov M, Ronneberger O, Tunyasuvunakool K, Bates R, Žídek A, Potapenko A, Bridgland A, Meyer C, Kohl SAA, Ballard AJ, Cowie A, Romera-Paredes B, Nikolov S, Jain R, Adler J, Back T, Petersen S, Reiman D, Clancy E, Zielinski M, Steinegger M, Pacholska M, Berghammer T, Bodenstein S, Silver D, Vinyals O, Senior AW, Kavukcuoglu K, Kohli P, Hassabis D. Highly accurate protein structure prediction with AlphaFold. Nature. 2021;596:583–589.

Schallmey M, Koopmeiners J, Wells E, Wardenga R, Schallmey A. Expanding the halohydrin dehalogenase enzyme family: identification of novel enzymes by database mining. Appl Environ Microbiol. 2014;80(23):7303–7315.

Yang Z, Rannala B. Molecular phylogenetics: principles and practice. Nat Rev Genet. 2012;13:303–314.

Zhang Z, Schäffer AA, Miller W, Madden TL, Lipman DJ, Koonin EV, Altschul SF. Protein sequence similarity searches using patterns as seeds. Nucleic Acids Res. 1998;26(17):3986–3990.

Optimization of Enzymes

8

Dominique Böttcher and Uwe T. Bornscheuer

What You Will Learn in This Chapter

This chapter describes the strategies for the optimization of enzymes for industrial applications using protein engineering methods. Firstly, the two main concepts—rational protein design and directed evolution—are introduced and the major differences are explained. Molecular biology methods for generating mutant libraries by site-directed mutagenesis or random mutagenesis techniques are described, as well as concepts for screening or selection to identify the desired enzyme variants. Lastly, successful first and recent examples of optimized biocatalysts are given.

When using enzymes as biocatalysts in industrial applications, one quickly encounters limitations due to their natural function in the living cell. This is because enzymes are often not compatible, for example, with the requirements for organic synthesis. The properties of enzymes such as substrate or product inhibition, stability, and catalytic efficiency are finely tuned by natural evolution to ensure survival of the organism. However, this is rather disadvantageous in an industrial application where high substrate concentration and complete turnover are required. Also, the naturally high substrate specificity, which is very important in the cell to prevent undesired side reactions, means that the enzyme can often be used only for the synthesis of a few products.

D. Böttcher · U. T. Bornscheuer (✉)

Department of Biotechnology and Enzyme Catalysis, Institute of Biochemistry, Greifswald University, Greifswald, Germany

e-mail: dominique.boettcher@uni-greifswald.de; uwe.bornscheuer@uni-greifswald.de

© The Author(s), under exclusive license to Springer Nature Switzerland AG 2024   165
K.-E. Jaeger et al. (eds.), *Introduction to Enzyme Technology*, Learning Materials in Biosciences,
https://doi.org/10.1007/978-3-031-42999-6_8

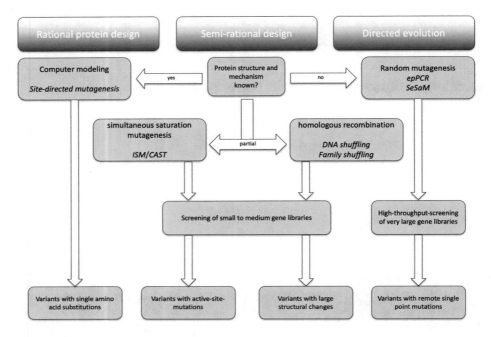

Fig. 8.1 Protein engineering strategies

The design of enzymes with tailor-made properties is therefore of great importance for industrial applications.

8.1 Strategies for the Optimization of Enzymes

Depending on the information available about the protein of interest, there are two different approaches for optimizing enzymes: rational protein design and directed evolution or in vitro evolution (Fig. 8.1).

8.1.1 Rational Protein Design

The rational protein design uses the methods of bioinformatics to gain knowledge about how a protein structure must be changed so that a biocatalyst has the desired properties such as activity, stability, and specificity.

For this purpose, the target protein must first be purified, characterized in biochemical detail, and then its three-dimensional structure needs to be elucidated by X-ray crystallography. If the protein structure cannot be elucidated, a homology model can be created based on the structural data of related (homologous) proteins stored in the Protein Data Bank (PDB). Once the location of the active site and the catalytic mechanism are known,

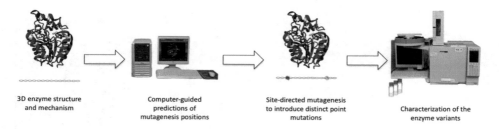

3D enzyme structure
and mechanism

Computer-guided
predictions of
mutagenesis positions

Site-directed mutagenesis
to introduce distinct point
mutations

Characterization of the
enzyme variants

Fig. 8.2 Rational protein design

substrate-binding modeling (*molecular modeling*) and planning of mutagenesis positions are performed. Computer programs (e.g., Yasara with AutoDock) or web server-based applications (Caver3.0, HotSpot Wizard) are often used for this purpose. Based on the predictions, individual amino acids are then exchanged by targeted mutagenesis, and the resulting protein variants are produced, purified, and biochemically characterized.

However, rational protein design alone rarely yields new enzyme variants with significantly improved properties (Fig. 8.2). Further rounds of optimization are often necessary.

8.1.2 Directed Evolution

Unlike rational design, the methods of directed evolution do not require detailed knowledge of the three-dimensional structure of the enzyme, the reaction mechanism, or the relationships between structures, sequences, and mechanisms. The prerequisites for the evolutionary approach are only the presence of the gene encoding the enzyme to be optimized, a suitable expression system (usually *Escherichia coli*; see also Chap. 9), an effective method to produce high-quality mutant libraries, and a powerful screening or selection system (Fig. 8.3). In directed evolution, only random mutagenesis methods are used. Here, a distinction is made between PCR-based in vitro methods and in vivo methods that mutate the entire genome in the growing bacterial cell in addition to the plasmid. In the case of in vitro random mutagenesis methods, a further distinction is made between non-recombining methods (e.g., *error-prone PCR*) and recombining methods such as DNA shuffling.

8.1.3 Semi-rational Design

Based on the findings of structural data and biochemical analyses, semi-rational methods combine the advantages of rational approaches and random mutagenesis methods to generate small "smart" libraries of enzyme variants (Fig. 8.4). Moreover, if the analysis of the crystal structure or the homology model does not provide concrete evidence for a targeted exchange of individual amino acids in the active site, several positions in the active site can alternatively be subjected to (simultaneous) saturation mutagenesis. A frequently

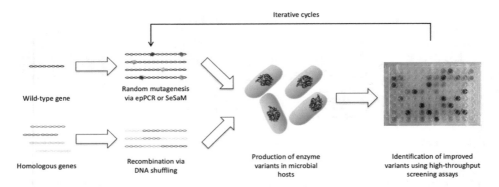

Fig. 8.3 Directed evolution

used method of semi-rational design is the iterative saturation mutagenesis (ISM), in particular the CAST method (Sect. 8.2.2).

Questions
1. Why is optimization of enzymes often required?
2. Which two concepts are used for enzyme engineering?
3. What is the main difference between these two methods?

8.2   Methods for Generating a Mutant Library

8.2.1   Site-Specific Saturation Mutagenesis

In this method, a mutant library is created in which a distinct amino acid is replaced by the other proteinogenic amino acids using so-called wobble primers, i.e., oligonucleotides with degenerate base triplets (codons). The selection of the codons largely determines the screening effort (Table 8.1). The standard codons NNN or NNK (with $N = A$, T, G, or C and $K = G$ or T) encode for all nucleotides and thus cover all 20 amino acids. However, to minimize the probability of a randomly inserted stop codon, only an NNK primer is used for a completely randomized amino acid position.

If one wants to reduce the effort for screening, special codons are used, in which the selection of encoded amino acids is restricted to the different amino acid classes. The codon NDT (with $D = A$, G, or T) encodes only for 12 amino acids that differ in charge, size, and polarity.

To simplify the design of degenerate primers, codons can also be generated using a web server-based application such as DYNAMCC (http://www.dynamcc.com). This tool eliminates unwanted amino acids, stop codons, redundancy, and codon bias (Sect. 8.2.4). The resulting compressed codons are also optimized for the codon usage of the corresponding host organism (Pines et al. 2015).

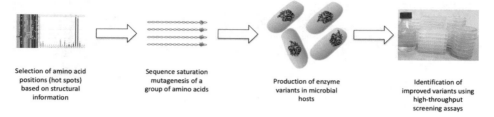

Fig. 8.4 Semi-rational protein design

Table 8.1 Screening effort for 95% coverage. (Reetz et al. 2008)

Number of positions	NNK codon		NDT codon	
	Codons	Number of colonies to screen	Codons	Number of colonies to screen
1	32	94	12	34
2	1028	3066	144	430
3	32,768	98,163	1728	5175
4	1,048,576	3,141,251	20,736	62,118
5	33,554,432	100,520,093	248,832	745,433

8.2.2   Iterative Saturation Mutagenesis (ISM)

For this method, selected amino acid positions that are thought to have a high probability of influencing a particular property of the protein, so-called hotspots, are saturated with all other amino acids. A region can consist of one, two, three, or even more amino acids, each of which is "saturated" in turn (Fig. 8.5). The best mutant of each region serves as the basis for the saturation mutagenesis of the next region. The amino acid positions are thus mutated one after the other, whereby the desired property should constantly improve. If a mutation leads to an inactive variant, one goes back a step and selects another position for mutagenesis.

The CAST method (*combinatorial active-site saturation test*; Reetz et al. 2005), a variant of ISM, is a frequently used method of semi-rational protein design and is primarily used to alter substrate scope or stereospecificity. Here, the focus is primarily on the amino acids in the first sphere (first shell residues) around the catalytic center, and they are subjected to position-directed saturation mutagenesis.

A variation of the CAST method is B-Fit (B-factor iterative test; Reetz et al. 2006) where the basis for the choice of positions for iterative saturation mutagenesis are the so-called B-factors of the amino acids in the crystal structure. This method is mainly used to stabilize and increase the thermostability of a protein.

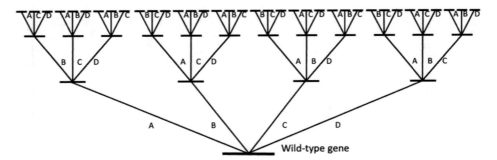

Fig. 8.5 Schematic representation of an iterative saturation mutagenesis as an example with four mutagenesis positions A, B, C, and D

8.2.3 ProSAR

Another example of a semi-rational method is ProSAR (*protein sequence activity relationships*; Fox et al. 2007). ProSAR is a learning system that evaluates the results of directed evolution and allows predictions for further experiments. All mutagenesis methods can be used. An algorithm enables the calculation of the influence of each individual mutation on activity, regardless of how many mutations the gene carries. In this way, mutations in genes that improve activity can be identified, even though the protein may have a lower activity than the parent protein.

8.2.4 Error-Prone PCR

In error-prone PCR (epPCR) (Cadwell and Joyce 1992), the conditions are selected in such a way that higher error rates occur during gene amplification (1–5%). The frequently used polymerase from *Thermus aquaticus* already has an increased error rate due to its non-proofreading function. This can be increased by using $MnCl_2$ instead of $MgCl_2$, an unbalanced nucleotide ratio, or an increased concentration of nucleotides. The mutation rate can also be increased by the concentration of the gene template or the use of special polymerases, e.g., Mutazyme (GeneMorphII Random Mutagenesis Kit, Agilent Technologies) or nucleotide triphosphate analogs (JBS dNTP Mutagenesis Kit, Jena Bioscience), to an error rate of up to 20%.

A major disadvantage of error-prone PCR is that not all theoretical mutations are accessible and thus not all possible protein variants can be generated. The probability of exchanging more than one base per triplet (codon) is very low with this method. Due to the degenerate genetic code, only 6 of the 20 amino acids of the theoretically possible amino acid exchanges are thus experimentally accessible on statistical average (codon bias). This in turn leads to a limited diversity of the mutant library.

In addition, polymerases have the property of preferring certain nucleotide exchanges, causing some mutations to occur more often than others (polymerase bias). By combining several polymerases or using commercial kits with different polymerases, one can reduce the error in the library. Amplification bias arises from the exponential amplification of mutations from the first PCR cycles. Thus, a mutation that arose in the first amplification ends up being 25% overrepresented. This can be circumvented by running several different reaction approaches in parallel and combining them at the end to then generate the mutant library. With the help of commercial kits with special polymerases, the problem of nonstatistical exchange of nucleotides can be minimized as far as possible. Setting an optimal mutation rate and minimizing *bias* are thus the most important parameters for a high-quality gene library.

An alternative to error-prone PCR is the sequence saturation mutagenesis (SeSaM), which generates a pool of DNA fragments of different lengths to which the universal base deoxyinosine is subsequently added enzymatically. The fragments are then amplified in a subsequent PCR, in which the parent gene serves as a template, to produce a now modified full-length gene sequence. In this process, the deoxyinosine is replaced by one of the four standard bases (Wong et al. 2004).

8.2.5 DNA Shuffling

Recombining methods produce changes in the gene sequence by separating and recombining different DNA.

In DNA shuffling, one of the most important methods of in vitro recombination, one or more related (homologous) genes with different positive mutations are cut with the endonuclease DNase I (Stemmer 1994). Overlapping DNA fragments randomly distributed throughout the gene are produced. The fragmented gene sequences are recombined, and gene variants from mutants with improved properties are used as starting points for further rounds of mutation and recombination. DNA shuffling can be performed with mutants of the same gene, but also with homologous genes of different species (family shuffling).

8.2.6 In Vivo Mutagenesis

For mutagenesis, there are also nonspecific in vivo methods, including traditional methods such as treatment with UV radiation, chemicals such as N-methyl-N'-nitro-N-nitrosoguanidine (MNNG) or ethyl methanesulfonate (EMS), radioactive radiation, and commercially available bacterial mutator strains such as *Epicurian coli* XL1-Red with defective DNA repair system. However, all these methods have the major disadvantage that the mutations are not only introduced in the gene of interest on the plasmid but also arise in the entire genome of the host cell and, moreover, may also be distributed anywhere on the plasmid. The first case sometimes leads to nonviable clones and the second usually to a large number of truncated and thus inactive protein variants. Furthermore, if the ribosome binding site (rbs), the promoter region, and the antibiotic resistance gene are affected by mutations, no recombinant protein is produced at all by the expression host. Similarly, setting a defined mutation rate is almost impossible.

Due to these drawbacks, few protein engineering applications have been described for this method.

In order to circumvent this problem, so-called mutator plasmids can be used as an alternative. For this purpose, a method was first described in 2001 by scientists from the company Genencor, in which the ε-subunit of DNA polymerase III, which is responsible for the proofreading function, was additionally equipped with a plasmid-encoded, inactive variant from E. coli (MutD5 protein) (Selifonova et al. 2001). Despite the loss of its catalytic activity, the MutD5 protein can still bind very effectively to polymerase III and compete with the functional chromosomal copy of the MutD protein, significantly increasing the mutation rate of the expression host at times. Subsequent selection allowed enrichment of clones that further developed a desired trait. To prevent the new phenotype from mutating further, loss of the mutator plasmid was induced by using a temperature-sensitive origin of replication after increasing the growth temperature.

With the help of this new method, the mutation rate could now be controlled to a large extent, but the disadvantage remained that the mutations affected the entire genome of the bacterium.

It was not until the use of a plasmid-encoded error-prone variant of DNA polymerase I, which performs error-prone replication of a target gene on a second plasmid, that mutator plasmids were able to insert mutations in a targeted manner (Camps et al. 2003).

The production of high-quality mutant libraries remains a time-consuming challenge despite the development of a wide variety of mostly reliable methods. Therefore, the use of commercial mutant libraries from, e.g., Genscript, Thermofisher, or Eurofins is an increasingly interesting alternative.

Questions
  4. What is the difference between in vivo and in vitro mutagenesis methods?
  5. How is the error rate of the DNA polymerase increased during the error-prone PCR?
  6. What are the polymerase bias and the codon bias?
  7. How can you reduce the screening effort during site saturation mutagenesis?

8.3   Screening and Selection

Rational protein design usually yields a manageable number of protein variants that can be analyzed by standard analytical methods such as gas chromatography (GC) or high-performance liquid chromatography (HPLC).

In contrast, in directed evolution, large libraries with a very large number of protein variants are often generated by the methods of random mutagenesis. Theoretically, 19^M

Table 8.2 Screening effort depending on sequence length

Number of amino acid substitutions (M)	Number of variants depending of the sequence length (N)	
	10	200
1	190	3800
2	16,245	7,183,900
3	823,080	9,008,610,600
4	27,367,410	8,429,807,368,950

[N!/(N - M)! M!] variants of a gene with N amino acids could be generated if M positions are exchanged for 19 other amino acids (Table 8.2).

In principle, two different methods are available for finding improved enzyme variants: selection or screening.

Selection can occur when the improved protein variant allows the bacterial cell to have a growth or even survival advantage over other cells. This can be, for example, antibiotic resistance, toxin degradation, or the use of an alternative energy or metabolite source. Selection can be used to screen large libraries ($>10^{10}$ clones) very quickly, usually by growth on selective agar media.

If no suitable selection system is available, the libraries must be screened (screening; Chap. 6). For this purpose, a fast, simple, and reliable high-throughput assay is the prerequisite for finding improved enzyme variants. This is mainly carried out in microtiter plate format and allows the screening of 10^3–10^6 enzyme variants.

The reaction in the assay should be as similar as possible to the true reaction to the real substrate. In addition, it is better to use the real substrate and not a test-specific substrate, as otherwise only variants that accept the (artificial) test substrate better ("you get what you screen for") may be identified.

Questions
   8. What is the difference between selection and screening?
   9. What is the meaning of the phrase "You get what you screen for"?
   10. How many variants are usually screened using an MTP assay?

8.4   Successful Examples of Optimized Biocatalysts

The first example of enzyme optimization using directed evolution methods was the adaptation of an esterase to organic solvent reaction conditions for the synthesis of the antibiotic loracarbef (Moore and Arnold 1996). After several rounds of random mutagenesis by error-prone PCR and recombination of the best variants by DNA shuffling, a variant of *p*-nitrobenzyl esterase was found that exhibited 150-fold increased activity in 15% dimethylformamide (DMF).

In another example, scientists from the company Codexis and the University of Delft were able to demonstrate in an impressive way what is possible with protein engineering when the appropriate effort is made—a total of almost 600,000 clones were screened. For the production of a starting material for the cholesterol-lowering drug atorvastatin (Lipitor), a drug that lowers cholesterol levels, a halohydrin dehalogenase was to be optimized. By combining various methods of random mutagenesis and a complex statistical analysis of the influence of amino acid exchanges on the activity, using the ProSAR algorithm, a variant was finally generated whose volumetric productivity was increased by a factor of 4000 (see Chap. 7; Fox et al. 2007).

Another particularly impressive example is the development of a biocatalyst for the production of sitagliptin, a drug used to treat type 2 diabetes mellitus. The parent enzyme, a transaminase, initially showed no measurable activity toward the sterically demanding ketone substrate. Planned mutations were introduced following rational design and substrate docking studies, which progressively generated more active-site space. After several rounds of saturation mutagenesis, combination of the best mutations, and subsequent random mutagenesis, an enzyme variant (with 27 mutations) with 40,000-fold increased activity and excellent stereoselectivity was finally generated under process conditions (Savile et al. 2010).

These examples demonstrate that the process of optimizing enzymes has itself undergone a kind of "evolution" and has developed from a method of "naïve" random mutagenesis followed by simple selection and/or screening to a well thought-out, strategically planned approach within just a few years. Consequently, this method is also successfully used for the industrial development of tailor-made biocatalysts (Bornscheuer et al. 2012; Buller et al. 2023). This is largely due to the development of sophisticated computer programs and web server-based applications, as well as access to vast genome, DNA, protein sequence, and protein structure databases. These enable the production of high-quality mutant libraries, and modern techniques such as microfluidic systems or flow cytometry (e.g., FACS) allow the rapid and successful screening of huge libraries of protein variants.

Take-Home Message
- Enzymes often need to be optimized to fit the needs for industrial application.
- Based on the knowledge about the enzyme candidate's structure and mechanism, methods of rational protein design or directed evolution are applied.
- Enzyme libraries can be created by introducing either random or site-directed (point) mutations within the gene sequence.
- Optimized enzyme candidates can be found by screening or selection methods.
- Depending on the library size, low-, medium-, or (ultra)high-throughput assays are applied.

Answers

1. Enzymes are often not compatible with the requirements for industrial applications. Properties of enzymes such as stability, catalytic efficiency, and substrate specificity often need to be optimized.
2. The two concepts for enzyme (or protein) engineering are called rational protein design and directed evolution.
3. The main difference of both methods is the degree of information needed. For (semi) rational design, a three-dimensional structure or at least a reliable homology model is needed. For applying methods of directed evolution, only the gene sequence and a suitable expression host are needed.
4. In vitro mutagenesis methods introduce mutations into the nucleotide sequence via PCR using the isolated plasmid DNA. In vivo methods are based on living cells that introduce mutations during replication cycles due to mutagenic conditions.
5. The error rate of the DNA polymerase in an error-prone PCR can be increased up to 5% by the following factors: $MnCl_2$ instead of $MgCl_2$, unbalanced nucleotide ratio or an increased concentration of nucleotides, variation of the concentration of the gene template, the use of special polymerases, and an increase of the number of PCR cycles.
6. The polymerase bias means that some polymerases have preferred nucleotide exchanges because they have a higher tendency to exchange A into T rather than G to C which limits the diversity of the library. The codon bias is based on the degenerate genetic code, meaning that 1 amino acid is encoded by 2 to 6 codons and that in average only 5.7 of all 20 amino acids are accessible upon single site mutagenesis.
7. The screening effort can be reduced by using special codons (like NDT, encoding 12 out of the 20 proteinogenic amino acids), in which the selection of encoded amino acids is restricted to the different amino acid classes.
8. Selection can be applied if the improved protein variant allows the bacterial cell to have a growth or even survival advantage over other cells. This can be, for example, antibiotic resistance, toxin degradation, or the use of an alternative energy or metabolite source. Screenings are usually based on microtiter plate assays using fluorogenic or chromogenic substrates in which the reaction product is then detected by absorption or fluorescence measurements.
9. This phrase, which is also called the first law of directed evolution, means that by using artificial chromogenic substrates in the screening assay instead of real substrates, there might be the chance that the improved variants accept the (artificial) test substrate better than the real substrate.
10. In microtiter plate format, usually between 10^3 and 10^6 enzyme variants are screened.

References

Bornscheuer UT, Huisman G, Kazlauskas RJ, Lutz S, Moore J, Robins K. Engineering the third wave in biocatalysis. Nature. 2012;485(7397):185–94.

Buller R, Lutz S, Kazlauskas RJ, Snajdrova R, Moore JC, Bornscheuer UT. Harnessing enzymes for biocatalysis. Science. 2023;382(6673):eadh8615.

Cadwell RC, Joyce GF. Randomization of genes by PCR mutagenesis. PCR Methods Appl. 1992;2: 28–33.

Camps M, Naukkarinen J, Johnson BP, Loeb LA. Targeted gene evolution in *Escherichia coli* using a highly error-prone DNA polymerase I. Proc Natl Acad Sci U S A. 2003;100(17):9727–32.

Fox RJ, Davis SC, Mundorff EC, Newman LM, Gavrilovic V, Ma SK, Chung LM, Ching C, Tam S, Muley S, Grate J, Gruber J, Whitman JC, Sheldon RA, Huisman GW. Improving catalytic function by ProSAR-driven enzyme. Nat Biotechnol. 2007;25(3):338–44.

Moore J, Arnold FH. Directed evolution of a *para-nitrobenzyl* esterase for aqueous-organic solvents. Nat Biotechnol. 1996;14(4):458–67.

Pines G, Pines A, Garst AG, Zeitoun RI, Lynch SA, Gill RT. Codon compression algorithms for saturation mutagenesis. ACS Synth Biol. 2015;4:604–14.

Reetz MT, Bocola M, Carballeira JD, Zha D, Vogel A. Expanding the range of substrate acceptance of enzymes: combinatorial active-site saturation test. Angew Chem Int Ed. 2005;44(27):4192–6.

Reetz MT, Carballeira JD, Vogel A. Iterative saturation mutagenesis on the basis of B factors as a strategy for increasing protein thermostability. Angew Chem Int Ed. 2006;45(46):7745–51.

Reetz MT, Kahakeaw D, Lohmer R. Addressing the numbers problem in directed evolution. Chembiochem. 2008;9(11):1797–804.

Savile CK, Janey JM, Mundorff EC, Moore JC, Tam S, Jarvis WR, Colbeck JC, Krebber A, Fleitz FJ, Brands J, Devine PN, Huisman GW, Hughes GJ. Biocatalytic asymmetric synthesis of chiral amines from ketones applied to sitagliptin manufacture. Science. 2010;329(5989):305–9.

Selifonova O, Valle F, Schellenberger V. Rapid evolution of novel traits in microorganisms. Appl Environ Microbiol. 2001;67(8):3645–9.

Stemmer WPC. Rapid evolution of a protein *in vitro* by DNA shuffling. Nature. 1994;370(6488): 389–91.

Wong TS, Tee KL, Hauer B, Schwaneberg U. Sequence saturation mutagenesis (SeSaM): a novel method for directed evolution. Nucleic Acids Res. 2004;32(3):e26.

Enzyme Production

9

Fabienne Knapp, Andreas Knapp, and Karl-Erich Jaeger

What You Will Learn in This Chapter

The choice of a host organism is crucial for enzyme production. We provide an overview of the limiting factors in protein biosynthesis that influence the biotechnological production of enzymes. The advantages and disadvantages of the prokaryotic and eukaryotic (micro)organisms commonly used for enzyme production are described, and we explain how to handle an organism to produce a heterologous protein or enzyme. Finally, we briefly list problems in biotechnological enzyme production and their solutions and possibilities for optimization.

F. Knapp
Institute for Molecular Enzyme Technology, Heinrich Heine University Düsseldorf at Forschungszentrum Jülich, Jülich, Germany
e-mail: f.knapp@fz-juelich.de

A. Knapp
Institute for Molecular Enzyme Technology, Heinrich Heine University Düsseldorf at Forschungszentrum Jülich, Jülich, Germany

Castrol Germany GmbH, Mönchengladbach, Germany
e-mail: a.knapp@fz-juelich.de

K.-E. Jaeger (✉)
Institute for Molecular Enzyme Technology, Heinrich Heine University Düsseldorf at Forschungszentrum Jülich, Jülich, Germany

Institute for Bio- and Geosciences IBG-1: Biotechnology, Forschungszentrum Jülich GmbH, Jülich, Germany
e-mail: karl-erich.jaeger@fz-juelich.de

© The Author(s), under exclusive license to Springer Nature Switzerland AG 2024
K.-E. Jaeger et al. (eds.), *Introduction to Enzyme Technology*, Learning Materials in Biosciences,
https://doi.org/10.1007/978-3-031-42999-6_9

9.1   Choice of Host Organism

9.1.1   Homologous or Heterologous Protein Production

For the production of an enzyme, one can use the organism that naturally synthesizes the enzyme. Examples include a lipase produced by the Gram-negative bacterium *Burkholderia glumae* or a phytase produced by the filamentous fungus *Aspergillus niger*. Lipases are used both as biocatalysts in chemical syntheses and in detergents, where they help to remove grease stains. Phytases are used as additives in animal feed, where they make phosphates in plant food more digestible. After the industrial potential of both organisms became clear, the enzyme yields were significantly increased by genetic optimization of the two organisms, making their use as enzyme producers more profitable.

However, it is estimated that only about 1% of all microorganisms can be cultivated under laboratory conditions, and even these often have disadvantages such as a slow cell division rate or cultivation conditions that are difficult to implement. Filamentous fungi, for example, naturally produce a number of interesting enzymes, e.g., cellobiohydrolase I from *Trichoderma reesei* for the utilization of cellulosic waste. However, cultivation of filamentous fungi is usually more expensive and laborious than that of prokaryotic organisms. In cases where homologous enzyme production is not feasible or does not make sense from an economic point of view, enzymes can also be produced in another host organism.

Advantages of such heterologous production can be a cheaper and faster cultivation of the selected host organism but also an increased enzyme yield. The handling of well-characterized organisms is also easier, since a large number of established protocols are available for this purpose. It also provides the opportunity to produce enzymes from potentially dangerous organisms or organisms that cannot be cultured in the laboratory. Furthermore, the desired enzyme can be modified and optimized during heterologous production by altering the corresponding gene sequence. An overview of the original organisms producing industrially relevant proteins and enzymes and their production host strains can be found in Table 9.1.

Thanks to intensive research, many different organisms from the various phyla of the phylogenetic tree of living organisms are available for heterologous enzyme production. Each of them has particular advantages and disadvantages, which are to be weighted differently depending on the desired target protein as described below (see also Tables 9.2, 9.3, and 9.4).

Advantages of heterologous/recombinant enzyme production are as follows:

– Inexpensive and fast cultivation
– Easy handling of the heterologous host organism
– Increased yield due to molecular biological optimization of the host organism
– Handling of safe production strains
– Modifications of the target enzyme possible

Table 9.1 Sources and production hosts of industrially relevant proteins and enzymes (Selection; after Maurer et al. 2013)

Enzyme	Origin	Production hosts
Proteases	Various *Bacillus* species (prokaryotes)	Various *Bacillus* species (prokaryotes)
Amylases	*Bacillus licheniformis, B. stearothermophilus* (prokaryotes)	*Bacillus licheniformis, B. amyloliquefaciens, B. stearothermophilus* (prokaryotes)
Amylases	*Aspergillus niger, Trichoderma reesei* (eukaryotes)	*Aspergillus niger, A. oryzae, Trichoderma reesei* (eukaryotes)
Phytases	*Escherichia coli* (prokaryote)	*Trichoderma reesei, Pichia pastoris, Schizosaccharomyces pombe* (eukaryotes)
Phytases	*Aspergillus niger, Peniophora lycii* (eukaryotes)	*Aspergillus niger* (eukaryote)
Cellulases	*Trichoderma reesei* (eukaryote, homologous production host)	
Cellulases	*Clostridium thermocellum, C. cellulolyticum* (prokaryotes)	*Escherichia coli, Bacillus subtilis* (prokaryotes)
Cellulases	*Acidothermus cellulolyticus* (prokaryote)	*Nicotiana tabacum* (tobacco), *Zea mays* (maize), and *Oryza sativa* (rice)
Lipases	*Candida antarctica* (eukaryote)	*Escherichia coli* (prokaryote), *Pichia pastoris, Saccharomyces cerevisiae* (eukaryotes)
Xylanases	Various *Bacillus* species (prokaryotes)	*Bacillus subtilis* (prokaryote)
Xylanases	*Actinomadura* species (prokaryote), *Trichoderma* species (eukaryote)	*Trichoderma reesei, T. longibrachiatum* (eukaryotes)
Insulin	Human	*Escherichia coli* (prokaryote), *Saccharomyces cerevisiae* (eukaryote)
Antibodies	Mouse, rabbit (homologous production host)	
Antibodies	Human	*Escherichia coli* (prokaryote), *Saccharomyces cerevisiae* (eukaryote), *CHO,* and *HEK*

CHO: Chinese hamster ovary, ovarian cell line isolated from Chinese hamster ovaries. HEK: human embryonic kidney, cell line derived from human embryonic kidney cells

Table 9.2 Advantages and disadvantages of different expression hosts. (Tripathi and Shrivastava 2019)

Prokaryotes	Eukaryotes
(+) Simple handling (cultivation, processing of the target enzyme)	(−) More complex handling
(+) Simple genetic accessibility (cloning, insertion of DNA and mutations)	(−/+) More complex genetic accessibility
(−) Only few protein modifications possible	(+) Posttranslational protein modifications possible

Table 9.3 Characteristics and advantages of commonly used industrial prokaryotic host organisms. (Yin et al. 2007; Demain and Vaishnav 2009; Fernandez and Vega 2013)

	Escherichia coli (Gram-negative)	*Bacillus* species (Gram-positive)
Generation time	0.2–0.5 h	0.6–1.0 h
Time required to isolate an active enzyme[a]	1–2 weeks	1–3 weeks
Advantages	– Well characterized – Rapid growth/enzyme production – High yields of target protein – Variety of molecular biology tools available – Many different strains commercially available – Inexpensive cultivation	– Well characterized – Rapid growth/enzyme production – High secretion capacity without formation of inclusion bodies – Variety of molecular biology tools available – GRAS (generally recognized as safe) status – Inexpensive cultivation

[a]Ideally needed by an experienced experimenter

9.1.2 Limiting Factors in Protein Biosynthesis

The genetic code determines the sequence of amino acids in each protein with only four different information units, the nucleobases adenine, thymine, guanine, and cytosine. Three bases in a row (triplet) code for a particular amino acid. Since this coding is universal, i.e., basically applies to every living cell, in principle "foreign" (heterologous) proteins can also be produced in any organism, as long as it is ensured that the underlying DNA sequence contains all the features necessary for successful protein biosynthesis (Fig. 9.1). These include a promoter sequence at which an RNA polymerase can begin transcribing the gene encoded on the DNA into mRNA (transcription) and a terminator sequence at which transcription ends. The transcription rate, i.e., the frequency and speed of mRNA production, as well as the transcript stability has an influence on the amount of mRNA present in the cell and thus also significantly determines the amount of the target protein formed. In prokaryotes, transcription is directly followed by translation of the mRNA sequence into an amino acid sequence, whereas in eukaryotes, modifications of the mRNA are carried out before translation. These include transport from the cell nucleus into the cytoplasm, modifications at the sequence termini, and splicing, i.e., removal of certain regions of the mRNA (introns) from the transcript so that their genetic information is no longer translated into the amino acid sequence. Since only eukaryotes have mechanisms to remove introns, they must be removed in advance if eukaryotic genes are to be heterologously expressed in prokaryotic hosts. Translation, i.e., the synthesis of proteins, then takes place at the ribosomes. Here, specific parts of the mRNA sequence are recognized by the ribosomes as binding sites, as well as the start and stop codons of the coding sequence. The speed of

Table 9.4 Characteristics and advantages of eukaryotic host organisms. (Yin et al. 2007; Demain and Vaishnav 2009; Fernandez and Vega 2013)

	Yeasts (e.g., S. cerevisiae, P. pastoris)	Filamentous fungi (e.g., A. niger)	Cell cultures	Living transgenic insects, mammals, plants
Generation time	1.3–3.0 h	3.0–4.0 h	14–36 h (mammalian cell culture like CHO)	18–24 h (insects), weeks to years for mammals and plants
Time required to isolate an active enzyme[a]	2–6 weeks	2–6 weeks	3–8 weeks	Weeks (insects) to years (for mammals and plants)
Advantages	– High protein yields and cell densities – Relatively fast growth – Protein modification (folding, glycosylation) similar to mammals – GRAS status – Inexpensive cultivation	– Complex PTM – High secretion capacity – GRAS status – Relatively inexpensive cultivation	– PTMs similar to human PTMs	– PTMs similar to human PTMs – High protein yields – Accumulation of target enzyme in defined tissue (plants) or organs/secreted fluids like milk (animals)

CHO: Chinese hamster ovary, ovarian cell line isolated from Chinese hamster ovaries. PTM: posttranslational modification

[a]Ideally needed by an experienced experimenter

translation (translation rate) directly influences the yield of the desired protein and is determined by factors such as tRNA availability or the frequency of the codons to be translated. The mature protein can be further modified with the type and variety of these posttranslational modifications (PTMs) depending on the respective organism. More complex protein modifications occur preferentially in eukaryotes, such as glycosylation, acetylation, and phosphorylation of specific amino acids. Some proteins fold spontaneously into their active conformation during or after protein biosynthesis; others need the help of other proteins (so-called chaperones) for proper folding. Some proteins can be transported into specific cell compartments or even out of the cell, for which complex transport mechanisms exist.

Questions

1. Can you name three enzymes that are of high industrial importance?

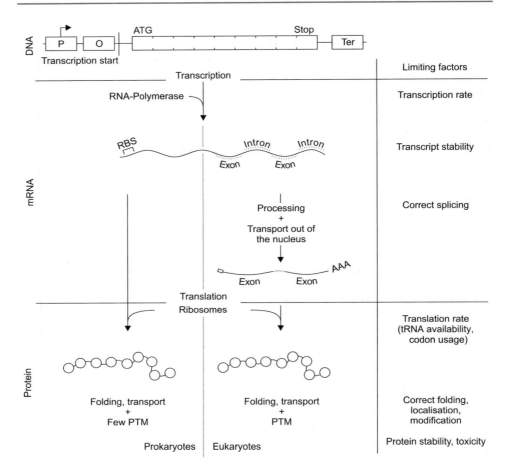

Fig. 9.1 Protein biosynthesis (PBS) and limiting factors. Upstream of a gene are a promoter (P) and optionally an operator (O) sequence. Transcription begins with a start codon (often ATG) and ends with a stop codon followed by one or more terminator sequences (Ter). Additionally, a ribosome binding site (RBS) is present on the mRNA. During both mRNA (transcription) and protein (translation) formation, rate-limiting factors can affect the amount of protein synthesized. Transcript and protein stability also determine the yield of the target protein. Depending on the chosen expression host, the ability to process the mRNA (e.g., splicing) and the protein (folding, localization, PTM) is important for protein yield. Toxic proteins can be harmful to the host organism and thus negatively affect their own production

2. Which prokaryotic and eukaryotic microorganisms are commonly used for enzyme production?
3. Which factors do you know that can limit the biosynthesis of proteins?

9.2   Production in Eukaryotes or Prokaryotes

If one decides to produce a target enzyme as a recombinant protein in a heterologous host, not only the protein yield is important but also the desired enzymatic activity (bioactivity). Both factors can be influenced by the chosen host organism and the cultivation conditions. For example, not all organisms are able to carry out certain posttranslational modifications to proteins that are required for their enzymatic activity (Sect. 9.1.2). Furthermore, it is possible that a produced enzyme cannot develop its bioactivity due to the physical or chemical conditions in the cultivation medium (temperature, pH, salinity, etc.). The host organism as well as the cultivation conditions must therefore be selected in such a way that not only the enzyme is produced but also its bioactivity is maintained. Economic considerations often also play a role in the choice of host organism, since the costs of cultivation can vary significantly, and enzyme production must remain profitable in industrial enzyme technology. Depending on the application of the enzyme, safety regulations must also be considered during production, which further limits the choice of a host organism.

Genetic accessibility as well as the knowledge of the complete genome sequences represents a clear advantage of some prokaryotic and eukaryotic microorganisms used for heterologous enzyme production. Furthermore, they are characterized by rapid and inexpensive cultivation as well as easy handling (e.g., purification of the target enzyme). These advantages result mainly from the relatively simple structure of the microorganisms. The production of eukaryotic enzymes may require posttranslational modifications (PTMs), which mostly cannot be carried out in prokaryotes (Rettenbacher et al. 2022). Here, eukaryotic yeasts can be used, which are easy to handle and genetically accessible. Higher eukaryotes such as insects, plants, and mammals or corresponding cell cultures are more complex to handle and thus considerably more expensive but are also used for the production of valuable proteins. The basic advantages and disadvantages of prokaryotes and eukaryotes as expression hosts are summarized in Table 9.2. Figure 9.2 shows schematically the cell structure of pro- and eukaryotes as well as important steps of protein biosynthesis.

In general, it is not possible to give a universally valid answer to the question which host organism will provide the best results for the production of a target enzyme. The requirements of the target enzymes for the respective protein biosynthesis (PBS) and modification mechanisms are too diverse. It is thus recommended to start with simple prokaryotic organisms and only switch to more complex hosts in case protein production or bioactivity fails.

9.2.1   Prokaryotes

Prokaryotes can be divided into three subgroups, the Gram-positive and the Gram-negative bacteria and the archaea. Gram-positive bacteria have only a single cytoplasmic membrane,

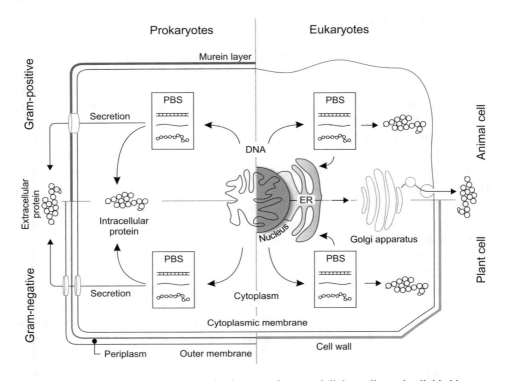

Fig. 9.2 Cell structure of prokaryotes and eukaryotes. In general, living cells can be divided into eu- (= with) and prokaryotes (= without nucleus). They all have in common a lipid bilayer, the cytoplasmic membrane, which surrounds the cytoplasm. The prokaryotes (bacteria) additionally possess a peptidoglycan or murein layer. In Gram-negative bacteria, an additional outer membrane is present which is missing in Gram-positive bacteria. Eukaryotic cells, which can be divided into plant (with a cell wall) and animal cells, possess an endoplasmic reticulum (ER) and a Golgi apparatus, both of which are required for the secretion of proteins, in addition to other organelles

overlaid by a murein (peptidoglycan) layer. Gram-negatives also have a second, outer cell membrane; the space between the two membranes is called the periplasm (see also Fig. 9.2). Although some members of archaea were exploited for biotechnological applications in the recent years, the majority of industrial processes focus on Gram-positive and Gram-negative bacteria. Thus, the two most commonly used representatives of these groups, namely, *Escherichia coli* and *Bacillus subtilis*, are presented in the following, and their advantages and characteristics are summarized in Table 9.3.

9.2.1.1 *Escherichia coli*

The Gram-negative bacterium *Escherichia coli*, which colonizes the human intestine, is probably the best-known and most frequently used microorganism. Since its discovery and description in the nineteenth century by the eponymous Theodor Escherich, *E. coli* has been the subject of worldwide research as a model organism and is also referred to as the "workhorse" in microbiology and biotechnology. More than 90% of the three-dimensional protein structures known today have been generated with proteins produced in this organism (Lübben and Gasper 2015).

Escherichia coli is particularly characterized by its short generation time; under optimal growth conditions, this is about 20 min per cell division, so that a high cell density can be achieved after only a few hours of growth. *Escherichia coli* theoretically reaches a maximum cell density of $1-10^{13}$ cells mL^{-1}, which corresponds to about 200 g cell dry weight per liter of cultivation medium. Such cell densities are not achieved in practical applications, but with about $1-10^{10}$ cells mL^{-1}, very high cell densities are still possible. In addition, *E. coli* can be cultivated in inexpensive media, offering the opportunity to produce recombinant enzymes as economically as possible (Rosano et al. 2019). Once cultivated, the target protein can constitute up to 80% of the dry mass of *E. coli* (Demain and Vaishnav 2009). Together with its very good genetic accessibility, these properties predestine *E. coli* as a standard host for heterologous enzyme production.

Once the target enzyme has been produced, it must often be available for further work in a purified form, i.e., free of other proteins. Since enzymes formed intracellularly are contaminated by a large number of other cellular proteins, the transport of the target enzyme out of the cytoplasm can be advantageous. Some expression systems use transport mechanisms in *E. coli*, which recognize the protein to be transported by means of a signal peptide at its N-terminus and transport it into the periplasm (see also Fig. 9.2). The yields of periplasmically or extracellularly localized target enzymes in *E. coli* are lower than those of cytoplasmically localized ones but can still be up to approx. $5-10$ g L^{-1} (Adrio and Demain 2014; Kleiner-Grote et al. 2018).

Many different strains are now available, most of which are based on *E. coli* strain B, in which the two proteases Lon and OmpT are deleted in order to minimize proteolytic degradation of the target protein. The target gene can be expressed with an RNA polymerase derived from bacteriophage T7 that specifically recognizes the T7 promoter behind which the target gene has been cloned (Sect. 9.4.2). Other *E. coli* strains have defined genetic modifications, e.g., strain *E. coli* BL21 (DE3) Rosetta, which contains additional genes for tRNAs rarely found in *E. coli*, so that heterologous proteins can also be expressed that require many of these tRNAs for correct translation.

9.2.1.2 *Bacillus* Species

Gram-positive bacteria in contrast to Gram-negative do not possess a second, outer membrane. The desired enzyme is therefore already present in the culture medium after transport across the cytoplasmic membrane. In contrast to the cytoplasm, relatively few other proteins are present in the culture supernatant, which simplifies subsequent isolation and purification of the target enzyme. The rod-shaped spore-forming bacteria of the genus *Bacillus* are therefore used as an alternative to *E. coli*. Primarily, *B. subtilis* must be mentioned, but other representatives of this genus such as *B. licheniformis* or *B. pumilus* are also coming into focus. Due to its ability to secrete relatively large amounts of homologous proteins into the culture medium (about 25 g L^{-1}; Rettenbacher et al.

2022), *Bacillus* is also used for the production and secretion of recombinant enzymes. For example, the production rate of the enzyme α-amylase, which is used as a starch-cleaving enzyme in the food and textile industries, could be increased 2500-fold in *B. subtilis* by introducing a heterologous gene from *B. amyloliquefaciens* (Adrio and Demain 2010). In addition, many compounds produced industrially by *B. subtilis* are classified by the FDA (US Food and Drug Administration) as *generally recognized as safe* (GRAS) for the food market, since this organism does not produce any endo- or exotoxins.

For a long time, one of the disadvantages of *Bacillus*, especially in comparison with *E. coli*, was the poorer genetic accessibility, e.g., due to plasmid instability issues (a brief explanation of plasmids can be found in Sect. 9.3.1). This could be circumvented by a stable integration of the corresponding gene into the genome of *B. subtilis* or by applying special plasmids with improved stability characteristics for these organisms. *Bacillus subtilis* has a natural mechanism for DNA uptake, which greatly simplifies its genetic modification. Additionally, strains are available in which different numbers of intracellular and extracellular proteases are deleted in order to prevent proteolytic degradation of target proteins.

9.2.1.3 Alternative Prokaryotic Hosts

In addition to *E. coli*, *B. subtilis*, and other members of the genus *Bacillus*, other prokaryotic host organisms are also established that fulfill specific requirements for the production of special enzymes. For expression of GC-rich genes, host organisms with a similar relatively high GC content of their chromosomal DNA may be used, such as *Pseudomonas putida* (Fernandez and Vega 2013). For the production of proteins and enzymes that need to be embedded in the cytoplasmic membrane, *Rhodobacter* species represent an interesting alternative, as those strains can form a very large membrane surface under certain environmental conditions and thus provide sufficient space for the incorporation of additional membrane proteins (Heck and Drepper 2017).

9.2.2   Eukaryotes

A general disadvantage of prokaryotes as host organisms for recombinant enzyme production is their lack of capacity for posttranslational biochemical modification of the proteins produced. Proteins of eukaryotic origin often require such modifications for their stability and bioactivity, for example, the addition of sugar residues to certain amino acids (glycosylation—almost 50% of all known eukaryotic proteins are glycosylated), which cannot easily be carried out by prokaryotes. Accordingly, eukaryotic hosts that are more closely related to the donor of the target gene are more promising for this task. Some characteristics and advantages of the presented host organisms are summarized in Table 9.4.

9.2.2.1 Yeasts

Yeasts represent a relatively simple form of eukaryotic life and comprise a diverse group of unicellular fungi. They are capable of many posttranslational modifications to proteins and enzymes known from higher eukaryotes and are therefore particularly suitable for the production of, for example, plant or human proteins. Yeasts can be divided into methylotrophs (utilizing methanol as a carbon source) and non-methylotrophs. The non-methylotrophic yeasts include the beer or baker's yeast *Saccharomyces cerevisiae* whose role in the production of alcohol has been known since antiquity, while the methylotrophs include *Pichia pastoris* (also: *Komagataella phaffii*) and *Hansenula polymorpha*. All three yeasts, or many of the products they produce, are classified as GRAS by the FDA because they do not produce toxic substances.

 S. cerevisiae, *P. pastoris*, and *H. polymorpha* can be cultivated easily and at low costs similarly to *E. coli*. With a generation time of a few hours, they are significantly slower in biomass production but still achieve high biomass yields (Fernandez and Vega 2013; Yin et al. 2007; Tripathi and Shrivastava 2019). They are also characterized by high enzyme yields; for example, *S. cerevisiae* and *P. pastoris* produce and secrete directly into the culture medium up to 9 and 20 g L^{-1}, respectively, of a recombinant glucose oxidase from *Aspergillus niger* (Demain and Vaishnav 2009; Ergün et al. 2021). The bioactivity of the target enzyme can be enhanced by posttranslational modifications such as glycosylation, acetylation, or phosphorylation. Also, correct folding can be significantly improved as compared to production in prokaryotes, if the target enzyme is derived from a higher eukaryote. Although phylogenetically relatively closely related, different yeasts perform different types of PTM on proteins. For example, *H. polymorpha* modifies proteins with shorter oligosaccharides than *P. pastoris* or the baker's yeast *S. cerevisiae* (Demain and Vaishnav 2009).

 Yeasts, like the higher eukaryotes, have cell compartments such as the endoplasmic reticulum and the Golgi apparatus, through which recombinantly produced enzymes can also be secreted into the external medium. This is advantageous for further purification of the enzyme and for its production, e.g., if the enzyme exerts toxic effects that can be avoided by secretion into cellular compartments.

9.2.2.2 Filamentous Fungi

Filamentous fungi such as *Aspergillus niger* can be cultivated in liquid cultures, even though their reproduction times are even slower than those of yeasts (3–4 h) and cultivation is more complex to handle (Adrio and Demain 2010). Filamentous fungi form filamentous hyphae whose growth occurs mainly at the tip; proteins are also secreted there. *Aspergillus niger* has long been used for food production and is considered a safe host organism also for heterologous enzyme production. However, the introduction of foreign DNA, as required for the production of recombinant enzymes, is proving more difficult than in prokaryotes and yeasts, even though increasingly easy-to-perform protocols have been published (Yin et al. 2007). An advantage of *A. niger* and other filamentous fungi is their high natural secretion capacity for homologous proteins of about 100 g L^{-1} under optimal fermentation conditions and up to 5 g L^{-1} for heterologous enzymes. The fungus *Trichoderma reesei*, which also grows filamentously, produces up to 35 g L^{-1} of heterologous enzymes (Demain and Vaishnav 2009). Filamentous fungi also posttranslationally modify proteins and enzymes and produce, for example, glycosylation patterns very similar to those of mammalian proteins. However, some PTMs required for human proteins cannot even be performed here.

9.2.2.3 Other Eukaryotic Hosts

Recombinantly produced human proteins or enzymes, such as those used in various medical therapies, must possess posttranslational modifications that are as "true to nature" as possible. Here, other and even more complex host organisms are needed. Genetically modified insect cells, for example, are able to produce mammalian-like glycosylation patterns if transformed with heterologous DNA to produce recombinant proteins (McKenzie and Abbott 2018). Yields of up to 30% of the target protein relative to the total amount of protein in the insect cell are possible (Demain and Vaishnav 2009). Furthermore, cell lines from mammals exist (such as the CHO cell line: Chinese hamster ovary, cells isolated from ovaries of Chinese hamsters), for which yields of up to approx. 4 g L^{-1} target protein are described. However, the cost and effort of production in mammalian cells is relatively high. The use of transgenic plants and animals is also possible and allows, among other things, the production of the enzyme only in certain tissues or organs. For example, it is possible to produce the human enzyme α-glucosidase recombinantly in rabbits, which secrete the enzyme directly into breast milk and excrete it in yields of up to 8 g L^{-1}. However, ethical concerns, complex handling, long reproduction times, and husbandry costs are clear disadvantages here. Transgenic plants, on the other hand, are more cost-effective in comparison, as they only require water, dissolved salts, air, and light to grow (Demain and Vaishnav 2009; Tripathi and Shrivastava 2019).

9.2.3   Cell-Free Production Systems

Cell-free protein synthesis represents another alternative, which, however, is not suitable for long-term enzyme production on a large scale (production yield of several mg mL^{-1} protein). Here, transcription and translation are carried out in vitro, i.e., in a test tube without a living host organism. The components required (RNA polymerase, ribosomes, etc.; see Sect. 9.1.2) are either provided by cell lysates prepared in advance or produced separately in purified form. Cell-free systems are often used in enzyme production when production in vivo is impaired by other components of the cell, for example, because the target enzyme forms aggregates with other proteins and thus loses its activity, or the target enzyme exerts a toxic effect on the cell (Zemella et al. 2015).

Questions
  4. Which cellular constituents are present in prokaryotic and in eukaryotic cells?

5. What are the major differences in the cellular structure of Gram-positive and Gram-negative bacteria?
6. Which factors are required for cell-free protein production?

9.3   Choice of Expression and Regulatory System

9.3.1   Episomal or Chromosomal Recombinant Genes

In order to express a target gene in an organism other than the original one, the heterologous gene must be genetically stable in this organism. It must therefore be duplicated with each cell division and be present again in both daughter cells after cell division. For this purpose, the gene can either be integrated into the genome of the host, or it is introduced into the cell on an autonomous genetic element, a vector or plasmid, where it is then present episomally, i.e., outside the chromosomal DNA.

Plasmids are used for the heterologous expression of episomal genes (Fig. 9.3). They are circular, double stranded DNA molecules that were originally found in bacteria, where they are replicated independently of the chromosomal DNA. Depending on the number of plasmids per cell, they are named low-, medium-, and high-copy plasmids (up to 700 copies per cell or per chromosome). The number of plasmids depends on the frequency of replication of the plasmid, which in turn is determined by the origin of replication (ORI). The ORI is recognized by the replication machinery of the host and represents the starting point for plasmid duplication. However, certain ORIs are only recognized in certain organisms. Plasmids that can be used in several, not closely related organisms, therefore, often carry several ORIs and are referred to as shuttle vectors.

If a gene is to be integrated into the genome of the host, it must first be introduced into the cell; this is done with the aid of a plasmid. After integration into the host chromosome, a heterologous gene usually remains stable and is automatically replicated together with the chromosome, whereas plasmids are subject to their own replication cycle. However, due to the higher copy number of the plasmid per cell, plasmid-encoded genes are significantly more abundant than genes integrated into the genome, which often also leads to a larger amount of corresponding mRNA and thus higher yield of the recombinant enzyme.

9.3.2   Stable Replicating Plasmids Require Selection Pressure

In order to maintain a plasmid in a dividing organism, selection pressure is necessary, as the plasmid can otherwise be lost in the course of many cell divisions. For this purpose, (expensive) antibiotics are often used as an additive to the culture medium and a corresponding resistance gene encoded on the plasmid, so that only those cells that contain the plasmid survive (Fig. 9.3). Alternatively, strains lacking an essential (vital) gene on the chromosome (such as for the production of an essential amino acid) are used, which is then

Fig. 9.3 The genetic elements of a plasmid. Optional elements are shown in dashed boxes. Abbreviations are promoter (P), operator (O), ribosome binding site (RBS), transcription terminator (Ter), and the origin of replication (ORI). The tag region codes for amino acid sequences that can be fused to the target enzyme

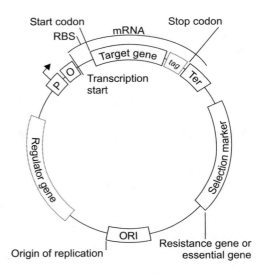

episomally reintroduced with the plasmid allowing only those cells to survive that carry the plasmid; this is also referred to as plasmid addiction (Rosano et al. 2019).

Question
7. Which genetic elements are required on a plasmid for expression of a heterologous gene?

9.4 Protein Production Can Be Modified and Optimized at any Level of Protein Biosynthesis

The decision whether a target gene should be chromosomally integrated or be present episomally affects the copy number of the gene and thus often the production rate of the target enzyme. In addition, numerous other regulatory elements are required (Sect. 9.1.2 and Fig. 9.3). In particular, the properties of the promoter are of fundamental importance, as discussed in more detail below.

9.4.1 Constitutive Promoters

Promoters are DNA regions to which the RNA polymerase binds and begins transcription. The RNA polymerase synthesizes an mRNA complementary to the gene, which is then translated into a protein. This basic mechanism applies to all genes in a cell but can be controlled by different regulatory processes. Promoters that lead to the permanent transcription of a gene are called constitutive promoters. In biotechnology, constitutive promoters are often used for the expression of a recombinant gene due to their simplicity,

Table 9.5 Frequently used expression hosts, promoters, and inducers

Organism	Promoter	Inducer
Escherichia coli	P_{lacUV5}	Lactose/IPTG
	P_{T7}, recognized by T7 RNA polymerase, which is produced under control of P_{lac} promoter	Lactose/IPTG
	P_{tac}, combining P_{lac} and P_{trp}	Lactose/IPTG
	P_{araI}	Arabinose
Bacillus subtilis	P_{HpaII}	Constitutive
	P_{T7} (see E. coli)	Lactose/IPTG
Saccharomyces cerevisiae	P_{TEF1}	Constitutive
	P_{Gal10}	Galactose
Pichia pastoris	P_{GAP}	Constitutive
	P_{AOX1}	Methanol
Aspergillus niger	P_{pkiA}	Constitutive
	P_{glaA}	Starch or maltose
Trichoderma reesei	P_{pdc}	Constitutive
	P_{cbh1}	Cellulose
Mammals (e.g., CHO)	P_{CHEF1}	Constitutive
	P_{MMTV}	Dexamethasone (artificial glucocorticoid)

CHO: Chinese hamster ovary, ovarian cell line isolated from Chinese hamster ovaries. IPTG: isopropyl β-D-thiogalactopyranoside

as enzyme production is then continuous. Such promoters are available for both Gram-positive and Gram-negative bacteria as well as for eukaryotic cells (Table 9.5). However, the disadvantage can be a relatively low expression rate of the target gene and at the same time a low biomass yield, as the cells have to continuously produce the target protein in parallel to the biomass. This can be prevented by using controllable promoters, which allow for target gene expression only under defined conditions.

9.4.2 Controllable Promoters

A cell does not have to express every gene at every point in its life. Therefore, regulatory mechanisms exist that can induce or repress the expression of certain genes only under certain (environmental) conditions. These naturally existing regulatory mechanisms are also used in biotechnology to control recombinant protein production in the host. In this way, biomass production can first be ensured and the expression of the target gene subsequently induced. A selection of frequently used inducible promoters can be found in Table 9.5. Among them, the *lac* promoter probably belongs to the best-known regulatory system, which is also used biotechnologically.

9.4.2.1 The *Lac Promoter*

The regulatory mechanism of the *lac* operon enables *E. coli* to produce lactose-degrading enzymes only when it makes sense to do so, i.e., when lactose is present. Without lactose in the medium, the structural genes of the *lac* operon (*lacZ, lacY, lacA*) are not transcribed because a repressor protein (LacI) binds between the *lac* promoter and the structural genes at the so-called *lac* operator and prevents the RNA polymerase from reaching the structural genes. Lactose, however, interacts with the repressor and causes a conformational change in the repressor, which can no longer bind to the operator. RNA polymerase can then transcribe the structural genes. This mechanism is called substrate induction. If the structural genes are replaced by a heterologous target gene and this construct is introduced into a host organism, e.g., *E. coli*, the same mechanism can be used to control its expression. Most of the inducible promoters function according to this or to similar principles (see also Fig. 9.4). In the case of the *lac* operon, however, the structural analog IPTG (isopropyl β-D-thiogalactopyranoside) is often used instead of the natural inducer lactose, which also inhibits the repressor LacI but cannot be metabolized by the host organism. A very elegant alternative is the use of a so-called autoinduction medium. This contains the inducer lactose but also glucose as a further carbon source. Glucose is preferentially metabolized by *E. coli* and simultaneously represses the *lac* operon, so that no target gene expression takes place as long as glucose is still present. In an initial growth phase of the culture, biomass is produced with glucose consumption. When the glucose has been consumed, lactose acts as an inducer and the cells, which are already present in large numbers, begin to express the target gene.

9.4.2.2 The T7 Expression System

The expression system based on T7 RNA polymerase is often used in prokaryotes such as *E. coli* and *B. subtilis* and is also sold commercially. It makes use of the RNA polymerase from the bacteriophage T7 which binds exclusively and with a high affinity to the promoter P_{T7}. In most T7 expression systems, the gene encoding T7 RNA polymerase is present in the genome of the host organism under the control of a *lac* promoter, whereas the target gene is introduced into the cell subsequently on a plasmid under the control of the P_{T7} promoter. Induction by IPTG now leads in the first step to the formation of T7 RNA polymerase, which then binds to the P_{T7} promoter resulting in target gene transcription (Fig. 9.4). Since there is also a *lac* operator between P_{T7} and the target gene, which only allows transcription of the target gene after the addition of IPTG or lactose, the system is regulated at multiple levels, minimizing basal expression (expression despite the absence of the inducer) of the target gene. The T7 expression system usually results in high enzyme yields due to the high transcription rate of T7 RNA polymerase.

9.4.2.3 Other Inducible Promoters

Inducible promoters are also used for enzyme production in eukaryotes (Table 9.5). In *S. cerevisiae*, the addition of galactose leads to the induction of gene expression under the control of the P_{Gal10} promoter; in *P. pastoris*, methanol induces the expression of

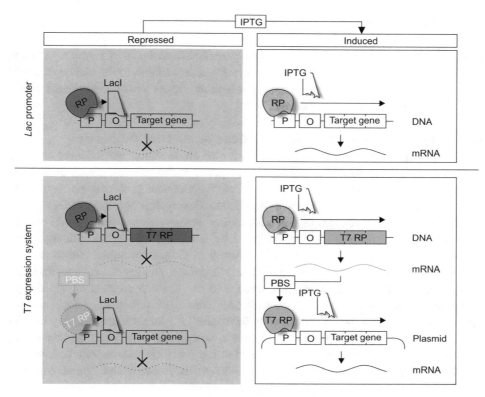

Fig. 9.4 Regulation of gene expression. Many inducible promoters function according to a similar principle as the *lac* promoter. In the absence of an inducer (in this case, lactose or IPTG), a repressor (in this case, LacI) binds to a region called operator (O) between the promoter (P) and the target gene. This prevents the RNA polymerase (RP) from transcribing the target genes. Binding of the inducer leads to a structural change in the repressor so that it can no longer bind to the operator, allowing expression of the target gene. The T7 expression system can also be induced by IPTG or lactose. Here, the target gene is cloned downstream of a T7 promoter which is recognized exclusively by a genomically encoded T7 RNA polymerase (T7 RP). Protein biosynthesis (PBS) of T7 RP occurs after transcription of the respective gene from the *lac* promoter and can therefore be induced by IPTG

recombinant genes regulated by the P_{AOX1} promoter. However, since *P. pastoris* belongs to the methylotrophic yeasts, methanol is also metabolized, and biomass production cannot be uncoupled from target enzyme production (Ergün et al. 2021). The most commonly used promoters in filamentous fungi can also be induced by addition of different carbon sources (Table 9.5). In higher eukaryotes such as plants, it is possible to produce a recombinant enzyme specifically in a certain tissue, organ, or only specific stage of the life cycle via the use of specific promoters. In addition to chemical induction, physical effects can also be used for induction, e.g., light, temperature, or pH (Rosano et al. 2019; Hughes 2018).

9.4.3 Modifications at Gene and Protein Level

In this chapter, only a few examples could be presented from the rapidly growing number of host organisms and expression systems. However, in addition to the choice of a host organism and the regulatory system used, many other factors must be considered on the way to a successfully produced enzyme.

9.4.3.1 Adaptation of a Recombinant Gene to the Host Organism

Heterologous production hosts are often used for the production of recombinant enzymes due to advantages in handling and costs. However, due to differences between the original and production hosts, some modifications of the recombinant gene may be required to enable or optimize efficient heterologous production. Eukaryotic genes contain regions that code for the target enzyme (exons) but have further regions (introns; see Fig. 9.1) that are removed from the mRNA after transcription by a process known as splicing. Since prokaryotes are not able to carry out this modification, the corresponding introns must be removed if these genes are to be expressed in prokaryotes. The genetic code itself, i.e., a DNA base triplet encoding a specific amino acid, is considered universal but is used differently by different organisms. Several base combinations can code for the same amino acid (e.g., six different triplets code for valine, two for cysteine, but only one for tryptophan), but not every triplet combination is used equally often in every organism. This so-called codon usage can determine the yield of target protein because rare codons may lead to slower protein biosynthesis and thus lower yields. Hence, codon optimization can be achieved by using the most abundant base triplet of the host organism for each amino acid to increase the yield, at least in some cases. Typically, genes are amplified from their organism of origin by PCR and subsequently inserted into a plasmid, which is then transferred into the host organism. Although the PCR method allows modifications of the amplified DNA fragment, complex modifications such as the removal of several introns or an adjustment of the codon usage for the entire gene would be very cost- and time-intensive. In the meantime, however, the technology of DNA synthesis has been developed to such an extent that desired sequences can be produced at prices of below € 0.10 per base pair.

In addition, some enzymes require additional proteins for correct folding and formation of enzymatic activity. These "helper" proteins can be chaperones, which assist the correct folding of an enzyme, or proteases that cleave a specific peptide resulting in enzyme activation. As these accessory proteins are present in the organism of origin but may be absent in the heterologous host, they must be identified and coproduced in the heterologous host to ensure the production of active enzymes. So-called chaperone toolboxes are available that allow to combine a number of known chaperones in an expression host to identify a combination that is necessary for the bioactivity of the desired target enzyme.

9.4.3.2 Secretion of a Recombinant Enzyme

Proteins secreted into the culture medium can be isolated more easily and cost-effective than those produced in cytoplasm. Hence, protein secretion finds application in biotechnology. For *E. coli*, a transport system is known that can be used to secrete recombinant proteins into the culture medium by fusion to a natural transport signal (Pourhassan et al. 2021). The secretion of a target enzyme by *Bacillus* can be enhanced by fusing it to different signal peptides located at the N-terminus of the secreted enzyme. Screening of a signal peptide library can identify the signal peptide that provides the most effective transport into the culture medium for a particular target enzyme. In plants, mammals, and fungi, which secrete proteins via the endoplasmic reticulum and the Golgi apparatus, "unconventional secretion" is known to bypass both compartments and thus the posttranslational modifications that take place therein.

9.4.3.3 Toxicity and Solubility of a Recombinant Enzyme

The bioactivity of an enzyme may be toxic to the host. An example is phospholipases which can degrade phospholipids as components of the host membrane, thereby inhibiting the growth of the host organism. In addition, especially with high transcription rates of the target gene (e.g., in the T7 expression system; Sect. 9.4.2.2), a large part of the energy available in the cell is rerouted into the production of the recombinant enzyme, which also restricts cell growth. Another side effect of (too) high protein production in bacteria is the aggregation of a recombinant enzyme in water-insoluble inclusion bodies. Since this aggregation is usually due to misfolding of the enzyme leading to loss of bioactivity, the formation of inclusion bodies is generally considered a disadvantage. However, an advantage may be that the recombinant enzyme in the form of inclusion bodies can be easily separated and purified from the cells and from other proteins. New methods have been developed allowing to isolate the protein aggregates and subsequently convert them into an enzymatically active conformation. It is also possible to immobilize enzymes in their catalytically active form in inclusion bodies and to introduce these so-called CatIBs (catalytically active inclusion bodies) directly into the desired reaction mixture (Ölçücü et al. 2021).

The formation of inclusion bodies can be avoided with several experimental strategies. Lowering the cultivation temperature leads to slower cell growth and reduced protein biosynthesis. Hence, the formed enzyme has more time to fold into its active conformation and does not exclusively form aggregates. Likewise, the transcription rate can be decreased by using a weaker promoter. Another possibility is the fusion with another protein, which can increase the solubility of the target enzyme (protein tag). For example, fusion of a target enzyme to a maltose-binding protein (MBP) can drastically increase the solubility of the target enzyme, although the reasons are not yet fully known. Such protein tags can also be used to purify target enzymes after production, taking advantage of their high affinity for certain substances (Rosano et al. 2019). In the case of a fusion to MBP, cellular proteins are separated from the target enzyme by affinity chromatography using an immobilized maltose matrix. The fusion protein binds specifically to the matrix, while the remaining proteins can be washed out. The pure fusion protein can then be eluted from the column. More often, shorter peptide tags are used to purify the target enzyme, the best known being the His-tag which consists of 6–10 histidine residues. These histidines interact with high affinity with a matrix consisting of nickel-nitriloacetic acid (Ni^{2+}-NTA). The corresponding affinity chromatographic methods are described in more detail in the chapter on enzyme immobilization (Chap. 11).

Questions
8. How does a constitutive promoter differ from an inducible promoter?
9. Which chemical and physical methods can be used to control gene expression in bacteria?
10. How can the genetic code affect the yield of protein production?

Take-Home Message
- Enzymes are produced at larger scale using microorganisms.
- Prokaryotic production hosts are Gram-negative and Gram-positive bacteria (e.g., *Escherichia coli* and *Bacillus subtilis*); eukaryotic production hosts are yeasts (*Pichia pastoris*) and filamentous fungi (*Trichoderma reesei*).
- Proteins can be produced in cell-free systems containing all enzymes and factors required for transcription, translation, and folding.
- Expression systems for enzyme production consist of a suitable host organism and a target gene cloned into a plasmid or integrated into the chromosome.
- The expression of a target gene can be controlled by a constitutive or a regulated promoter.
- A widely used expression system is based on *E. coli*, which produces bacteriophage T7 RNA polymerase and contains a plasmid with the target gene cloned behind a T7 promoter.

Answers
1. Enzymes of high industrial importance are proteases (e.g., as detergent additives), amylases (e.g., in the food industry), lipases (e.g., for the synthesis of enantiopure pharmaceuticals), etc.
2. Prokaryotic microorganisms include, among others, *Escherichia coli* and *Bacillus subtilis*; eukaryotic microorganisms include, among others, *Trichoderma reesei* and *Pichia pastoris*.
3. Limiting factors include rates of transcription and translation, stability of the transcript, availability of tRNAs and codons, correct folding, and stability of the protein.
4. All cells are surrounded by a cytoplasmic membrane and contain ribosomes and DNA. Eukaryotic cells possess as membrane-enclosed organelles a nucleus, an endoplasmic reticulum, a Golgi complex, mitochondria, and chloroplasts; such organelles are missing in prokaryotic cells.

5. The cell envelope of Gram-positive bacteria contains a thick layer of peptidoglycan (15 or more layers), whereas the Gram-negative cell wall contains a thin layer of peptidoglycan (1 layer). The Gram-negative cell envelope additionally contains an outer membrane, i.e., a second lipid bilayer consisting of phospholipids and lipopolysaccharide.

6. Factors required for cell-free protein production include the target DNA, RNA polymerase, ribosomes, amino acids, aminoacyl-tRNA synthetases, cofactors, and energy sources.

7. Genetic elements required on a plasmid include an origin of replication, a promoter and a ribosome binding site, the target gene with a start and a stop codon, and a selection marker (usually a gene encoding an antibiotic resistance).

8. A constitutive promoter is unregulated and allows continuous transcription of a target gene, whereas an inducible promoter allows control over transcription under inducing or repressing conditions.

9. Chemical compounds that bind to repressor or inducer proteins are often sugars or their derivatives; the most prominent ones are lactose and isopropyl β-D-thiogalactopyranoside (ITPG). Physical methods to control gene expression in bacteria can include illumination with light of specific wavelengths and alteration of pH or temperature.

10. The codon usage is different in different (micro)organisms. Rare (i.e., missing) codons in the expression host may result in slow protein biosynthesis and low yields of target protein.

References

Adrio JL, Demain AL. Recombinant organisms for production of industrial products. Bioeng Bugs. 2010;1(2):116–131. https://doi.org/10.4161/bbug.1.2.10484.

Adrio JL, Demain AL. Microbial enzymes: tools for biotechnological processes. Biomol Ther. 2014;4(1):117–139. https://doi.org/10.3390/biom4010117.

Demain AL, Vaishnav P. Production of recombinant proteins by microbes and higher organisms. Biotechnol Adv. 2009;27(3):297–306. https://doi.org/10.1016/j.biotechadv.2009.01.008.

Ergün BG, Berrios J, Binay B, Fickers P. Recombinant protein production in *Pichia pastoris*: from transcriptionally redesigned strains to bioprocess optimization and metabolic modelling. FEMS Yeast Res. 2021;21(7):foab057. https://doi.org/10.1093/femsyr/foab057.

Fernandez FJ, Vega MC. Technologies to keep an eye on: alternative hosts for protein production in structural biology. Curr Opin Struct Biol. 2013;23(3):365–73. https://doi.org/10.1016/j.sbi.2013.02.002.

Heck A, Drepper T. Engineering photosynthetic α-proteobacteria for the production of recombinant proteins and terpenoids. In: Hallenbeck P, editor. Modern topics in the phototrophic prokaryotes. Cham: Springer; 2017. https://doi.org/10.1007/978-3-319-46261-5_12.

Hughes RM. A compendium of chemical and genetic approaches to light-regulated gene transcription. Crit Rev Biochem Mol Biol. 2018;53(5):453–74. https://doi.org/10.1080/10409238.2018.1487382.

Kleiner-Grote GRM, Risse JM, Friehs K. Secretion of recombinant proteins from *E. coli*. Eng Life Sci. 2018;18(8):532–50. https://doi.org/10.1002/elsc.201700200.

Lübben M, Gasper R. Prokaryotes as protein production facilities. In: Kück U, Frankenberg-Dinkel N, editors. Biotechnology. Berlin/Bosten: Walter de Gruyter GmbH; 2015.

Maurer K-H, Elleuche S, Antranikian G. Enzymes. In: Sahm H, Antranikian G, Stahmann K-P, Takors S, editors. Industrial microbiology. Berlin/Heidelberg: Springer; 2013. https://doi.org/10.1007/978-3-8274-3040-3.

McKenzie EA, Abbott WM. Expression of recombinant proteins in insect and mammalian cells. Methods. 2018;147:40–9. https://doi.org/10.1016/j.ymeth.2018.05.013.

Ölçücü G, Klaus O, Jaeger K-E, Drepper T, Krauss U. Emerging solutions for *in vivo* biocatalyst immobilization: tailor-made catalysts for industrial biocatalysis. ACS Sustain Chem Eng. 2021;9 (27):8919–45. https://doi.org/10.1021/acssuschemeng.1c02045.

Pourhassan NZ, Smits SHJ, Ahn JH, Schmitt L. Biotechnological applications of type 1 secretion systems. Biotechnol Adv. 2021;53:107864. https://doi.org/10.1016/j.biotechadv.2021.107864.

Rettenbacher LA, Arauzo-Aguilera K, Buscajoni L, Castillo-Corujo A, Ferrero-Bordera B, Kostopoulou A, et al. Microbial protein cell factories fight back? Trends Biotechnol. 2022;40: 576–90. https://doi.org/10.1016/j.tibtech.2021.10.003.

Rosano GL, Morales ES, Ceccarelli EA. New tools for recombinant protein production in *Escherichia coli*: a 5-year update. Protein Sci. 2019;28:1412–22. https://doi.org/10.1002/pro.3668.

Tripathi NK, Shrivastava T. Recent developments in bioprocessing of recombinant proteins: expression hosts and process development. Front Bioeng Biotechnol. 2019;7:420 https://doi.org/10.3389/fbioe.2019.00420.

Yin J, Li G, Ren X, Herrler G. Select what you need: a comparative evaluation of the advantages and limitations of frequently used expression systems for foreign genes. J Biotechnol. 2007;127(3): 335–347. https://doi.org/10.1016/j.jbiotec.2006.07.012.

Zemella A, Thoring L, Hoffmeister C, Kubick S. Cell-free protein synthesis: pros and cons of prokaryotic and eukaryotic systems. Chembiochem. 2015;16:2420. https://doi.org/10.1002/cbic.201500340.

Enzyme Purification

10

Sonja Berensmeier and Matthias Franzreb

What You Will Learn in This Chapter

The growing importance of enzymes for the food, biotech, pharmaceutical, and cosmetics industries, among others, has increased the need for efficient methods to isolate and process them to the purities required for a particular application. With the protein purification techniques available today, most enzymes can be purified to the required quality and homogeneity. However, the costs associated with multi-stage separation processes are sometimes very high and account for a considerable proportion of the total production costs. This chapter introduces the general aspects of enzyme purification, an indispensable prerequisite for the technical and medical use of enzymes. Accordingly, industrial, analytical, and therapeutic uses of enzymes will be distinguished because the application strongly influences the required purity. Following a short overview of common separation technologies and the underlying physicochemical principles, examples of the most important processes will be described in more detail. Finally, the reader will be guided through the multi-step purification process of an important therapeutic enzyme.

S. Berensmeier (✉)
TUM School of Engineering and Design, Chair of Bioseparation Engineering, Technical University of Munich, Garching, Germany
e-mail: s.berensmeier@tum.de

M. Franzreb
Institute for Functional Interfaces (IFG), Department of Bioengineering and Biosystems, KIT – Karlsruhe Institute of Technology, Eggenstein-Leopoldshafen, Germany
e-mail: matthias.franzreb@kit.edu

© The Author(s), under exclusive license to Springer Nature Switzerland AG 2024
K.-E. Jaeger et al. (eds.), *Introduction to Enzyme Technology*, Learning Materials in Biosciences,
https://doi.org/10.1007/978-3-031-42999-6_10

10.1   Introduction

The raw materials necessary for the isolation of enzymes are microorganisms, animal organs/cell cultures, and plant materials. The choice of purification method and the number of steps depend on the localized enzyme. Isolating intracellular enzymes often entails separating them from a complex biological mixture after disruption of the starting material. Extracellular enzymes, on the other hand, have already been released into less complex media, which greatly simplifies downstream processing.

Enzymes are very complex proteins, and their high specific activity can only be maintained in their native form, which is strongly influenced by environmental conditions such as pH, temperature, and ionic strength. As a result, only mild, specially designed methods can be used for enzyme purification.

For efficient separation, different physicochemical properties can be exploited. These properties include surface charge and hydrophobicity, pI value, molecular weight, and affinity for biospecific ligands (e.g., dyes and metal ions), as outlined in Table 10.1 with corresponding separation methods.

Questions
1. Identify at least four physicochemical properties of an enzyme that can be used for purification and the corresponding technologies.
2. Two enzymes of comparable size, but different pI values, must be separated. Which separation method is commonly applied?

10.2   Parameters

Meeting parameters such as purity, yield, and concentration factor is essential in downstream processing to ensure the quality of a process. Purity P is defined by:

$$P = \frac{\text{Amount product}}{\text{Amount product} + \text{Amount impurities}} \qquad (10.1)$$

Enzyme purity is usually expressed in terms of specific activity (unit/mg total protein), as determined by activity measurements (Chap. 4). The purer an enzyme is, the greater the specific activity becomes and approaches a constant maximum value during the purification process. The ratio of the specific activities before and after purification determines the purification factor PF.

When calculating the yield, which is given as a percentage, a distinction is made between the yield of a single step and the total yield of the entire process. Here, the total amount of the target enzyme is related to the amount of the enzyme at the beginning of the overall process or at the beginning of each separation step under consideration. The yield of

Table 10.1 Classical separation methods

Physicochemical property	Separation process
Charge	Ion exchange chromatography
	Electrodialysis
	Aqueous two-phase extraction
	Reverse micelle extraction
Hydrophobicity	Hydrophobic interaction chromatography
	Reverse phase chromatography
	Precipitation
	Aqueous two-phase extraction
Specific binding	Affinity chromatography
Size	Gel filtration
	Ultrafiltration
	Dialysis
Sedimentation rate	Centrifugation
Surface activity	Adsorption
	Foam fractionation
Solubility	Crystallization
	Extraction
	Extraction with supercritical gases

each step can be very different and depends on many parameters. Yields of individual steps Y_i can vary between 50 and 99%, so as the number of purification steps n required increases, the overall yield Y_{total} of a process can become very low (Eq. 10.2, Fig. 10.1).

$$Y_{total} = \prod_{i=1}^{n} Y_i \tag{10.2}$$

In addition to the purities to be achieved with the highest possible yields, the concentration factor CF is also of high importance for industry. The volume to be processed should be kept as low as possible in order to be able to use lower mass flows and smaller apparatus. In addition, the final concentration of the target enzyme in solution should be as high as possible for the user; if necessary, it should even be available in solid form.

10.3 Process Development

In enzyme purification, there is not one purification process; instead, a multi-stage process is usually employed individually for each target product. In addition to the physicochemical properties of the enzyme mentioned above and the separation processes derived from them (Table 10.1), the required purity and quality of the target molecule must be defined, depending on the application and the associated guidelines. On this basis, a distinction is made between three main categories:

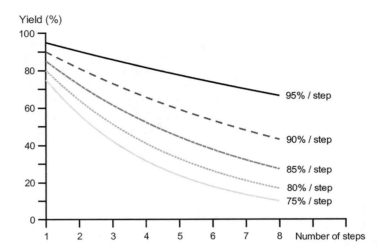

Fig. 10.1 Influence of yields and number of purification steps on the total yield of a process

- Technical (bulk) enzymes, e.g., hydrolases, such as amylases, cellulases, pectinases, and proteases
- Enzymes used in the diagnostic field and whose activities are frequently evaluated by photometric measurements, e.g., glucose oxidase, peroxidase, and cholesterol esterase (see Chap. 18)
- Therapeutic enzymes, e.g., α-glucosidase, arylsulfatase, iduronidase, and proconvetin (see Chap. 19)

In the diagnostic and therapeutic areas, the requirements for purity are very high in order to be able to prevent undesirable interactions. However, if the enzymes are used later, e.g., in detergents or for bioethanol production, low purities are often sufficient (Fig. 10.2).

The higher the requirements with regard to purity and legal regulations, the more process steps and analyses are needed, which affects the price. Depending on the enzyme and the application, the organism in which the enzyme is produced differs (Chap. 9). This in turn determines the localization of the enzyme (intracellular and extracellular) and its concentration before the start of the purification process.

Depending on the localization, the enzyme is first made accessible in a soluble form. For intracellular production, the processing can be done by mechanical disruption of the cells, e.g., by means of high-pressure homogenization. If poorly soluble *inclusion bodies* are formed, a further refolding step may be required. In contrast, extracellular products can be separated directly from the cell material by solid–liquid separation (e.g., centrifugation and microfiltration). The first goal is to considerably reduce the volume flow and to concentrate and stabilize the product. Depending on the required purity, a more or less multi-stage purification process follows and, depending on the application, the formation of the enzyme (Fig. 10.3).

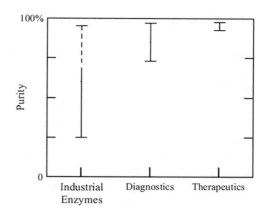

Fig. 10.2 Purities required for different products. (Adapted from Harrison 1994)

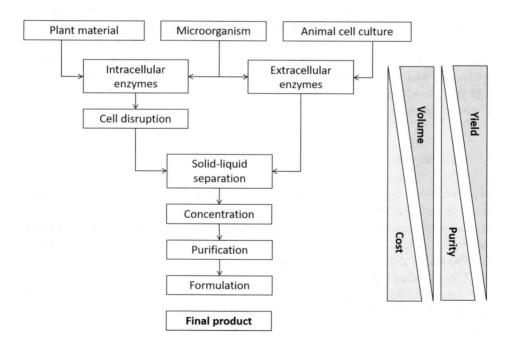

Fig. 10.3 General process for enzyme purification

Questions
  3. Identify the range of commonly required purities for industrial, diagnostic, and therapeutic enzymes. Briefly explain the differences in the demand.
  4. What are the main four steps in a downstream processing scheme of an extracellular enzyme? Please refer to the superordinate terms, not individual technologies.

10.3.1 Technical Enzymes

Industrial bulk enzymes (e.g., detergent enzymes and starch-degrading enzymes) are produced in large quantities (several tons per year; capacities of individual fermenters up to 100 m^3) and sometimes have low purity requirements (25–90%; Chap. 17). Enzymes used industrially on a large scale are often impure preparations that are sufficient to meet customer requirements in terms of activity and stability. Often, they contain many other enzymes, which in some applications are even useful for their overall performance.

In the food and feed sector, the industry prefers established GRAS (*generally regarded as safe*)-certified production strains, such as *Bacillus, Aspergillus*, and *Saccharomyces* (Chap. 9). Their biology is well understood, and in addition to their safe classification, they grow very rapidly with little effort and produce high space-time yields of the target molecule. Microbial enzymes are primarily produced in liquid cultures in batches, although traditional solid-phase fermentations for extracellular production using fungi are also common, depending on the capabilities and cultures of other countries. Many technical enzymes are secreted directly into the medium so that after separation of the cell mass by centrifugation or filtration usually only a concentration by means of precipitation, crystallization, or filtration is required. Depending on the application and formulation, stabilizing agents may be added or enzyme powder may be produced after a drying process. Due to the high costs involved, attempts are made to dispense with high-resolution chromatography processes for technical enzymes.

10.3.2 Diagnostic Enzymes

Many analytical methods in medicine and food analysis are based on enzyme-catalyzed reactions to detect metabolites, such as glucose and cholesterol, among others. Enzymes used in the diagnostic field must be very specific and sensitive. To ensure that no other enzymatic activities or other interferences occur during the diagnostic test, high purities (>95%) and quality must be guaranteed.

Glucose oxidase (GOD) is an example of a potential interference. It is typically produced industrially in GRAS-listed strains of *Aspergillus* and *Penicillium* and is available for a wide range of applications. In the diagnostic field, it is often used in combination with a peroxidase (POD) to detect glucose. GOD oxidizes glucose to gluconate with the formation of H_2O_2 (Chap. 18). For colorimetric evaluation, a reduced colorless chromogen is oxidized in the indicator reaction by H_2O_2 through peroxidase (POD). Commercially available preparations contain the additional enzyme mutarotase, which converts α-D-glucose into β-D-glucose.

$$\text{Mutarotase}: \alpha - D - \text{glucose} \rightarrow \beta - D - \text{glucose}$$

$$\text{Glucose oxidase}: \beta - D - \text{glucose} + H_2O + \frac{1}{2}O_2 \rightarrow \text{gluconolactone} + H_2O_2$$

$$\text{Peroxidase}: \text{Dye} - H_2 \ (\text{red.}) + H_2O_2 \rightarrow \text{Dye(ox.)} + 2H_2O$$

Depending on the production organism, GOD preparations may also contain carbohydrases, such as saccharase and maltase, as well as catalase, as accompanying enzymes (De Baetselier et al. 1991). To measure the glucose content in blood or urine, the former does not interfere since disaccharides are practically absent. However, to analyze the glucose content in the ever-growing field of food and food controls, carbohydrases must be eliminated to guarantee a high-quality result. Catalase can also indirectly influence the analytical result when H_2O_2 is detected by *horseradish peroxidase* (HRP) (Simpson et al. 2007).

The problem of the different interactions clearly shows the need for selective methods, such as chromatography and ultrafiltration, to purify diagnostic and therapeutic enzymes (Chap. 19). Similarly, product quality must be closely monitored to be able to make qualitative statements in diagnostics.

10.3.3 Therapeutic Enzymes

In addition to diagnostic purposes, enzymes are also increasingly used as biopharmaceuticals (Chap. 19; Walsh and Shanley 2005). Such therapeutic enzymes are traditionally extracted directly from natural sources, such as serum or animal tissue, and purified to greater than 99% in a subsequent multi-stage process (Walsh and Murphy 1999). Due to rapid advancements in molecular biology methods, production using recombinant microorganisms and cell cultures is becoming increasingly important, although the purification process required usually varies greatly with the type of source. Due to the injection or oral intake of the corresponding preparations, the highest purities of 99.9% and above are essential to avoid undesired immune reactions and side effects from impurities. Table 10.2 contains a short list of known therapeutic enzymes, the corresponding medical field of application, and the enzyme source used. The annual production of individual biopharmaceuticals is normally only a few kilograms, with low product titers usually present in the starting mixture.

With the purities required by the regulatory authorities, therapeutic enzymes are in the same range as therapeutically used monoclonal antibodies (mAb). In contrast to mAb, however, there is no established platform technology for therapeutic enzymes that enables the rapid and standardized development of a purification process. Purification schemes for therapeutic enzymes often contain purification steps of high affinity that exploit specific properties of the enzyme in question. In addition, the purification schemes for enzymes from serum or tissue and for enzymes from cell cultures differ significantly, at least with regard to initial product isolation.

Table 10.2 Examples of known therapeutic enzymes, their clinical field of application, and the enzyme source used for production

Therapeutic enzyme	Field of application	Enzyme source
Factor VIIa	Hemophilia	Cell culture (BHK)
Factor IX	Hemophilia	Plasma, cell culture (CHO)
Tissue-specific plasminogen activator (tPA)	Dissolution of blood clots, e.g., in heart attacks	E. coli, cell culture (CHO)
Urokinase	Dissolution of blood clots, e.g., in myocardial infarction	Cell culture (CHO)
DNase	Cystic fibrosis	Cell culture (CHO)
Activated protein C	Inhibition of blood clotting, sepsis	Cell culture (HEK 293)
Asparaginase	Cancer (leukemia)	E. coli, Erwinia chrysanthemi
Glucocerebrosidase	Hereditary disease Gaucher's disease	Cell culture (CHO)
α-Galactosidase	Hereditary Fabry disease	Cell culture (CHO)

Figure 10.4 shows typical process sequences for the purification of therapeutic enzymes from natural sources (plasma, tissue) or from animal cell cultures. These multi-step process sequences are shown in simplified form and omit, for example, the necessary solid–liquid separations following a precipitation step as well as possible volume reductions and re-buffering between the chromatography steps. The general conditions and tasks of each step in the process are briefly described below:

10.3.3.1 Cell Harvesting or Cell Separation and Solid/Liquid Separation

To produce therapeutic enzymes by means of a cell culture, the enzyme is usually present extracellularly. The cell separation or cell harvesting step is therefore normally used to remove the cells and clarify the supernatant from suspended solids. The main technologies employed are separators and microfiltration systems. The most common ones are continuously operating disc separators and cross-flow microfiltration (*tangential flow microfiltration*). With separators, clarification of the process solution is usually followed by depth filtration. Positively charged filter materials also reduce the DNA concentration and potential viruses in the filtrate. For reliable and complete clarification, the final step is generally direct flow filtration (*dead-end filtration*) with a filter membrane pore size of approx. 0.2 µm to avoid blockages in subsequent chromatography and membrane processes.

The Disc Separator

Disc separators enable the efficient separation of suspensions or emulsions. The separation principle is based on the different centrifugal forces experienced by the suspension or emulsion components due to their different densities. In a disc separator, the feed is via a

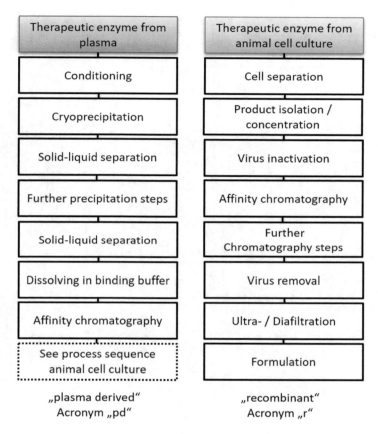

Fig. 10.4 Typical process sequences for the purification of therapeutic enzymes from natural sources (plasma and tissue) and from animal cell cultures

hollow shaft in which the suspension is transported to the separator bottom and gently brought up to full speed (Fig. 10.5). The main component of the separator is a pack of conical discs spaced a few millimeters apart. When flowing through the disc pack under rotation, the substance with higher density, i.e., the cells during cell separation, accumulates on the underside of the disks. Centrifugal force then directs the cells downward at an angle into a sediment collection chamber, where they are discharged at cyclic intervals by briefly lowering the bottom of the bowl. The clarified fermentation liquid flows inwards on the top of the plates, where it is collected by a so-called gripper and conveyed out of the plate separator.

10.3.3.2 Precipitation Steps

For a long time, human serum was one of the main sources of therapeutic enzymes and still plays an important role. Serum has a very high protein concentration of approx. 70 g L^{-1}, with albumin and immunoglobulins making up by far the largest proportion. In the run-up

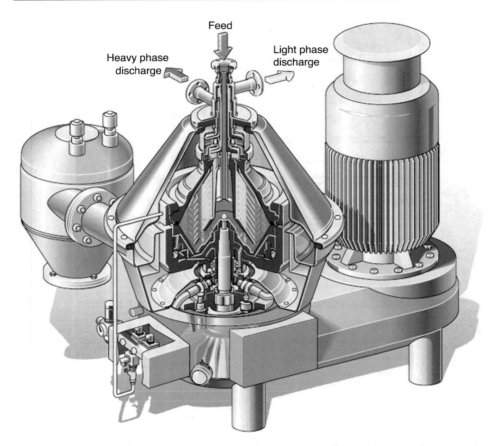

Fig. 10.5 Disc separator with automatic solid discharge from GEA (GEA Westfalia Separator Group, with kind permission)

to purification methods for therapeutic enzymes, such as factor VIIa and Factor IX, which are also present in serum, the proportion of these main proteins in the serum is first greatly reduced by fractionation. This fractionation is carried out by staggered precipitation steps with increasing ethanol concentration, whereby the temperature and pH at which the precipitation takes place also have an important influence on the solubility of various serum proteins. The best-known method for this stepwise precipitation, also known as cryoprecipitation, is the Cohn process, named after E. J. Cohn (Cohn et al. 1946). The main steps of the Cohn process are:

  (i) One precipitation at 8% ethanol, −3 °C and pH 7.2
 (ii) One precipitation at 25% ethanol, −5 °C, pH 6.9
(iii) One precipitation at 18% ethanol, −5 °C, pH 5.2
(iv) One precipitation at 40% ethanol, −5 °C, pH 5.8

(v) One precipitation at 40% ethanol, $-5\,°C$, pH 4.8

Cooled centrifuges are required for the solid–liquid separation following the precipitation process since the selected temperature level must also be maintained precisely during this separation process. Special tube or chamber centrifuges are quite commonly used for this process.

10.3.3.3 Product Isolation and Concentration
In addition to stabilizing the target enzyme, its rapid concentration is one of the primary goals of a purification process. The concentration reduces the volume to be processed and thus the necessary size of downstream process equipment. In addition to the precipitation processes previously described, chromatographic steps are particularly suitable for concentration. Due to the high loading capacities that can be achieved, ion exchange chromatography is generally used, applying cation or anion exchange resins depending on the charge of the target enzyme. The operational capacities and throughputs are up to 100 g enzyme and $10\ L\ h^{-1}$ per L column volume. A typical ion exchange chromatography procedure includes equilibration, loading, washing, elution, and regeneration steps (Carta and Jungbauer 2010). Chromatography is therefore a time- and resource-intensive process and often one of the main cost items of a purification process. An alternative to a separate chromatography step for product isolation is concentration and purification in the course of affinity chromatography.

Chromatography Columns
Chromatography is an umbrella term for physicochemical separation, which is based on the distribution of the substances to be separated between a mobile and a stationary phase. In column chromatography, the stationary phase is located as a compact bulk in a cylindrical column (Fig. 10.6). When the mobile phase with the solutes flows through the column, the substances interacting more strongly with the stationary phase tend to be retained (retarded) and accordingly appear later in the run. Despite the supposedly simple principle, chromatography media, i.e., the stationary phase, and chromatography columns are complex and often expensive components of a purification process. Chromatography media for industrial applications must have high capacities and selectivities for the target component and at the same time must be chemically and mechanically robust as well as highly porous. A chromatography column requires a very efficient liquid distributor in the inlet to achieve uniform flow through the packed chromatography media. In addition, the packing must be kept under a defined pressure via a movable plunger that prevents fluidization or channel formation. The design and material of the column must be able to withstand higher pressures and allow complete cleaning of the system with, e.g., 1 M NaOH.

10.3.3.4 Further Chromatography Steps and Affinity Chromatography
In addition to concentration, chromatography steps are also employed to remove the main contaminants (other proteins of the production strain or of the starting medium, DNA) and

Fig. 10.6 AxiChrom™
chromatography column
(GE Healthcare, with kind
permission)

to remove unwanted variants (e.g., dimers) of the target enzyme. A wide range of chromatography variants (e.g., ion exchange, hydrophobic interaction, *mixed-mode,* affinity chromatography, etc.) can be used for separation. Affinity chromatography is of particular importance as it facilitates a strong increase in the purity of the target enzyme in one step and will therefore be discussed more thoroughly in the next section. However, the chromatography methods reviewed thus far are largely unsuitable for removing undesired variants of the target enzyme because, for example, a dimer of an enzyme has many of the same physicochemical properties as the corresponding monomer. An exception is the size of the macromolecule, for which the variants can be separated by size exclusion chromatography.

10.3.3.5 Affinity Chromatography

Affinity chromatography is a special method that uses highly selective, biospecific interactions between the ligands of the chromatography medium and the enzyme to be purified (Janson 2011). Ligands can be distinguished between group-specific ligands, which have a high affinity for a whole group of proteins, and monospecific ligands, which only have a high affinity for a specific protein. Among the best known representatives of group-specific ligands are complexing agents, such as NTA (nitrilotriacetic acid) or IDA (iminodiacetic acid), which, after loading with divalent metal ions such as Cu^{2+}, Ni^{2+}, or Co^{2+}, have a high affinity for proteins with several histidine or cysteine residues in a suitable arrangement. In order to use this process, which is referred to as immobilized metal ion affinity chromatography (IMAC), polyhistidine groups are often fused to *C-* or *N-terminally* to the enzyme of interest via molecular genetic methods. For therapeutic enzymes in particular, however, working with so-called fusion tags is generally used only for research purposes since a molecular genetic modification of the enzyme is undesirable in a medical application. For cofactor-dependent enzymes, other group-specific ligands are suitable, which are similar in structure to the particular cofactor, and do not require additional tags. A well-known example is *Blue Sepharose* produced by Cytiva. In this chromatography material, the reactive dye Cibacron Blue 3G is coupled to a Sepharose matrix. The structure of Cibacron Blue 3G mimics that of the cofactor NAD and can be used to purify, e.g., dehydrogenases or kinases. Due to the high added value of therapeutic enzymes, the special development of monospecific ligands is often worthwhile. These ligands are usually monoclonal antibodies that recognize and bind with the target enzyme as an antigen. The antibodies are produced in a dedicated cell culture, purified, and covalently bound to an activated chromatography material, namely NHS-Sepharose. A correctly prepared chromatography column results in a highly selective capture of the target enzyme directly from the clarified process solution and concentrates the final product at the same time.

10.3.3.6 Virus Removal

In addition to the removal of all proteins (except the target enzyme) and DNA, a production process for therapeutic enzymes must also ensure the complete removal of any viral contamination from the serum-containing starting medium. For this purpose, the process must contain at least two steps for virus removal, which differ in their physical principle. The most commonly used methods are virus filtration and membrane chromatography. For both processes cartridge systems are employed, which are often designed for a single use *(disposable)*. However, the function of the methods differs. While virus filtration uses membranes with a nominal cut-off of 20 nm, which achieves virus depletion mainly by a mechanical sieving effect, virus depletion by membrane chromatography is based on charging effects. A membrane with a thickness of a few millimeters and a pore structure in the micrometer range has positive charge groups on the surface of the membrane fibers. Since viruses usually have a negative surface charge, this leads to the effective attachment of the viruses to the fibers. However, pore size and membrane thickness make it clear that this is not a surface filtration but a special form of depth filtration. Since the potentially occurring virus mass is low even in the case of contamination, the systems for membrane chromatography are not optimized in terms of their loading capacity but with regard to their throughput, which is in the range of up to 30 bed volumes per min.

Cartridge Systems

Modern cartridge systems for virus depletion are sterile-packed and directly insertable hollow fiber modules (Fig. 10.7). During manufacture, the hollow fibers composed of polyethersulfone are embedded directly in a closed housing, which can no longer be opened. The cartridges are therefore designed for single use only. In addition to the main

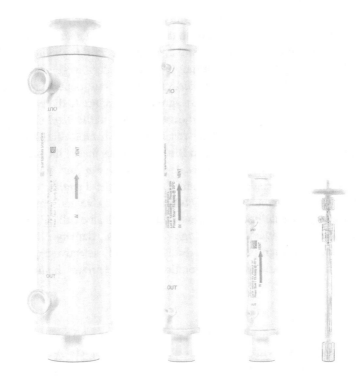

Fig. 10.7 Disposable filtration cartridges Virosart® HF from Sartorius for virus removal. (©Sartorius Stedim Biotech, with kind permission)

connections for the inlet and outlet, there are further connections for venting and flushing. At an operating pressure of 2 bar, such cartridge systems have a throughput of approx. 5 L min^{-1} m^{-2}.

10.3.3.7 Formulation and Freeze-Drying

The target of the formulation is to convert the purified therapeutic enzymes into a stable, transportable, and storable state (Parkins and Lashmar 2000). The steps necessary for this are the adjustment of pH and ionic strength to values favorable for the enzyme by transferring the enzyme into an appropriate storage buffer, which is typically accomplished by a combination of ultrafiltration and diafiltration (UF/DF). The ultrafiltration step first serves to further concentrate the enzyme and then saves storage buffer in the subsequent diafiltration step. The diafiltration step is used for the buffer exchange, the conditioning for the next process step. The enzyme is retained by the filtration membrane, while the target buffer flows across the membrane in a cross-flow direction and buffer components pass through the membrane driven by concentration differences. The continuous or cyclic replenishment of fresh storage buffer results in a complete exchange of the buffer. Other substances, such as polysorbates, are usually added to the solution to prevent aggregation of the active ingredient, i.e., the therapeutic enzyme. The stabilized enzyme solution is finally frozen or stored in a cool place. A better alternative for long-term storage or transport is freeze-drying. For this purpose, the frozen active substance solution is gently dried under a vacuum, during which the moisture sublimates from the frozen state.

10.3.3.8 Requirements for Purified Therapeutic Enzymes

For approval of a purification process and for routine quality control, therapeutic enzymes must meet numerous high requirements for the purity of the active substance solution and for the unique conformation of the enzyme itself. Table 10.3 shows an overview of the most important requirements.

One of the most important requirements for the purity of therapeutic agents is being free from viruses since, depending on the type of virus, even a single virus could cause an infection. Therefore, several independent stages for virus separation are required during the purification process. The effectiveness of the individual steps is usually 4–6 log levels, i.e., a reduction of the virus concentration by a factor of 10^4 to 10^6. Overall, a purification process must have, for example, a theoretical virus reduction of 14 log levels so that even with a virus contamination of 10^6 in the initial volume of cell culture corresponding to a subsequent dose, the probability that a purified dose still contains a virus is less than 10^8. That is, even under these unfavorable assumptions, only one in a hundred million doses would contain a virus. For host cell proteins and DNA, reduction to zero is impossible. However, strict limits exist, often in the range of 0–100 ng host proteins per milligram therapeutic enzyme, i.e., the required purity is >99.99%. For DNA, a common limit is <10 ng of DNA per dose. With pure aggregates of the therapeutic enzyme, no correspondingly low limits can be set because aggregates, such as dimers, can only be incompletely separated from their monomers and they can also be reconstituted from the highly concentrated enzyme in the ready-to-use formula. On the other hand, correctly folded dimers usually also possess the desired therapeutic effect. The permissible limits for aggregates are therefore relatively wide, e.g., 0–5%. The question of how critical other conformational variants, e.g., variants in glycosylation, are cannot be answered in general terms and must be investigated separately for each therapeutic enzyme.

10.3.4 Application Example: Industrial Purification of Factor VII

Coagulation factor VII (proconvertin), a vitamin K-dependent enzyme, is a glycoprotein with a molecular weight of 59 kDa. In the absence of injury, factor VII circulates in the bloodstream as a proenzyme in an inactive monomeric form. In the case of injury, factor VII is converted into its active dimeric form, factor VIIa, and the coagulation cascade is activated (Per et al. 2005). In patients with coagulation disorders, administration of a factor VIIa preparation may be necessary in case of serious bleeding or surgery. In the 1950s and 1960s, concentrates derived from blood plasma were used, but they were a mixture of

Table 10.3 Requirements for therapeutic enzymes

Purity requirements		Conformity requirements Demonstration of correct enzyme compliance with regard to:	
Virus-free	Free of microorganisms	Aggregates	Fractions
DNA <10 ng per dose	Proteins (HCPa) <100 ng per mg enzyme	Conformation	Glycosylation
Defined, serum-free cell culture medium	Defined buffer components	Oxidation	Disulfide bridges

ª Host cell protein

several clotting factors. In the 1980s, recombinant factor VIIa was successfully developed, and clinical trials for the use of this therapeutic enzyme began in 1988. In 1996, the drug NovoSeven (rFVIIa) was finally approved as a product of the company Novo Nordisk in the EU, followed by approval in the USA (1999) and Japan (2002).

Figure 10.8 shows a simplified schematic representation of a purification process for factor VIIa. Batch production begins with the preparation of the inoculum, starting from frozen starting cells of a strictly controlled cell bank. Recombinant *baby hamster kidney* (BHK) cells, which possess the gene for human factor VII, serve as production cells. The BHK cells are propagated through several scaling steps and finally used as an inoculum for cell culture in a bioreactor. With regard to the purification process and possible contamination, fetal calf serum is still occasionally used in the cell culture medium. In addition to the higher amount of different proteins from a second organism (hamster and bovine), the risk of viral contamination is increased, so special attention must be paid to virus removal in the subsequent purification process. Since the resulting enzyme FVII is extracellular, the first step after cell culture is cell separation, e.g., by a disc separator and a microfilter, followed by conditioning of the clarified solution for the subsequent chromatography steps. The first chromatographic step is anion exchange chromatography to concentrate the target enzyme and to remove other proteins. A subsequent addition of surfactants chemically inactivates a virus. The main purification step is immunoaffinity chromatography using an FVII-specific monoclonal antibody as a ligand, followed by two further anion exchange chromatography analyses. These chromatography analyses remove any leached ligands of the chromatography resin, and completely convert FVII into the active form FVIIa. As a second, independent step of virus removal, membrane filtration with a pore size of 20 nm is applied. Finally, various formulation steps are required, such as ultrafiltration and diafiltration, to transfer the product into the storage buffer, for sterile filtration and freeze-drying.

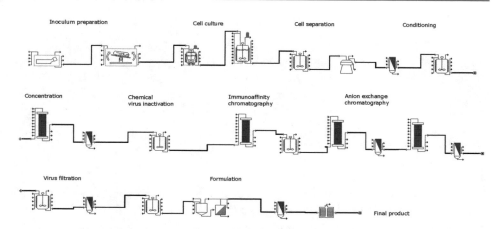

Fig. 10.8 Schematic purification process of recombinant coagulation factor VIIa

The simplified purification process described above illustrates special features of the purification of therapeutic enzymes. First, complex purification procedures with a large number of steps are often required. Second, the purification procedure does not follow a platform process but contains enzyme-specific steps that usually cannot be transferred to other products. For example, the process described for FVIIa contains three anion exchange chromatographs. Such repetition of purification steps with a comparable mechanistic principle should be avoided for standard processes but serves here, among other things, to transform the product into its active form. Third, the lack of a universally applicable highly selective purification step, comparable to the use of Protein A affinity chromatography for monoclonal antibodies, requires special purification steps.

Questions
5. Which type of chromatography is usually selected to separate unwanted dimers of an enzyme from its monomeric form?

Take Home Message
In conclusion, the following basic rules can be summarized for the development of a downstream processing procedure:

1. Determine the required product purity, product activity, and quantity of total product.
2. Determine the physicochemical properties of the target enzyme and critical impurities for simplified method selection.

(continued)

3. Develop analytical detection methods to detect both the target molecule and its activity as well as the critical contaminants.

4. If possible, drastically reduce the volume at the beginning of the work-up and then reduce it again and again in subsequent steps in order to avoid long processing times and large apparatus. A strong reduction of buffer solutions and cleaning agents can also be achieved in this way.

5. If possible, employ separation methods that differ with regard to the type of separation (e.g., size, charge, ligand specificity, and hydrophobicity).

6. Avoid adding additives (e.g., detergents, protease inhibitors, and lysozyme for cell disruption) when possible, as they have to be separated afterward and often cause problems in the analysis.

7. Remove proteases that can denature the enzymes at the beginning of the work-up process. In addition to avoiding the use of protease inhibitors, working at 4–8 °C can be omitted if the proteins are not thermally unstable under process conditions (usually room temperature).

8. Reduce the number of steps required as much as possible and combine them appropriately to optimize the overall yield and process time required.

Answers to the Questions
1. Charge: Ion exchange chromatography and electrodialysis

Hydrophobicity: Hydrophobic interaction chromatography, precipitation, and aqueous two-phase extraction
Size: Filtration, dialysis, and size exclusion chromatography
Bioaffinity: Affinity chromatography, e.g., IMAC

2. Ion exchange chromatography at a pH between the two pI values.
3. Industrial enzyme: 25–90%; diagnostic enzyme: 85–99%; therapeutic enzyme: >98%. In industrial applications, e.g., detergents, impurities such as remaining host cell proteins usually do not interfere with the required catalytic effect of the target enzyme. In diagnostic applications, impurities may lead to unwanted side-effects and must be reduced to an acceptable level. Therapeutic enzymes that are administered to patients must be extremely pure in order to avoid unwanted immune responses.
4. Solid–liquid separation, concentration, purification, and formulation
5. Size exclusion chromatography (also called gel filtration).

References

Carta G, Jungbauer A. Protein chromatography: process development and scale-up. Weinheim: Wiley-VCH; 2010.

Cohn EJ, Strong LE, Hughes WL, Mulford DJ, Ashworth JN, Melin M, Taylor HL. Preparation and properties of serum and plasma proteins. IV. A system for the separation into fractions of the protein and lipoprotein components of biological tissues and fluids 1a,b,c,d. J Am Chem Soc. 1946;68(3):459–75. https://doi.org/10.1021/ja01207a034.

De Baetselier A, Vasavada A, Dohet P, Ha-Thi V, De Beukelaer M, Erpicum T, De Clerck L, Hanotier J, Rosenberg S. Fermentation of a yeast producing A. Niger glucose oxidase: scale-up, purification and characterization of the recombinant enzyme. Nat Biotechnol. 1991;9(6):559–61.

Harrison RG, editor. Protein purification process engineering bioprocess technology. New York: Marcel Dekker, Inc; 1994.

Janson J-C, editor. Protein purification: principles, high resolution methods, and applications. 3rd ed. Hoboken: Wiley; 2011.

Parkins DA, Lashmar UT. The formulation of biopharmaceutical products. Pharm Sci Technol Today. 2000;3(4):129–37. https://doi.org/10.1016/S1461-5347(00)00248-0.

Per R, Elisabeth E, Nikolai B, Niels Kristian K, Egon P. Recombinant Factor VIIa. In: Directory of therapeutic enzymes. CRC Press; 2005. p. 189–207. https://doi.org/10.1201/9781420038378.ch10.

Simpson C, Jordaan J, Gardiner NS, Whiteley C. Isolation, purification and characterization of a novel glucose oxidase from Penicillium sp. CBS 120262 optimally active at neutral pH. Protein Expr Purif. 2007;51(2):260–6. https://doi.org/10.1016/j.pep.2006.09.013.

Walsh G, Murphy B, editors. Biopharmaceuticals, an industrial perspective. Springer; 1999. https://doi.org/10.1007/978-94-017-0926-2.

Walsh G, Shanley N. Applied enzymology. In: Directory of therapeutic enzymes. CRC Press; 2005. p. 1–15. https://doi.org/10.1201/9781420038378.ch1.

Enzyme Immobilization

11

Marion B. Ansorge-Schumacher

What You Will Learn in This Chapter

Immobilization is a comparatively old approach to make enzymes technically usable. The aims are both to increase productivity and enzyme lifetimes and to improve practical handling and facilitate enzyme recovery. This chapter gives an overview of the main, practically relevant immobilization principles for enzymes and explains their most important molecular and physicochemical effects on catalytic performance. Key parameters of the immobilization process and the resulting immobilizates are described, and the choice of method is considered against the background of a technical application.

11.1 Importance, Definition and Goals

Immobilization is a comparatively old approach to make enzymes technically usable. It was first described in 1916 (Nelson and Griffin 1916), and the first immobilized enzymes were used in practice in 1953 (Grubhofer and Schleith 1953). Today, immobilized enzymes are widely used in production processes in the chemical and food industries, in diagnostics and biosensors, or in detergents, agricultural products, and cosmetics.

Enzyme immobilization is generally defined as "physical confinement or localization of enzymes in a confined space while maintaining catalytic activity." This was settled in 1973

M. B. Ansorge-Schumacher (✉)
Department of Molecular Biotechnology, TU Dresden, Faculty of Biology, Dresden, Germany
e-mail: marion.ansorge@tu-dresden.de

© The Author(s), under exclusive license to Springer Nature Switzerland AG 2024
K.-E. Jaeger et al. (eds.), *Introduction to Enzyme Technology*, Learning Materials in Biosciences,
https://doi.org/10.1007/978-3-031-42999-6_11

during the International Enzyme Engineering Conference in Henniker, New Hampshire (USA) (Katchalski-Katzir 1993). Noteworthily, the definition only refers to artificially induced confinement, not to natural compartmentalization. The primary aim of immobilization is to broaden the basis for a technical use of isolated enzymes. Immobilization can achieve a local concentration of enzymes and consequently a higher productivity (space–time yield). At the same time, the separation of the enzymes from the reaction medium is facilitated, which on the one hand prevents product contamination with possibly pyrogenic enzyme molecules and on the other hand enables the use of the enzymes in several reaction cycles. The latter has an enormous effect on the economic efficiency of enzyme-catalyzed processes, since enzyme costs can still account for a disproportionately high share of total process costs, even in the age of production with recombinant high-performance strains. In addition, immobilization can be accompanied by an intrinsic stabilization of the enzymes, so that storage capability and process life are significantly improved compared to the free enzyme. Last but not least, immobilization can increase the motivation for the technical use of enzymes, since practical handling is simplified and health risks that arise from the allergenic potential of enzyme-containing preparations are reduced.

Questions
1. *Would a catalytic active enzyme located in a microbial whole cell be termed immobilized?*

11.2   Methodological Principles

By definition, approaches to the immobilization of enzymes are versatile. Basically, carrier binding, entrapment, and carrier-free cross-linking can be distinguished as methodological principles (Fig. 11.1). The most important subcategories of these principles are explained in more detail in the following sections and illustrated with selected examples of commercially available enzyme immobilizates.

11.2.1 Noncovalent Bonding to Support Materials

Due to the strong functionalization of their molecular surface, enzymes tend to attach to the surfaces of various materials spontaneously. Hydrogen bonding and hydrophobic, polar, electrostatic, or chelating interactions can play a role (Fig. 11.2). Affinity and binding strength depend individually on the type and number of interactions between an enzyme and material surface. As a result, almost any material can potentially function as an enzyme carrier. Correspondingly, the repertoire of materials with which noncovalent immobilization of enzymes has already been described is vast (Jesionowski et al. 2014; Zucca and Sanjust 2014). It ranges from poorly defined materials such as wood chips to natural or synthetic hetero- and homopolymers (e.g., agarose and polystyrene, respectively) to

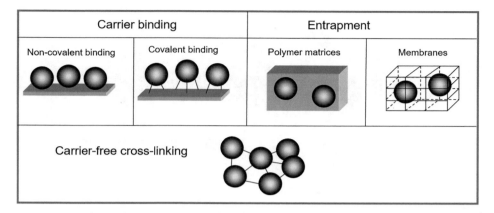

Fig. 11.1 Basic methodological principles and subcategories of enzyme immobilization

activated carbon, minerals (e.g., clays), and metal compounds (e.g., alumina and metal-organic networks).

Commercially available **carrier materials** for the noncovalent immobilization of enzymes predominantly rely on natural or synthetic, highly functionalized organic polymers such as agarose, cellulose, and dextran or poly(acrylate), poly(methacrylate)- poly(propylene), and poly(styrene), respectively, as basic materials. They are widely known from chromatographic protein purification. The mechanical and chemical stability of these materials as well as their pore structure can be modified specifically by internal cross-linking, e.g., with epichlorohydrin. By introducing additional hydrophobic groups, such as octyl chains, or ionizable groups, such as amino- or carboxyl functions, the interaction between a carrier material and enzyme can be controlled via the pH value. Increasingly, materials are getting commercially available that specifically allow the adsorption of tagged recombinant enzymes (i.e., enzymes terminally extended with specific amino acid sequences), but these are usually rather expensive. Enzyme immobilization via chelation of matrix-embedded metal ions, e.g., through the interaction of histidine-tagged enzymes with nickel nitrile acetate matrices (Fig. 11.2) or Cu^{2+}-activated silicates, is also not advisable from a practical point of view, since the continuous, initially hardly detectable desorption of the metal ions from the matrix can lead to contamination and destabilization of synthesis products. In particular for further use in pharmaceutical formulations, this must be avoided by all means.

Among the inorganic materials, **silicates** (SiO_2-based materials) in particular are relevant for the adsorptive binding of enzymes. The silicate surface is highly functionalized with free hydroxyl functions and therefore allows both direct interaction with the enzyme molecule and the introduction of alternative functional groups. The latter is achieved by the process of silanization, during which functionalized trialkoxysilane derivatives form a surface coating that is covalently bonded via siloxane bridges. Amino-, epoxy-, cyano-,

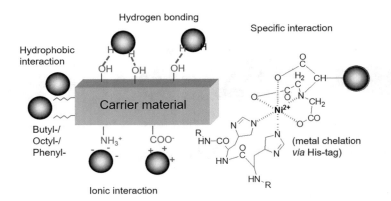

Fig. 11.2 Mechanisms of adsorptive enzyme binding. Hydroxyl functions are variously exposed by natural polymers such as agarose or cellulose and inorganic silicates. Amino and carboxyl functions are found in high density on ion-exchange materials such as diethylaminoethyl (DEAE) and carboxymethyl cellulose, respectively

and sulfhydryl-functionalized silanes as well as phenolic triethyoxy- or trimethoxysilanes are available. Most commonly used samples are amino-functionalized trialkoxysilanes such as APTES (3-aminoethoxysilane). Commercially available basic silicates are simple glass beads or particles in micro- and nano-sizes, zeolites, and diatomaceous earth. Particularly promising for the adsorptive immobilization of synthetically active enzymes are mesoporous silicates such as SBA-15 or MCM-41 or their numerous evolutions (Hartmann and Kostrov 2013). Their adjustable pore volumes, ordered pore structure, uniform pore size distribution, high specific pore volume, and large internal surface area make it possible to efficiently introduce enzymes into the matrix of this material while making them readily accessible to the inflow or outflow of substrates and products. In principle, direct granulation of enzymes and silicates into catalytically active materials is also possible, a process established for the production of water-soluble detergent preparations. For synthetic purposes, however, such immobilizates are only relevant for reactions in nonaqueous media as they readily dissolve in water.

11.2.1.1 Novozym® 435

The probably most prominent commercially traded preparation of an adsorptively bound enzyme for synthetic-technical application is Novozym® 435 from Novozymes S/A (Denmark). It binds the catalytically very versatile lipase B (EC 3.1.1.3) from *Candida antarctica* (reannotated as *Moesziomyces antarcticus*) to Lewatit® VP OC 1600 from Lanxess (Germany). The porous poly(methyl methacrylate) carrier has an average particle size of 0.3–0.9 mm and a large specific surface area of 80 m^2 g^{-1}. The lipase penetrates the pore system of the carrier to a depth of about 100 µm so that only the outer shell is occupied by the active enzyme. Novozym® 435 is characterized by excellent catalytic properties and a broad reaction spectrum. On a large scale, it is used, for example, for the synthesis of emollient esters.

11.2.2 Covalent Bonding to Carrier Materials

Reactive functional groups on the enzyme surface enable covalent linkage with the surface of functionalized carrier materials. In particular, the sulfhydryl group of the amino acid cysteine, the free amino functions of lysine, and the N-terminal amino acids as well as the free carboxyl functions of glutamate, aspartate, and the C-terminal amino acids exhibit sufficiently high reactivity (nucleophilicity). However, due to the favorable relationship between reactivity and average abundance on the enzyme surface, the majority of enzymes are covalently bound via the ε-amino function of lysine.

In principle, all organic and inorganic materials that are suitable for adsorptive immobilization (Sect. 11.2.1) can also be used as carriers for covalently bound enzymes, if free hydroxyl, amino, and carboxyl functions are present (Novick and Rozzell 2005; Zucca and Sanjust 2014). Usually, these must be activated again before reacting with the enzyme so that a nucleophilic attack by the amino acid side chains of the enzyme can be achieved under conditions agreeable with the enzyme (low temperatures and moderate pH values) (Fig. 11.3).

Free **hydroxyl functions** of unmodified natural polymers or inorganic silicates can, for example, be converted by cyanogen bromide (BrCN) to form reactive cyanate esters and cyclic imidocarbonates, which form iso-urea (or imidocarbonate) bonds with amino functions of the enzyme. The latter exhibits only low resistance to hydrolysis, and BrCN itself is characterized by a high hazard potential (volatility, toxicity, and explosiveness). However, alternative cyanylating agents such as 4-nitrophenyl cyanate, N-cyanotriethylammonium bromide, and 1-cyano-4-diemethylaminopyridinium bromide, which exhibit lower toxicity and lead to more hydrolysis-stable bonds, have not been established yet. The activation of **amino functions** occurs predominantly by reaction with bifunctional aldehydes. Here, a prominent position takes glutardialdehyde, which reacts rapidly and almost irreversibly at neutral to acidic pH values and produces thermally and chemically stable bonds more efficiently than other reagents. For the activation of **carboxyl functions,** carbodiimides such as 1-ethyl-3-(3-dimethylaminopropyl)carbodiimide (EDC) are used. However, all activation approaches share the disadvantage that the stability of the generated reactive groups in an aqueous solution is poor. Thus, the stockpiling of activated carriers is hardly possible, and the subsequent linkage with the enzyme requires very rapid working.

A functionalization that enables the covalent binding of enzymes without further activation and is stable over a long time is offered by materials with **oxiran- (epoxy) functions**. Organic carriers with corresponding modifications are commercially distributed by various manufacturers; inorganic carriers can be functionalized by silanization (Sect. 11.2.1), for example using glycidoxypropyltrimethoxysilane. Covalent binding of enzymes to epoxy-functionalized materials occurs spontaneously at room temperature and neutral pH. However, due to an overall low reactivity, satisfactory loading requires large enzymes in excess and long reaction times (typically ≥48 h). Higher efficiency is achieved by exposing hydrophobic domains or additional functional groups (e.g., amino functions and

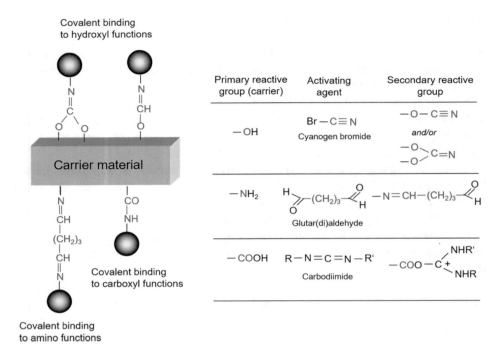

Fig. 11.3 Mechanisms of covalent enzyme binding to hydroxy-, amino- and carboxyl-functionalized carriers using free amino functions of the enzyme

iminodiacetyl functions) on the support (Mateo et al. 2007). These bring the reactive functions of carrier and enzyme into spatial proximity by rapid adsorption of the enzyme and thereby support the formation of the covalent bond.

The activity of covalently bound enzymes often benefits from the introduction of longer, non-reactive molecular chains between the enzyme and the carrier, so-called **spacers**. Predominantly, these are simple bifunctional molecules such as 1,6-diaminohexane or 1,8-diaminooctane, which form Schiff bases with the enzyme and carrier that are subsequently reduced to stable bonds. To prevent enzymes from being cross-linked to each other in the same course, the spacer is introduced in two separate steps, first at the carrier, and then at the enzyme.

11.2.2.1 Plexazyme® AC

An example of a commercially available preparation with covalently bound enzymes is Plexazyme AC from Röhm GmbH & Co. KG (Germany), which is provided for the chiral synthesis of amino acids. It consists of L-amino acylase I (EC 3.5.1.14) from *Aspergillus* sp. that is linked via oxirane functions to the carrier from the same company. The macroporous, hydrophilic carrier consists of a copolymer of methacrylamide, glycidine methacrylate, and allyl glycidine ether, which is cross-linked via *N,N'*-methylenebis (methacrylamide). It has an average particle size of 150 μm and a dense occupancy of active oxirane functions. Carriers with similar properties but different porosity and particle size are currently commercially available, for example, under the trade names Amberzyme™ (Röhm GmbH & Co. KG, Germany), Sepabeads FP-EP (Resindion, Italy), and Immobead 150P (ChiralVision, The Netherlands).

11.2.3 Entrapment in Polymer Matrices

Polymeric materials with the ability to form three-dimensional networks are, in principle, capable of accommodating enzymes in their network cavities (interstices). In contrast to the binding of enzymes to porous materials, this occurs in the course of network formation and not on the finished carrier. A direct interaction between the enzymes and carrier material is therefore not necessary but cannot be excluded generally. In principle, a distinction must be made between network formation based on monomers and network formation by the cross-linking of polymer molecules. Customary polymers of both types are shown in Fig. 11.4.

The **monomer-based entrapment of** enzymes predominantly involves the process of polymerization or polycondensation. Through polymerization, mainly poly(acrylates) and poly(acrylamides) are obtained. In this process, the reaction solution consists of a mixture of monomers (acrylates, acrylamides) and cross-linkers (e.g., N,N'-methylenebis(acrylamide)) in defined concentration ratios. The chemical properties of the monomers determine overall properties such as functionalization, hydrophilicity, and hydrophobicity of the finished matrix, while the amount of cross-linker determines the network density. Chemical decomposition or UV radiation radicalizes a starter molecule (e.g., ammonium persulfate or riboflavin), which in turn initiates the radical polymerization of the monomers. TEMED (N,N,N',N'-tetramethylenediamine) is usually added to stabilize free radicals. Good efficiency of matrix formation via radical polymerization basically requires the use of oxygen-free media. The reaction process is characterized by a high heat of reaction. The extent to which the enzyme also is covalently incorporated into the polymer network during the process has not been clarified fully to date.

Polycondensation reactions mainly occur during the inclusion of enzymes in sol-gels based on inorganic silicates (Pierre 2004). Reactive species of the general form Si-O-Si are built from monomeric alkyl silanes such as $Si(OC_2H_5)$ by acid hydrolysis and then form polymeric three-dimensional networks. The hydrophilicity/hydrophobicity of this network is predominantly determined by the alkylsilane precursor, in particular the length and degree of branching of the alkyl radical. Thus, it can be adapted to the needs of the immobilized enzyme. The addition of enzyme-stabilizing additives such as isopropyl alcohol, crown ether (18-crown-6), Tween 80®, methyl-β-cyclodextrin, or KCl can have a favorable effect on immobilization. For more sensitive enzymes, the generation of a hybrid nanocomposite of tetrakis(2-hydroxyethyl)orthosilicate (THEOS) and natural polysaccharides (xanthan gum and locust bean gum) has also been demonstrated as favorable. In the composite, both the silicate network and enzymes are embedded in a

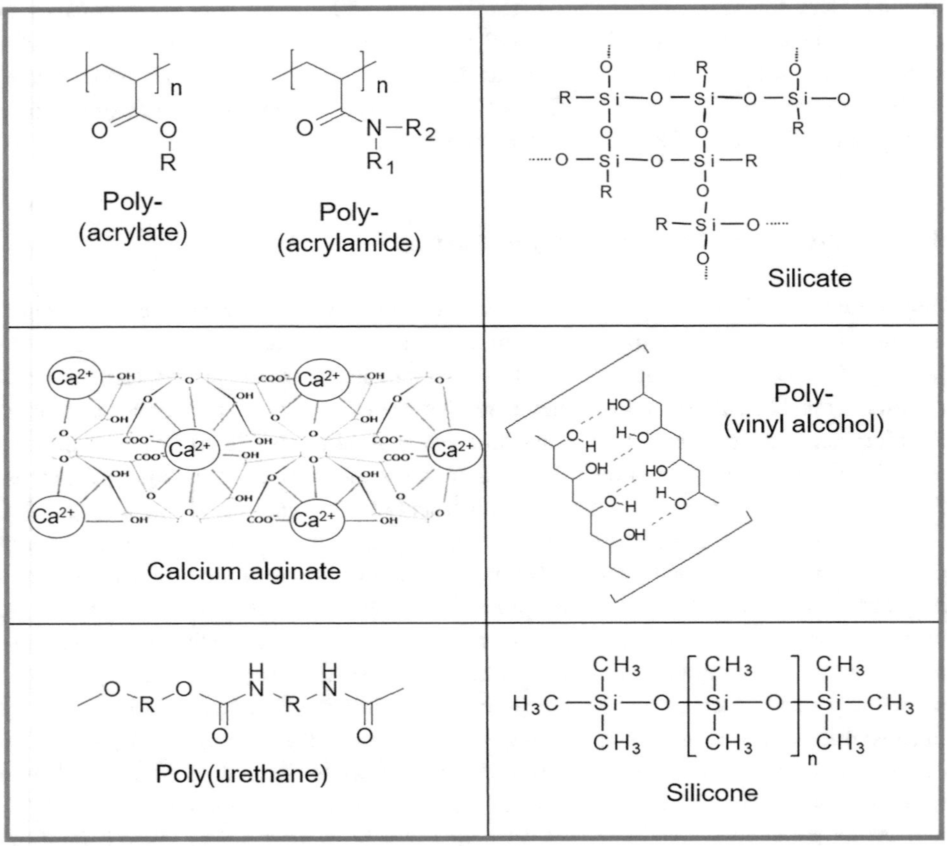

Fig. 11.4 Customary polymers for inclusion immobilization of enzymes

matrix of the polysaccharides, leading to an increase in the stability of the enzyme and a decrease in the fragility of the whole material. The presence of the polysaccharide during the polymerization process also avoids the use of organic solvents, reduces heat generation, and allows a higher pH.

Polymer-based entrapment of enzymes resorts to the three-dimensional cross-linking of natural or synthetic, usually linear polymers by inducing physical interactions or by forming covalent bonds. Entrapment in networks of natural polymers is usually achieved by thermally inducing the rearrangement of molecular interactions between polymer chains (agar, gellan, κ-carrageenan, and gelatin) or by their ionotropic cross-linking via multivalent ions (alginates, pectate). The result is a gel-like matrix with a low polymer ($\leq 4\%$) and a high water content (so-called hydrogels).

While thermogelling requires, at least initially, the application of comparatively high temperatures (80–100 °C) and long times for gel formation depending on the content of foreign ions, ionotropic gel formation proceeds very quickly and efficiently at room temperature. With this in mind, alginate is popularly used as an immobilization matrix for biological components. Alginate is a natural polymer of block-arranged units of β-*D*-mannuronic acid (M) and α-*L*-guluronic acid (G), which forms gels via interactions with divalent cations (preferably Ca^{2+}). The physical properties of the resulting gels vary with the (natural) G and M content of the alginates employed, the sequence and extent of the GM blocks in the polymer, the polymer chain length, and the final concentration of the polymer in the gel. In addition, the type and concentration of the cross-linking ion play a decisive role (Smidsrod and Skjak-Braek 1990).

Enzyme entrapment into three-dimensional networks of synthetic polymers occurs predominantly through the formation of covalent bonds between polymer chains. An exception is the use of poly(vinyl alcohol) (PVA). PVA is capable of forming partially crystalline structures via hydrogen bonds, especially in the fully saponified state (i.e., in the presence of free hydroxyl groups to almost 100%). Based on this, the repeated freezing and thawing (up to five cycles) of a concentrated aqueous solution of the polymer (80–100 g L^{-1}) leads to the formation of so-called cryogels (Lozinsky 1998). In the equimolar mixture with poly(ethylene glycol) (molecular weight up to 1000 g mol^{-1}), even a single cycle of freezing and slow thawing leads to the formation of a stable macroporous matrix.

Polyurethane matrices for enzyme entrapment are formed from short-chain polyurethane polymers with free isocyanate functions (molecular weight 500–5000 g mol^{-1}). The addition of water to these polymers leads to the formation of primary amines at isocyanate end groups, which generate covalent bonds with further isocyanate functions. Depending on the nature of the polyurethanes chosen, gel-like matrices or foams are generated (Romaskevic et al. 2006). The high reactivity of the isocyanate functions makes that enzymes are covalently bound into the polyurethane network via accessible free amino functions.

Formation of silicone matrices by Pt^0-catalyzed hydrosilylation of vinylsiloxanes and linear poly(hydroxymethylsiloxane-ω-dimethyl-siloxane) proceeds without direct interaction with the enzyme. Here, the number and distribution of functional groups (vinyl and SiH functional ions) involved in the cross-linking determine the density of the resulting network. Silicones allow the entrapment of free enzymes, the stabilization of enzyme-containing emulsions, and the formation of stable composites with porous enzyme carriers, so-called *silCoat* immobilizates. The hydrophilicity of the matrix can be modified by incorporating poly(ethylene glycol)400 instead of a siloxane. A direct interaction of the silicone network with the incorporated enzymes does not occur.

Generally, when immobilizing enzymes in polymer networks, the **shape and size of** the immobilized particles must be defined in the course of formation. For technical use, distinct particles with small diameters (micrometer range) have the greatest importance since they can be used efficiently both as a bulk in fixed-bed and dispersed in stirred reactors.

Irregular particles are easily obtained by first creating larger blocks of the polymeric matrix and then grinding them to the desired size. Soft polymers such as poly(acrylamide) gels can also be cut with knives or pressed through metal screens. Generally, the resulting particles have a wide size distribution. The formation of very small particles can result in a considerable loss of material (rejects).

More clearly defined, usually approximately **spherical particles** are obtained by portioning the entrapment material before network (matrix) formation using suspension or drop-in processes (Fig. 11.5). In the **suspension process**, the entrapment material is emulsified as a liquid phase in an immiscible second liquid phase. With agitation, distinct droplets are formed, which harden into stable spheres during matrix formation. The size of the droplets is significantly influenced by the viscosity of the continuous phase, the interfacial tension, the phase ratio, and the stirring intensity (energy input). The latter, in turn, is a function of the shape of the stirrer and reactor as well as the stirring speed. Under constant conditions, spherical matrices with a broadly scattered size distribution are formed in a dynamic equilibrium between coalescence and redispersion. In principle, the process can be applied to any entrapment material. In the **drop-in process**, the entrapment material is introduced in defined portions into a liquid container by pressing it in a pulse-like manner through a nozzle or by splitting a thin material jet with fine wires or by vibration. In the air, droplets are formed and fixed in shape by rapid formation of a surface network on the way into the liquid container or directly on contact with the container liquid. Dispersed in the liquid, complete curing of the matrix occurs. The size of the droplets is mainly determined from the diameter of the droplet nozzle and optionally by the pulse or vibration frequency or by the speed of the material jet and cutting wires. Thus, the droplet size can be well defined, and a comparatively narrow size distribution is obtained. Limitations lie in the fact that the method can only be applied to matrices with very fast network formation. Thus, it is used particularly frequently in connection with ionotropic network formation.

11.2.3.1 Lentikats®: β-Galactosidase

One of the few examples of commercially available enzyme preparations obtained from a direct entrapment in polymer matrices is the lentikats®-β-galactosidase from LentiKat's Biotechnologies (Czech Republic). Lenticular preparations with an average diameter of 3–4 mm and a maximum thickness of 200–400 µm are generated at room temperature using a special solution of poly(vinyl alcohol) (LentiKat® Liquid) developed by geniaLab (Germany). They are formed by placing drops of a solution mixed with the enzyme on a solid surface, partially drying and finally reswelling. Lentikats®-β-galactosidase can be used, for example, for the hydrolysis of lactose and the synthesis of galactooligosaccharides. The immobilization process can be transferred to different enzymes.

11.2.4 Inclusion in Membranes

Membranes, i.e., semi-permeable thin material layers, can be used for enzyme immobilization as small capsules (diameter in the millimeter to micrometer range) or as sheets incorporated into prefabricated technical modules. The prerequisite is that the molecular weight cut-off (MWCO) of the membrane structure prevents the passage of enzymes, while low-molecular compounds (substrates and products) diffuse as freely as possible.

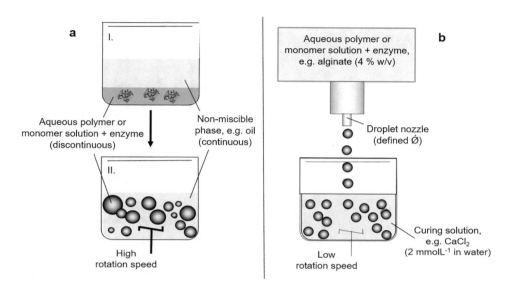

Fig. 11.5 Entrapment of enzymes in polymer matrices by suspension (**a**) and drop-in (**b**)

Due to the fully closed structure of **membrane capsules,** enzymes must be included directly during capsule formation. Most commonly applied are coacervation (phase separation), interfacial polymerization, or liquid drying (Park and Chang 2000). In coacervation, an aqueous solution of the enzyme is dispersed in a water-immiscible solvent, in which the membrane-forming polymer is dissolved. By adding a second solvent that is immiscible with water, the polymer is separated, deposits on the surface of the emulsified water droplets, and then fuses to form a continuous membrane. It is mandatory that the polymer forms a concentrated solution without precipitating. Polymers that can be used for membrane formation in this way include modified celluloses (cellulose nitrate, ethyl cellulose, nitrocellulose), poly(styrene), poly(ethylene), poly(vinyl acetate), poly(methyl methacrylate), and poly(isobutylene). Membrane formation by interfacial polymerization starts from an aqueous mixture of the enzyme with hydrophilic monomers that is dispersed in a water-immiscible solvent. A hydrophobic monomer dissolved in the same water-immiscible solvent is added to the emulsion. Upon contact at the interface between the aqueous and nonaqueous phases, the hydrophilic and hydrophobic monomers polymerize. This method is typically used for the production of nylon membranes but can be applied to the production of other polymer membranes as well, e.g., polyurethanes and polyesters. For membrane formation by liquid drying, a polymer, e.g., poly(styrene) or ethyl cellulose, is dissolved in a water-immiscible solvent that has a boiling point below that of water (e.g., benzene or chloroform). An aqueous solution of the enzyme is emulsified in this solvent, and the resulting emulsion is in turn dispersed in an aqueous phase containing stabilizers (e.g., gelatine). By evaporation of the nonaqueous solvent, membranes are formed on the surface of the enzyme-containing aqueous phase.

Fig. 11.6 Basic design of a
filtration module for enzyme
retention

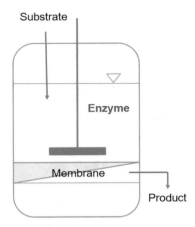

The retention of enzymes in prefabricated **technical modules** is usually accomplished with commercially available membrane sheets made of poly(sulfone)cellulose acetate or acrylic copolymers, which have a MWCO between 200 Da and 100,000 Da. The technical modules follow the principles of common filtration or aeration techniques; hollow fibers or ultrafiltration units are predominantly used (Fig. 11.6). Enzymes are inserted on one side of the membrane, usually the lumen or main chamber, prior to operation. The retention of low molecular weight cofactors can be improved by their covalent modification with higher molecular weight compounds such as poly(ethylene glycol).

11.2.4.1 Amicon®: Stirring Cell

Technical modules for the immobilization of enzymes by membranes are commercially available, especially on a small scale (working volume <1 L). A much-used example is the Amicon®—Stirring cell from Merck Millipore (Germany), which operates according to the dead-end filtration principle. The cell consists of a cylindrical vessel with a working volume between three and 400 mL, in the lower part of which the selected membrane is located. The effective membrane area ranges from 0.9 to 41.8 cm². Mixing is accomplished by using a magnetic stir bar on top of the membrane, and a positive pressure of up to 75 psi (5.3 kg cm^{-2}) can be applied to increase efficiency.

11.2.5 Carrier-Free Cross-Linking

Carrier-free insoluble agglomerates (immobilizates) of enzymes are obtained by direct or spacer-mediated covalent cross-linking of the individual protein molecules (Rössl et al. 2010). This involves the use of bifunctional reagents addressing the reactive functions of the amino acid side chains on the enzyme surface. As in the case of covalent carrier binding (Sect. 11.2.2), the ε-amino functions of lysine residues primarily serve as reaction partners. Suitable   cross-linking   reagents   are   dialdehydes   (e.g.,   glutardialdehyde   and dextrandialdehyde), diisocyanates (e.g., hexamethyldiisocyanate), and diisothiocyanates (e.g., p-phenylenediisothiocyanate). Cross-linking is possible both on the basis of dissolved enzymes or starting from freeze-dried or spray-dried preparations, enzyme crystals, and pre-aggregated enzymes. Especially cross-linked enzyme crystals (CLECs) and cross-linked enzyme agglomerates (CLEAs) have gained some importance.

CLECs, originally developed to facilitate X-ray structural analysis, have a solid micro-porous structure with uniform channels running through the crystal. The particle diameter can be adjusted flexibly between 1 and 100 μm via the ratio of the enzyme and cross-linking reagent or via the incubation time. The physicochemical properties of the CLEC are mainly determined by the crystallization conditions; additionally, the choice of cross-linking reagent has a minor influence. Suitable crystallization conditions must be determined individually for each enzyme. A general prerequisite for crystal formation is a high purity of the enzymes.

The generation of **CLEA** takes advantage of the spontaneous agglomeration and precipitation of proteins in the presence of various organic solvents, electrolytes, non-ionic polymers, and acids, which occur while preserving the three-dimensional structure, i.e., without denaturation. The resulting aggregates are cross-linked to form particles with a diameter of 5–50 μm and gel-like consistency. The particle diameter can be influenced by the ratio of enzyme and cross-linking reagent or by the incubation time. Similar to the generation of CLEC, the properties of the resulting CLEA depend on the cross-linking reagent and the precipitation conditions. Predominantly, ammonium sulfate and poly(ethylene glycol), which also give good results in protein purification, are used for precipitation.

11.2.5.1 CLEA® 301: Penicillin Acylase

An example of a commercially available, carrier-free cross-linked enzyme is the preparation CLEA® 301 from the company CLEA Technologies (Netherlands). It consists of aggregated and glutardialdehyde cross-linked molecules of penicillin acylase (EC 3.5.1.11). The preparation catalyzes the synthesis or hydrolysis of penicillin G in the same ratio as other commercially available immobilizates of penicillin acylase (e.g., PGA-50, polyacrylamide support) and exhibits high stability in organic solvents.

Questions
  2. *Why is it possible to noncovalently bind enzymes to the surface of a broad variety of materials?*
  3. *Why is it not advisable to use metal chelation for the immobilization of enzymes employed in the production of pharmaceutics?*
  4. *What makes silicates a good choice of carrier for noncovalent enzyme immobilization?*
  5. *Why is lysine a frequent target for covalent enzyme immobilization or for carrier-free cross-linking?*
  6. *Enzyme immobilization with poly(urethane) combines covalent binding to a carrier and entrapment. Why is this?*

11.3 Molecular and Physicochemical Effects

The catalytic performance of enzymes is largely determined by individual molecular properties such as the three-dimensional structure, the position and nature of the active site, and the molecular dynamics (Chap. 2). The achievement of the maximum specific activity (U mg_{prot}^{-1}) basically requires the constant occupation of the active site with substrates (Chap. 4). Consequently, both the alteration of molecular properties and the restriction of mass transfer (substrate supply) inevitably affect the catalytic performance of enzymes. Both are common effects of an immobilization-induced concentration of enzyme molecules, their contacting with catalytically nonactive materials, and transformation from a solute (homogeneous) to a solid or pseudo-solid (heterogenized) catalyst. During the immobilization process, loss or complete deactivation of a portion of the enzyme population may also occur. The extent and direction of the aforementioned immobilization effects are described in their entirety by the activity yield or residual activity of the immobilizates (Sect. 11.3.1). However, due to the manifold interactions and overlaps, the determination of the actual contribution of individual effects to the overall effect is at best approximate.

11.3.1 Molecular Modification

The interaction of components participating in the immobilization process with the target enzymes can have many effects on their structural and thus catalytic properties. Besides complete deactivation by loss of the catalytically active conformation or reactions at catalytically active amino acids, permanent "distortions" of the active conformation, changes in the charge and charge distribution, and changes in molecular dynamics occur.

The complete **deactivation** of enzymes is usually related to the reaction conditions of the immobilization process, such as temperature, pH, and the presence of organic solvents, or to reaction components involved, such as toxic reagents or by-products. Comparatively high incubation temperatures are required, for example, for inclusion in thermo-induced gels such as agar, gellan, κ-carrageenan, or gelatine. In the course of the formation of inclusion matrices by radical polymerization (e.g., poly(acrylamide) gel) or polycondensation (e.g., inorganic sol-gels), considerable heat is often generated (Sect. 11.2.3). The formation of simple sol-gels is additionally characterized by low pH values and the requirement of organic solvents, whereby enzymes of low stability only show sufficient residual activity upon the addition of stabilizing additives or in inorganic–organic composites of special composition. The formation of membrane capsules generally occurs at phase boundaries (predominantly organic-aqueous; Sect. 11.2.4), which can cause rapid denaturation of enzymes. During the formation of poly(acrylamides), the necessary reaction components, in particular the employed monomers, the starter, and the stabilizer, develop a highly toxic effect, especially for enzymes sensitive to oxidation.

Permanent **changes in catalytic properties** are predominantly due to the direct interaction of enzymes with a catalytically nonactive material. Thus, hydrophobic/hydrophilic interactions, ionic interactions, the formation of hydrogen bonds between the enzyme and support surface, and the formation of inter- and intramolecular bonds result in changes in the charge and charge distribution in the molecule, limitations of the molecular dynamics, and deviations of the molecular structure from the solute active conformation ("distortions"). Covalent bonding can also result in direct changes at the active site. The result of the carrier bond is often a low residual activity of the enzymes with a simultaneous significant increase in enzyme stability. Particularly, in connection with the immobilization of lipases, the immobilization process is also reported to increase the catalytic activity. In sol-gels, for example, lipases can show up to 100-fold higher activity than in the free state. Similar results have been reported for immobilization in static emulsions consisting of water in silicone. The reason for this behavior is probably the fixation of the enzyme molecules in the open conformation typical for active lipases. The effects of physicochemical interactions are significantly reduced by the introduction of low-functionalized spacer molecules between the carrier and enzyme.

11.3.2 Limitations of Mass Transfer

A direct consequence of the immobilization of enzymes is almost always the reduction of the rate of the catalyzed reaction due to the limitation of substrate supply to the catalytically active sites. This is due, on the one hand, to the distribution of substrates between the liquid phase and the support, which is influenced by the functionality and hydrophilic/hydrophobic properties of the material. On the other hand, depending on the particle size and pore structure of the immobilizates, the diffusive transport of the substrates to the enzyme is delayed. In addition, necessary substrate rotations and substrate access to the active site can be considerably restricted by an unfavorable orientation of enzymes in the immobilizate.

Mass transport limitations usually have a greater effect when the substrate molecules involved in the reaction are larger. Particularly, clear effects also arise if the enzyme kinetics are subject to product inhibition or if pronounced pH gradients are formed. The contribution of limited diffusion to activity reduction is determined by quantifying the influence of the flow rate on the conversion in a flow tube. If, for a defined residence time in the reactor, the conversion increases with increasing flow rate, the reaction proceeds in a diffusion-limited manner. If no influence of the flow rate can be determined, the reaction rate is predominantly subject to the kinetics of the catalyst.

Questions
   7. *Why is it important to distinguish between molecular and mass transfer effects of an immobilization approach?*

Table 11.1 Technically relevant parameters of the enzyme immobilization process and the resulting immobilizates

Characteristics of the immobilization process	
Immobilization yield [%]	$Y_{\text{Immo}} = \frac{EA_{\text{immo}}}{EA_{\text{total}}} \times 100 = \frac{(EA_{\text{total}} - EA_{\text{free}})}{EA_{\text{total}}} \times 100$
Immobilization efficiency [%]	$Eff_{\text{immo}} = \frac{EA_{\text{app}}}{EA_{\text{immo}}} \times 100 = \frac{EA_{\text{app}}}{EA_{\text{total}} - EA_{\text{free}}} \times 100$
Activity yield (residual activity) [%]	$Y_{EA} = Y_{\text{immo}} \times Eff_{\text{immo}} = \frac{EA_{\text{app}}}{EA_{\text{total}}} \times 100$
Characteristics of immobilizates	
Carrier loading [$g_{\text{enzyme}}/g_{\text{carrier}}$]	$m_{\text{sup}} = \frac{m_{\text{enzyme}}}{m_{\text{carrier}}}$
Carrier-specific activity [U/g_{carrier}]	$EA_{\text{sup}} = \frac{EA_{\text{app}}}{m_{\text{carrier}}}$
Activity per volume element [U/L]	$EA_V = \frac{EA_{\text{app}}}{V_{\text{carrier}}} = EA_{\text{sup}} \times \delta_{\text{carrier}}$
Specific catalyst productivity (turnover number) [$g_{\text{product}}/g_{\text{immobilizate}}$ h^{-1}]	$TN = \frac{m_{\text{product}}}{m_{\text{carrier}} \times t}$
Total turnover number [$g_{\text{product}}/g_{\text{immobilizate}}$]	$TTN = \frac{m_{\text{product}}}{m_{\text{carrier}}}$
Deactivation constant [h^{-1}]	$k_d = \frac{\ln(EA_0 - EA_t)}{t}$
Half-life [h]	$t_{0,5} = \frac{\ln 2}{k_d}$
Leaching (enzyme loss) [$g_{\text{enzyme}}/g_{\text{carrier}}$]	$m_{\text{desorb}} = m_{\text{sup, 0}} - m_{\text{sup, }t}$

δ_{carrier}: mass density of the immobilizate; EA: enzymatic activity related to a defined substrate and assay; EA_0: enzymatic activity at the start of the reaction; EA_{app}: actual (apparent) activity of the immobilizate; EA_{free}: residual activity in the immobilization approach after the immobilization process; EA_{total}: total activity in the immobilization approach before the immobilization process; EA_{immo}: theoretical immobilized activity; EA_t: enzymatic activity after defined time; m_{enzyme}: mass of the enzyme; m_{product}: mass of the product of the enzymatic reaction; $m_{\text{sup,0}}$: loading at reaction start; $m_{\text{sup,}t}$: carrier loading after defined time; m_{carrier}: mass of the carrier/immobilizate; t: time; V_{carrier}: volume of the carrier material; Y: yield

11.4   Performance Characteristics

The quality of an enzyme immobilization is characterized by a series of parameters that describe both the immobilization process itself and the performance of the resulting preparations. They are explained in detail below; an overview is given in Table 11.1.

11.4.1 Characteristics of the Immobilization Process

For characterization of the immobilization process of catalytically active enzymes, the immobilization yield, the immobilization efficiency, and the activity yield are frequently (but not exclusively) used as parameters.

The **immobilization yield** (in %) describes the percentage of the total enzymatic activity originally present in the immobilization mixture that has theoretically been incorporated into the immobilizates. The determination of the theoretically immobilized activity (EA_{immo}) is carried out by subtracting the activity remaining in the preparation after the end of the immobilization process (EA_{free}) from the total activity originally present in the preparation (EA_{total}). If necessary, the value obtained must be corrected for the enzymatic activity lost during the immobilization process due to deactivation. This is determined by a blank approach in which the free enzyme is exposed, as far as possible, to the reaction conditions of the immobilization process and the remaining activity is determined. For simplicity, the immobilization yield is often referred to as the amount of protein present in the preparation rather than to the enzymatic activity. However, especially when using enzyme solutions containing various proteins, this can lead to a distortion of results because different proteins can exhibit very different immobilization behavior. Indications for such a problem provide a difference between the protein-specific activity (enzyme activity per protein mass) of the immobilized and the free enzyme.

The **immobilization efficiency** (in %) refers to the percentage of the apparent, i.e., actually measurable, enzymatic activity of the immobilizate (EA_{app}) in relation to the theoretically immobilized activity (EA_{immo}, see above). Ideally, it is $\geq 100\%$.

The **activity yield** (in %), which often is also referred to as **residual activity**, indicates the percentage of the apparent enzymatic activity of the immobilizate (EA_{app}, see above) in relation to the total activity originally present in the immobilization mixture (EA_{total}, see above) and is the product of immobilization yield and immobilization efficiency.

All parameters result from the determination of absolute activities (units in $\mu mol\ min^{-1}$), not from volume- or protein-related values. Comparability is only given if the activity determination is carried out under identical conditions (substrate, concentration, temperature, pH, etc.) since the level of activity varies with the employed assay.

11.4.2 Characteristics of Immobilizates

Characteristics of enzyme immobilizates predominantly describe their catalytic performance and stability, which together determine the input quantities required for a process and thus the enzyme costs. Their actual values are very strongly influenced by the test conditions (temperature, pH, substrate and product concentrations, co-solvents, etc.), which is why these should be as identical as possible to the conditions of the envisaged application.

Key parameters of **catalytic performance** are the carrier-specific activity and the specific catalyst productivity or turnover number (TN). The carrier-specific activity describes the directly measurable catalytic activity of the immobilizate in relation to the immobilizate (or carrier) quantity. Taking into account the mass density ($g\ mL^{-1}$) of the immobilizate, it can be used to determine the maximum achievable catalytic activity per volume element ($U\ mL^{-1}$) of a reactor. The TN indicates the mass of product obtained per mass of immobilized material (or catalyst) in a defined period of time. It is determined from the direct quantification of the amount of product, i.e., independent of substrate consumption, taking into account the stoichiometric factor of the product in the catalyzed reaction. The total turnover number (TTN) describes the absolute amount of product that can be produced with a defined amount of the immobilizate and thus intrinsically takes into account the catalytic stability of the immobilizate during application.

The **catalytic stability of** enzyme immobilizates is described by the deactivation constant or the half-life derived therefrom. The deactivation constant is calculated as the slope of the measurable catalytic activity of the immobilizate, which decreases exponentially with time. The half-life is the period of time after which the immobilizate has only half of its original catalytic activity. Assuming a decrease in activity following a first order reaction, it is calculated from the quotient of the natural logarithm of two (ln2) and the deactivation constant. In the case of repeated use of an enzyme immobilizate in several reaction cycles (recycling), the half-life of the immobilizate is obtained by plotting the measurable activity at the beginning of each reaction cycle (if necessary as a percentage of the initial activity) against the accumulated reaction time. The observed loss of activity does not necessarily describe a decrease in enzyme activity, as is the case with free enzymes, but can also be caused completely or proportionally by leaching. Leaching refers to the loss of activity due to a physical loss of the enzyme from the immobilizate and is equivalent to a reduction of enzyme loading on the carrier. Leaching can be quantified directly by determining the amount of enzyme in the supernatant or in the run of a reaction mixture or indirectly by the loss of activity compared to the free enzyme under identical conditions. Both approaches are not trivial since, on the one hand, the quantification of enzyme amounts (especially when using protein mixtures for immobilization) causes problems and, on the other hand, the manifold possible effects of immobilization on the measurable catalytic activity of enzymes (Sect. 11.3) have to be considered.

As is generally the case in enzyme catalysis, a fundamental distinction must be made between the storage stability and the process stability of enzyme immobilizates. **Storage stability** describes changes in the catalytic or mechanical properties of the immobilizates before or between catalytic applications, whereas **process stability** refers to changes during catalytic use. Due to the absence of reactants and molecular dynamics during storage, storage and process stability usually differ significantly. In addition, reactor configuration, process control, and deviating ambient conditions have an effect on the respective parameters. Therefore, it is usually not possible to conclude from the storage stability of an enzyme immobilizate to its process stability.

Questions

  8. *Why is it important to consider both characteristics of the immobilization process and of the obtained immobilizates?*

11.5   Choice of Methods

Against the background of the wide range of available methods, which also includes the combination of different basic principles, the identification of the best method to immobilize an enzyme is not a trivial concern. The focus of considerations is usually on achieving a high activity yield (residual activity) during the immobilization process and the optimization of the catalytic performance of the resulting immobilizates (catalyst productivity and TTN; Sect. 11.4). However, these are not constant quantities, but vary with the requirements imposed by the specific application. The entire selection or development of the immobilization process must therefore be based on the envisaged application, including the reaction conditions and the intended process control. In addition, economic considerations regarding material availability and costs, as well as energy requirements and waste quantities, should be included.

As shown in Sect. 11.3, the outcome of an immobilization process is determined by the individual properties of the enzymes and materials involved and the resulting interactions. Completely different immobilization approaches can lead to the same activity yield due to very different effects. Based on the improved understanding of the structural properties of enzymes and the relation between the structure and catalytic activity, as well as on modern methods of molecular biology and polymer sciences, a rationally oriented immobilization that anticipates immobilization effects is increasingly pursued. The promising prospects of such an approach are illustrated by the progress made in the immobilization of lipases, where the characteristic surface properties of these enzymes are exploited for stable adsorption onto carriers and even for activation in sol-gels and silicones. For most other classes of enzymes, however, the relationship between the nature of the carrier and the performance of the resulting immobilizate is as yet poorly understood, so immobilization strategies are still predominantly developed by empirical screening.

For catalyst productivity (Sect. 11.4.2), apart from catalytic activity and stability of the enzymes, material-related properties such as stability against chemical and mechanical stress and catalyst retention play an important role. How important depends on the reaction conditions and the process control of the intended technical application. Relevant to the choice of the immobilization method are, above all, the type of medium (aqueous or nonaqueous), the type of reactor (stirred or fixed-bed), and the reaction control in (repeated) batch or continuous mode.

In aqueous media or in the presence of surface-active substrates or products, **enzyme leaching** (Sect. 11.4.2) is of considerable importance. It occurs particularly in connection with immobilization processes that do not involve covalent binding of the enzymes. When enclosed in polymeric matrices or membranes, the mesh size of the matrices or the pore diameters of the membranes determine the retention of the enzyme during the reaction. In principle, complete retention is possible but is usually associated with a considerable restriction of mass transfer between the reaction medium and immobilizate. Binding of enzymes to material surfaces via purely physical interactions is usually in thermodynamic equilibrium with an unbound enzyme. The position of this equilibrium depends on the type and number of interactions between the enzyme and carrier, which in turn are influenced by environmental conditions such as temperature, pH, and ionic strength, as well as by the presence of solubilizing compounds such as surfactants. Therefore, both transfer of the immobilizate into new reaction mixtures (recycling) or continuous operation and changes under ambient conditions during the reaction lead to gradual, more or less severe losses of the enzyme from the carrier material. In nonaqueous reaction media, the problem of leaching usually does not arise as the low solubility of enzymes in these media effectively prevents their diffusion. Under these conditions, advantageous properties of noncovalent immobilization, such as the recyclability of the carrier material or the shielding of the enzymes from damaging influences of the solvent, can have a favorable effect on the economy of the application.

The specific composition of the reaction medium and the reaction temperature cause a **chemical or physicochemical load** on the immobilizates, which they must withstand over a longer period of time. This is especially critical when enzymes are included in matrices of natural polymers. Thus, ionotropically produced gels like Ca-alginate are particularly sensitive to chelating components such as phosphates, citrates, EDTA, and lactates or cations that inhibit gel formation (e.g., Na^+ or Mg^{2+} ions), whereas gels produced by thermogelation remain stable only at comparatively low temperatures. In contrast, materials produced on the basis of synthetic building blocks or by covalent cross-linking of natural materials usually behave inertly under typical reaction conditions of biocatalysis.

The optimum physical properties of immobilizates are significantly determined by the reactor configuration. Stirred tank reactors and fixed-bed reactors have to be distinguished in particular. **Stirred tank reactors** are characterized by predominantly turbulent mixing of the reaction media. The immobilizates suspended therein are subject to mechanical stress due to the shear forces generated by the agitator and due to collisions with internals. As a result, abrasion and fractures in the carrier material occur. This can be visualized directly by (electron) microscopic images or indirectly by the change in the particle size distribution of immobilizates during use (increase in the proportion of smaller particles). It can be assumed that immobilizates in stirred tank reactors benefit from a particle size that is as small as possible, as this favors mass transport and thus requires less mixing and reduces the effects of shear forces. A high porosity can, on the one hand, have a negative effect on the resistance to abrasion and material fracture but, on the other hand, can lead to an improvement in mass transfer. In **fixed-bed reactors**, immobilizates are stacked on top of each other in a loose bed, and the reaction medium flows through the bed in a directed manner from one end to the other. There are hardly any shear forces worth mentioning in such a setup. However, the immobilizates are subjected to mechanical stress by the weight of the overlying bed. The weight load can cause material deformation up to fracture. This can be demonstrated by a simple experimental setup that records the resistance of the material as the compressive load increases. The fracture point is characterized by a sudden, momentary drop in material resistance. It is described by the compression force F_{krit} (in N cm^{-2}) required to reach the breaking point. With increasing compression of the particles in the reactor and decreasing particle size, the pressure drop in the fixed bed increases, resulting in

Fig. 11.7 Experimental setup for determining the breaking strength of immobilization materials. Care should be taken to ensure that the compression ram has a larger diameter than the particle to be measured. It should also be lowered at a slow, constant speed

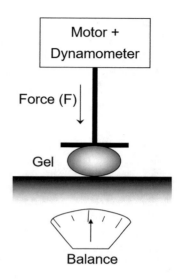

a less effective flow. A high porosity of the immobilization material can, on the one hand, increase the compressibility and, on the other hand, simplify the flow through the fixed bed (Fig. 11.7).

The possibility to rationally influence the physical properties of immobilizates by the choice of the carrier material leads to a superiority of carrier-bound over carrier-free preparations in both reactor configurations. In particular, CLEAs exhibit both low mechanical stability in the stirred tank reactor and high compressibility in the fixed-bed reactor due to their gelatinous consistency. CLECs have a somewhat higher stability against mechanical stress. Considering the fracture strength, a very low mechanical stability is also observed for most entrapment matrices made from natural polymers, whereby ionotropic gels predominantly perform better than gels generated by thermogelation. Synthetic materials such as poly(vinyl alcohol), on the other hand, have a very high fracture strength but are also highly compressible, so a high pressure drop can occur in fixed-bed reactors even at low pouring heights. Inorganic carrier materials such as sol-gels and other silicates are particularly unstable in stirred tank reactors. Even at low stirring power, they disintegrate into smallest particles. However, recent investigations on commercially available carriers have shown that the mechanical stability of various carrier materials can be improved by the formation of composites with silicones (*silCoat* technology). At the same time, the frequently observed leaching of porous enzyme carriers is reduced. The physical properties of carrier materials can thus be influenced quite well by rational material design. This should further promote the rational selection and development of suitable immobilization processes.

Questions

9. *Which considerations should lead the choice of immobilization method for implementation in a production process?*
10. *In nonaqueous media, enzyme leaching plays a minor role for the activity loss of an immobilizate over time. Why is this? Which parameter will probably play a major role?*

Take Home Message
- Immobilization often is a requirement for the synthetic use of enzymes on a production scale.
- Enzyme immobilization can be done in multiple ways. Theoretically, every enzyme can be immobilized. Immobilization affects both molecular features of enzymes and accessibility.
- The best way to immobilize an enzyme depends not only on the enzyme but also on the envisaged application.

Answers to Questions

1. *Strictly speaking: No, because cells are natural containers of enzymes.*
2. *As proteins, enzymes offer various functional groups on their surfaces, enabling interaction with differently functionalized materials.*
3. *Metals leach from the carrier (in small amounts) and thus contaminate products during catalysis. This is particularly detrimental if products are used as pharmaceutics.*
4. *Silicates are easily functionalized with different groups through silanization and thus can be adapted to the requirements of the adsorption of different enzymes. Additionally, they are available in different preparations.*
5. *The ε-amino function of lysine is highly reactive and abundant in many enzymes.*
6. *During the formation of poly(urethane) gels for enzyme entrapment, isocyanate function reacts not only with amino functions at poly(urethane) prepolymers but also with amino functions on proteins. Thus, enzymes are at the same time entrapped and covalently linked within the matrix.*
7. *Mass transfer effects might be counteracted by changes in the operation mode of a process, while molecular effects can only be eliminated by changes in the immobilization method. Thus, distinguishing between the two effects is required for rational optimization.*
8. *Process and preparation might be optimized separately. To identify potential for optimization, both must be investigated thoroughly.*
9. *The envisaged application/process, economics, energy requirements, and waste quantities should, if possible, be considered.*

10. *Enzymes are hardly soluble in nonaqueous media and therefore do not tend to leach into them. Such media, however, often have detrimental (molecular) effects and therefore deactivate the enzyme upon direct contact.*

References

Grubhofer N, Schleith K. Nat Sci. 1953;40:508.

Hartmann M, Kostrov X. Chem Soc Rev. 2013;42:6205.

Jesionowski T, Zdarta J, Krajewska B. Adsorpt. 2014;20:801.

Katchalski-Katzir E. Trends Biotechnol. 1993;11:471.

Lozinsky VI. Russ Chem Rev. 1998;67:573.

Mateo C, Grazu V, Pessela BCC, Montes T, Palomo JM, Torres R, Lopez-Gallego F, Fernandez-Lafuente R, Guisan JM. Biochem Soc Trans. 2007;35(6):1593.

Nelson JM, Griffin EG. J Am Chem Soc. 1916;38:1109.

Novick, Rozzell. In: Barredo JL, editor. Microbial enzymes and biotransformations, vol. 247. Totowa: Springer; 2005.

Park JK, Chang HN. Biotechnol Adv. 2000;18:303.

Pierre AC. Biocatal Biotransformation. 2004;22:145.

Romaskevic T, Budriene S, Pielichowski K, Pielichowski J. Chemija. 2006;17:74.

Rössl U, Nahalka J, Nidetzky B. Biotechnol Lett. 2010;32:341.

Smidsrod O, Skjaj-Braek G. Trends Biotechnol. 1990;8:71.

Zucca P, Sanjust E. Molecules. 2014;19:14139.

Enzymatic Reactions in Unusual Reaction Media

12

Christoph Syldatk

What You Will Learn in This Chapter

The performance of enzymatic reactions in unusual reaction media offers a wide range of possibilities. Polar and nonpolar organic solvents, supercritical fluids, ionic liquids, and deep eutectic solutions can be used as media for enzymatic reactions. The main motivations for this are to significantly increase the solubility of non- or low-water-miscible substrates and products compared to aqueous reaction systems, to simplify subsequent work-up, and to avoid microbial contamination as well as side and degradation reactions. Also, many enzymes exhibit different properties in unusual reaction media and catalyze different reactions than in aqueous systems. Hydrolytic enzymes can be successfully used for synthesis reactions, making specific use of their properties of regioselectivity, stereoselectivity, and substrate specificity. Therefore, the use of enzymes in unusual reaction media has gained great importance in recent years, e.g., for organic synthesis and in the chemical industry. This chapter describes the possibilities of carrying out enzymatic reactions in unusual reaction media. The different reaction media are presented in this chapter, and their influence on enzyme activity and stability is discussed. In all reactions, the consideration of water activity a_W is an essential prerequisite for successful performance. Due to the reaction product water formed in the

(continued)

C. Syldatk (✉)

Institute of Process Engineering in Life Sciences II – Electro Biotechnology, KIT - Karlsruhe Institute of Technology, Karlsruhe, Germany

e-mail: christoph.syldatk@kit.edu

© The Author(s), under exclusive license to Springer Nature Switzerland AG 2024
K.-E. Jaeger et al. (eds.), *Introduction to Enzyme Technology*, Learning Materials in Biosciences,
https://doi.org/10.1007/978-3-031-42999-6_12

respective synthesis reactions, an equilibrium is reached above a certain concentration and should either be removed from the reaction preparations or its formation should be avoided by selecting suitable reaction conditions or substrates. Inactivation of enzymes in unusual reaction media can usually be avoided by immobilization.

12.1   Introduction

Many interesting target products of organic synthesis are much more soluble in unusual reaction media than in aqueous solutions and often have a much higher stability in aqueous solutions. The absence of water also generally prevents side and degradation reactions as well as microbial contamination in the case of sensitive compounds, which eliminates the need to work under sterile conditions as a further advantage.

For these reasons, there was early interest in the history of biocatalysis in carrying out biotransformation reactions with the addition or in the presence of organic solvents, e.g., to increase the solubility of substrates and products in aqueous reaction media by adding water-miscible solvents or to simplify the conversion of water-insoluble substrates or the subsequent reactant and product separation by using non-water-miscible solvents in two-phase systems. However, it was also discovered early on that solvents can influence the activity and stability of enzymes quite differently depending on whether they are present in monophasic or biphasic systems.

A ground-breaking discovery in the 1980s was the demonstration that many enzymes of the EC 3 enzyme class (hydrolases) can catalyze the reverse synthesis reaction in almost anhydrous nonpolar solvents instead of the hydrolysis reaction (Fukui and Tanaka 1985; Zaks and Russell 1988). It was found to be further advantageous that the enzymes added in insoluble form can be easily separated from the reaction mixtures, as they are active but not soluble in the nonpolar solvents.

Subsequently, enzyme activity could also be detected in supercritical solutions as used in food technology, in ionic liquids, in deep eutectic solutions, in the gas phase, or even in solid phase systems.

In the meantime, the use of enzymes in unusual reaction media has also become an industrially established technology, although certain regularities and basic rules must be observed in their use, which will be dealt with in the following.

12.2   Enzymatic Reactions Using Organic Solvents

By definition, solvents are substances in which gases, liquids, or solids can be dissolved without undergoing a chemical reaction. Enzyme reactions usually take place in water as a solvent, in which, in addition to the enzymes themselves, salts and organic compounds such as sugars, organic acids, amino acids, lipids, and higher-level compounds built up from them are usually homogeneously dissolved. In contrast to the above definition, water is not only a solvent in many enzyme reactions but also a reaction partner and substrate at the same time.

Water is characterized by its pronounced dipole character due to its positive and negative partial charges; it can dissociate and, with the possibility of forming hydrogen bonds, contributes to the dissolution of salts and organic compounds up to the formation of micelle and membrane structures in the cell. The ability of water molecules to form hydrogen bonds with each other is also the reason for the unusually high melting and boiling points of water compared to organic solvents. This effect is not found in organic solvents.

Organic solvents contain carbon and, in contrast to the water molecule, have molecular skeletons of different sizes to which, in addition to further carbon atoms, a large number of different functional groups can be bonded, which then ultimately determines the solvent properties such as hydrophobicity, hydrophilicity, and polarity. Well over a hundred different organic solvents are described in the literature and are commercially available (Laane et al. 1987).

12.2.1 Classification of Organic Solvents

Besides classification into hydrophilic (water-miscible) and hydrophobic (non-water-miscible), polar and nonpolar solvents, or a classification according to their chemical composition, e.g., chlorinated and nonchlorinated solvents, there are many other possibilities of classification, which is important for the safe handling of solvents, such as flammability, toxicity, water hazard, or biodegradability. A number of these criteria already indicate possible interactions of solvents with biological systems. An important way of classifying solvents for practical applications is the eluotropic series. This groups the most common organic solvents according to their elution effect or elution selectivity in the separation or elution of substances in chromatography, whereby this depends strongly on the stationary phase used in each case.

While the denaturing effect of many solvents, such as acetone or ethanol, on proteins in the presence of higher concentrations has been known for a long time, both empirically and in the literature, and was even used specifically for the precipitation of proteins, for example, it was not until the 1980s that the first systematic investigations were carried out in order to identify possible regularities in the handling of enzymes in combination with organic solvents. In this context, selected enzyme reactions were experimentally investigated in a large number of organic solvents in order to determine the possible influence of the parameters polarity, dielectric constant, and $\log P$ value of the solvents used on the enzyme activity (Laane et al. 1987).

The $\log P$ value is a parameter used in physical chemistry, which describes the solubility of a substance X at 25 °C in a mixture of water and *n-octanol*. It is defined as the logarithm of the partition coefficient of the respective substance X in a two-phase system of *n-octanol* and water (Eq. 12.1).

Table 12.1 Log$P$ value and water miscibility of organic solvents and their effect on protein structure and enzyme activity. (Modified according to Faber 2011)

Log$P$ value	Examples	Water miscibility	Effect on protein structure and enzyme activity
−2.5 to 0	Acetone, ethanol, DMFA, DMSO	Completely water-miscible	Depending on the solvent, when added, increase the solubility of lipophilic compounds without inactivating the enzymes in them up to a certain concentration
0 to 2	Propanol, ethyl acetate, butanol, hexanol	Partially water-miscible	As a rule, rapid enzyme inactivation occurs in the presence of these solvents
2 to 4	Toluene, octanol, styrene, n-hexane	Slightly water miscible	Effects on protein structure and enzyme activity are possible depending on the enzyme but cannot be predicted
>4	Diphenyl ether, octane, dodecane, butyl oleate	Not water miscible	Usually no effect on protein structure and preservation of enzyme activity

(1) Fully water-miscible solvents (log$P$ −2.5 to 0). (2) Partially water-miscible solvents (log$P$ 0 to 2). (3) Slightly water-miscible solvents (log$P$ 2 to 4). (4) Non-water-miscible solvents (log$P$ > 4)

$$\log P = \log \frac{[X]_{octanol}}{[X]_{water}} \qquad (12.1)$$

With the aid of this parameter, the solvents can be divided into four large groups (Table 12.1; Faber 2011):

12.2.2 Effects of Organic Solvents on Enzyme Activity

While in the case of polarity and the dielectric constant of the respective solvents, no systematic correlation with the activity of enzymes was apparent, a clear influence was shown in the case of the log$P$ value of the solvents used (Fig. 12.1; Laane et al. 1987): While an addition of completely or partially water-miscible solvents (log$P$ −2.5 to 2) to aqueous reaction media up to a certain concentration was possible without loss of activity, but a direct use of the enzymes in it led to their inactivation, the use of enzymes in little (log$P$ 2 to 4) and not water-miscible solvents (log$P$ >4) was possible with preservation of activity in many cases and even at high temperatures.

Another important factor influencing the activity of the enzymes in these solvents was identified as the water activity aW (Zaks and Klibanov 1986), a parameter that also plays an important role, for example, in the preservation of foodstuffs and other sensitive or perishable products. The water activity aW is defined as the ratio of the water vapor partial pressure in the product or protein (p) to the saturation vapor pressure of pure water (p_0) at a certain temperature (Eq. 12.2).

Fig. 12.1 Initial activity of a lipase-catalyzed transesterification between tributyrin and heptanol in various almost water-free organic solvents as a function of the log$P$ value. (According to Laane et al. 1987)

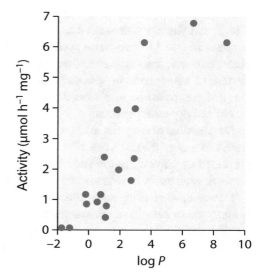

$$a_{\mathrm{w}} = \frac{p}{p_0} \tag{12.2}$$

The a_{w} value is therefore dimensionless and ranges between 0 (no water available) and 1 (formation of condensation water). In biochemical reactions, the aW *value* indicates the proportion of available water in the reaction system. It is of decisive importance for enzyme reactions and also for microbial growth: Bacteria generally require an aW *value* of at least 0.98 for growth, most fungi only 0.7. The aW *value* can be influenced by high salt or sugar concentrations, for example, which is used in the preservation and storage of food.

If we now consider the interaction of enzymes or, more generally, of proteins with the various groups of solvents mentioned above, completely water-miscible solvents (log$P$ − 2.5 to 0) can lead to a withdrawal of the hydrate shell (as in the precipitation of proteins when higher concentrations of acetone or ethanol are used in aqueous media), since the affinity of these solvents for water is higher than that of the proteins. A similar effect is also observed, for example, in the so-called salting out of proteins, as in precipitation with ammonium sulfate (Rupley et al. 1983).

Even if the course of the activity of enzymes in different solvents seems to be very different depending on their water content, this is almost identical in principle after normalization to the essential water content of the respective enzyme necessary for activity (Zaks and Klibanov 1986). However, different enzyme groups differ significantly with regard to the essential water content required for activity.

In contrast, non-water-miscible solvents (log$P$ >4) can stabilize the hydrate shell of the enzymes, which is necessary for activity—even at high temperatures—and thus prevent their denaturation (Zaks and Klibanov 1984).

In contrast, the influence on the activity of enzymes in partially water-miscible ($\log P$ 0 to 2) and slightly water-miscible solvents ($\log P$ 2 to 4) cannot be clearly predicted. If enzymes are used in two-phase systems consisting of an aqueous and an organic solvent phase, however, there is often an inactivation of the enzymes at the phase interface, while in the case of monophasic presence of these solvents with the addition of water until saturation and the formation of a second phase, similar to the non-water-miscible solvents, no loss of activity must yet occur.

Denaturation of enzymes in or by solvents can be avoided by various measures (Iyer and Ananthanarayan 2008). Very often, the denaturation of enzymes can be successfully prevented by immobilization and "freezing" in their active conformation, as was impressively demonstrated in fundamental work in the 1990s (Mozhaev et al. 1990).

However, when using anhydrous enzymes in a nonpolar, non-water-miscible solvent, it is important to note that a minimum water concentration in the reaction system is essential for their activity. As a rule, so much water should be present in the reaction system that each enzyme molecule is fully hydrated (Rupley et al. 1983).

Moreover, enzymes will be present in the solvent in the structure in which they were previously precipitated or dried from the aqueous reaction mixture at a certain pH (usually at the isoelectric point, IEP, which usually does not correspond to the pH optimum of the desired reaction), and will remember it, so to speak (*pH memory effect*) (Zaks and Klibanov 1986). Before use in the organic solvent, they should therefore be buffered and dried again, if necessary.

Thus, both the $\log P$ value and the water activity a_W are two crucial parameters that are essential to consider when working with enzymes in organic solvents, but also in the other unusual reaction media mentioned below.

Questions
1. What are the possible reasons for using enzyme reactions in unusual reaction media?
2. What are possible ways of classifying organic solvents?
3. How is the $\log P$ value defined? What is its importance for the classification of organic solvents?
4. How is the water activity aW defined? What is its importance for enzymatic reactions in completely water-miscible- and non-water-miscible organic solvents?

12.2.3 Enzymatic Reactions with Addition of Water-Miscible Solvents

Industrial processes for the biotransformation of poorly water-soluble steroids with the addition of completely water-miscible solvents such as methanol, ethanol, dimethyl sulfoxide (DMSO), or dimethyl formamide (DMFA) have been described in the literature since the 1940s. As a rule, growing or resting microbial cells are used, to which the poorly or not at all water-soluble starting substrate is then added and dissolved in the corresponding solvent in order to improve the bioavailability or to enable easier dosing. The subsequent processing of the products from the reaction mixtures is usually carried out by extraction with little or no water-miscible solvents.

Another field of application of water-miscible solvents is their addition to influence the reaction equilibrium in protease-catalyzed reactions such as the trypsin-catalyzed exchange of amino acids in modified insulin molecules.

12.2.4 Enzymatic Reactions in Two-Phase Systems with Little or Non-Water-Miscible Solvents

As an alternative method to the described procedure for improving the availability of poorly water-soluble substrates by the addition of water-miscible solvents (Sect. 12.2.2) is the use of solvents that are not or only slightly water-miscible, such as cyclohexane, methyl isobutyl ketone (MTBE), toluene, or ethyl acetate in a two-phase system. This has the advantage that the substrate and product are dissolved in the organic phase and thus significantly higher concentrations can be achieved than with the addition of water-miscible solvents. The enzymatic reaction takes place at the phase interface. Compounds that are sensitive in the aqueous phase can thus be stabilized, and possible product inhibition can be avoided. The separation of the product from the unreacted residual substrate can easily be carried out either at the end of the reaction after phase separation or even continuously during the reaction (in situ *product removal,* ISPR) from the organic phase, whereby the latter can also be used to avoid product inhibition.

However, frequently observed possible disadvantages can be a denaturation of the biocatalysts at the phase interface or inhibitory effects due to the solvent used. In order to avoid direct contact of the biocatalysts with the solvent phase, it is therefore advisable to use either special membrane systems or special immobilization techniques (Hudson et al. 2005).

The aqueous phase of a two-phase system can be reduced to a necessary minimum (so-called "micro-aqueous systems"). This can be of interest, for example, if water-soluble coenzymes are involved in the reaction. Avoiding direct contact of the enzyme(s) with the organic solvent can be essential for maintaining enzyme activity. One possibility for this is the use of so-called "*reversed micelles*" (Eggers and Blanch 1988). In addition to the aqueous micro- and organic phase, this reaction system also contains surfactants and co-surfactants to separate the aqueous and organic phase from each other, which, however, makes the subsequent work-up of reactants and products as well as the reuse of the enzymes present in the aqueous phase difficult. In addition, the transport of the reactant from the organic phase to the enzyme in the aqueous phase and of the product from the reverse micelle back into the organic phase must be ensured during the reaction.

Another possibility is the use of so-called *Pickering emulsions*. These are named after the British chemist Percival Spencer Umfreville Pickering, who in 1907 described the phenomenon that emulsions can be stabilized by the addition of solid particles which adsorb at the interface between the aqueous and organic phases. A natural example is homogenized milk, in which the emulsion is stabilized by casein molecules at the interface between the fatty and aqueous phases. In biocatalysis, such reaction systems may be suitable for microencapsulation of enzymes (Wei et al. 2016).

Basic criteria for the selection of suitable solvents for enzyme reactions in two-phase systems are (Faber 2011):

- Capacity of the solvent for reactant and product
- Their partition coefficients in the aqueous and organic phases
- Possible denaturing effects at the phase interface
- For large-scale industrial use, flammability, toxicity, and reuse of the solvents used

Reactions in two-phase systems can be of interest for biocatalysis, especially when a reactant and product are each preferentially present in different phases.

A natural reaction system for enzymatic reactions in two-phase systems is lipase-catalyzed reactions with fats or oils in water, whereby the reaction of highly viscous triglycerides or fats can be promoted by the addition of nonpolar solvents. In contrast to esterases, many lipases are only fully active when a phase interface is present in the reaction system after the critical micelle concentration (CMC) has been reached and therefore do not exhibit typical Michaelis–Menten kinetics.

Further examples of reactions in two-phase systems described in the literature are the recovery of water-soluble enantiomerically pure amino acids starting from racemic amino acid esters present in the solvent phase, the recovery of L-menthol from D, L-menthylacetate, or biotransformation reactions on or with poorly water-soluble steroids, where both the reactant and product are present in the solvent phase.

12.2.5 Enzymatic Reactions in Almost Anhydrous Organic Solvents

Enzyme activity in almost anhydrous nonpolar organic solvents was demonstrated for the first time in the mid-1980s, thus disproving a dogma that had been valid until then: First, for lipases, it was shown that these catalyze the synthesis of ester bonds instead of hydrolysis in almost anhydrous organic solvents, albeit with much lower activity (Eq. 12.3).

$$\text{Hydrolysis} : \text{Ester} + H_2O \rightarrow \text{Carboxylic Acid} + \text{Alcohol}$$

$$\text{Reverse Hydrolysis} : \text{Ester} + H_2O \leftarrow \text{Carboxylic Acid} + \text{Alcohol}$$

$$K = \frac{[\text{Carboxylic Acid}] \times \text{Alcohol}}{[\text{Ester}] \times [H_2O]} \tag{12.3}$$

Subsequently, this could soon be shown for other enzymes from EC class 3, the hydrolases, but it was also demonstrated that an essential water content is decisive for the respective enzyme activity, which differs significantly for the different enzyme groups (Zaks and Klibanov 1988).

Since the respective synthesis reactions produce water as a reaction product, an equilibrium is reached above a certain concentration. In order to avoid this, the reaction water should either be removed from the reaction preparations, or its formation should be avoided altogether by selecting suitable reaction conditions or substrates.

Examples from the literature of enzymatic reactions in nearly anhydrous organic solvents are as follows:

– Enantioselective synthesis of chiral esters starting from fatty acids, fatty acid methyl esters, or vinyl derivatives of fatty acids and alcohols, catalyzed by lipases and esterases
– Transesterification reactions catalyzed by lipases and esterases
– Enantioselective syntheses of amino acid esters and peptides with peptidases
– Synthesis of oligosaccharides and glycolipids with glycosidases

It was also shown that the enantioselectivity of the corresponding enzyme reaction may strongly depend on the type of solvent used.

Meanwhile, the use of enzymes not only in almost anhydrous nonpolar organic solvents but also in partially or even fully water-miscible solvents is an established technique in the chemical and pharmaceutical industries, for which evolved, stability-enhanced immobilized enzymes are usually used, which can then be used like a chemical heterogeneous catalyst (Hudson et al. 2005).

By immobilizing the enzymes using various techniques (use as *cross-linked enzyme crystals,* CLECs, as *cross-linked enzyme aggregates,* CLEAs, immobilization on carriers by covalent bonding, etc.), a "freezing" and thus a stabilization of the respective active conformation can be achieved, which in the meantime even allows direct use in completely water-miscible organic solvents such as ethanol without denaturation even at high temperatures, as could be impressively demonstrated using the example of the immobilization of chymotrypsin (Mozhaev et al. 1990). However, the dry CLECs and the gel-like CLEAs differ significantly in their water activity, which can be of great importance for corresponding applications of such immobilized enzymes.

Alternative strategies for carrying out enzymatic reactions in almost anhydrous organic solvents in the future could be to make enzymes soluble in organic solvents by appropriate modification, such as "PEGylation" (Chap. 19) (according to Castro and Knubovets 2003).

12.2.6 Basic Rules for Working with Enzymes in Almost Anhydrous Organic Solvents

If one wants to use almost anhydrous enzymes in organic solvents, the following basic rules should be observed (Faber 2011):

- Hydrophobic and non-water-miscible solvents ($\log P > 3$) are more compatible for enzymes than hydrophilic and partly water-miscible solvents ($\log P$ value < 3).
- The essential water layer around the enzyme molecules should be maintained in order to preserve the enzyme activity. In the case of hydrophobic solvents, this can be ensured, for example, by saturating them with water beforehand, which can be done, for example, by simply shaking out the solvents with water in the separating funnel beforehand. If CLEAs are used, their water content must also be taken into account.
- Before using the respective enzyme in an almost water-free solvent, it should be ensured that the "micro pH" corresponds to the pH optimum of the enzyme. If necessary, the enzyme must be "re-buffered" and dried again beforehand.
- Since the enzyme is no longer dissolved in the solvent, diffusion of the reactant to and of the product away from the enzyme surface should be ensured by stirring, shaking, or ultrasonication.
- Stabilization of the enzymes can be achieved either by addition of stabilizing agents such as nonactive proteins, sugars, polyalcohols, polymers, and salts or by immobilization.

Questions

5. What are the basic criteria for the selection of suitable solvents for enzyme reactions in two-phase systems?
6. What are the basic rules for working with enzymes in almost anhydrous organic solvents?

12.3 Ionic Liquids, Deep Eutectic Solutions, and Supercritical Fluids as Reaction Media for Enzymatic Reactions

12.3.1 Ionic Liquids

Ionic liquids (ILs) are salts that are liquid at temperatures below 100 °C without the addition of water. They are usually composed of organic cationic compounds such as imidazolium, pyridinium, pyrrolidinium, and guanidinium, which may be alkylated, and anionic compounds such as a halide, tetrafluoroborates, trifluoroacetates, imides, or amides (Fig. 12.2; Moniruzzamana et al. 2010). When these compounds are combined with each other, charge delocalization and steric effects hinder the formation of a stable crystal lattice, the solid crystal structure in the individual compounds is broken down, and a so-called "ionic liquid" is formed.

For the production of ionic liquids there are a large number of possible combinations by means of which their respective desired physicochemical properties can be adjusted and adapted to possible technical requirements, e.g., in order to influence the solubility of substances that are difficult to dissolve in water, which is of great interest for dissolving

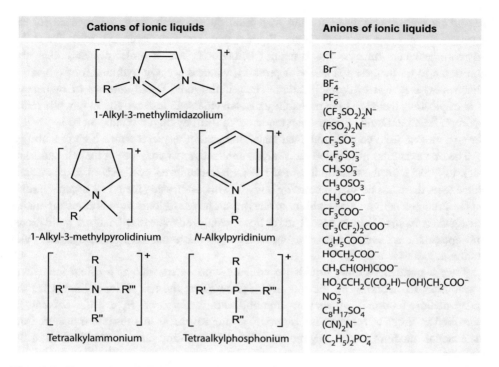

Fig. 12.2 Commonly used cations and anions of ionic liquids. (According to Moniruzzamana et al. 2010)

reactants and products when carrying out enzymatic reactions (Van Rantwijk and Sheldon 2007).

Compared to organic solvents, ionic liquids are also characterized by the fact that they are usually thermally stable and difficult to ignite, as well as having a very low, hardly measurable vapor pressure. Even if there is no danger of explosion or poisoning by inhalation due to their nonvolatility, as is the case with many organic solvents, IL-containing waste could possibly be problematic due to the unresolved biodegradability, which is why their recyclability can be an important criterion for applications.

Due to the large number of possible combinations, ionic liquids are highly interesting reaction media for biocatalysis (Potdar et al. 2015), and the aim is to achieve the desired physicochemical properties with the lowest possible toxicity through the right combination of components.

12.3.2 Enzymatic Reactions in Ionic Liquids

The interest in the use of enzymes in ionic liquids dates back to 1984, when the activity and stability of alkaline phosphatase in the cleavage of the model compound *p-nitrophenyl phosphate* were investigated for the first time. Subsequently, it was discussed for the interaction of enzymes with ionic liquids that strongly hydrated ions (so-called cosmotropes) can enhance the structural behavior of water molecules and thus also contribute to an increased stability of proteins, whereas weakly hydrated ions (so-called chaotropes) can lead to destabilization. Thus, if the ionic liquid consists of two rather chaotropic ions, there is a high probability that enzymes will lose activity in it, while rather hydrophilic ionic liquids increase the tendency to form hydrogen bonds, make it possible for enzymes to dissolve in it, and even maintain their activity in it with a high probability.

The first examples of targeted biocatalytic applications of enzymes in ionic liquids date back to 2000 with the description of the peptidase-catalyzed synthesis of *Z-aspartame*. Since then, there has been an increasing body of work addressing the effect of ionic liquids on the structure, activity, and stability of enzymes. The use of ionic liquids as an alternative to organic solvents has been described for lipase-catalyzed transesterifications, amidations, and epoxidations, among others, using both single-phase and two-phase systems (Van Rantwijk and Sheldon 2007).

Since a number of ionic liquids do not mix with ethers, this also offers interesting possibilities for working up the products. Furthermore, the use of ionic liquids for the encapsulation (*coating*) *of* enzyme immobilizates when used in organic solvents is described in order to achieve, for example, an improvement in substrate transport from the reaction medium to the enzyme in addition to avoiding enzyme inactivation at the interface.

There is a possibility of using ionic liquids to dissolve lignocellulose and its individual components with ionic liquids and thus to improve their availability for enzymatic reactions. Likewise, there are already successful applications for whole-cell biocatalysis in ionic liquids, e.g., for the production of chiral alcohols starting from prochiral ketones. Since ionic liquids have no vapor pressure, volatile compounds can be more easily separated from them by evaporation than from organic solvents (Van Rantwijk and Sheldon 2007).

In summary, it can be said that ionic liquids for enzyme catalysis offer a variety of interesting application possibilities as an alternative reaction medium to organic solvents. However, in addition to the still high costs, the main challenges still to be faced are their recyclability and their still unclear biodegradability and environmental compatibility.

12.3.3 Deep Eutectic Solutions

Closely related to ionic liquids, and also referred to by some authors as a "special class of biodegradable ionic liquids," are *deep eutectic solutions* (DES) or *natural deep eutectic solvents* (NADES; Paiva et al. 2014). For definition: A solution or alloy of different substances is called "eutectic," if its components are in such a ratio to each other that it becomes liquid or solid as a whole at a certain temperature. The corresponding point in the phase diagram is called a eutectic (Fig. 12.3). The prerequisite for this effect is that the

Fig. 12.3 Phase diagram of a deep eutectic solution. (According to Smith et al. 2014)

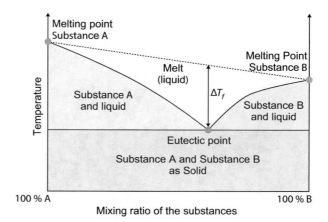

melting temperature of the mixing system is below the melting temperatures of the individual components and that a miscibility gap exists in the solid which remains until the melting temperature of the eutectic is reached. The "eutectic point" is the point in the phase diagram of a multicomponent system which is characterized by the concentration ratio of the eutectic and by its melting temperature (the "eutectic temperature").

In contrast to ionic liquids, eutectic solutions usually consist of an ammonium compound, namely choline or urea, in combination with organic compounds such as sugars, sugar alcohols, or carboxylic acids, which act as donors to form hydrogen bonds (Fig. 12.4; Smith et al. 2014). All building blocks can be obtained from renewable raw materials and, as individual components, have melting points above 100 °C. If mixtures of these are prepared in certain proportions and heated, solutions of varying viscosity are formed, which remain liquid even after cooling to room temperature.

Natural deep eutectic solutions are inexpensive reaction media and, like ionic liquids, are nonflammable and liquids. Unlike ionic liquids, however, they are usually harmless from a toxic point of view and are generally readily biodegradable. However, as a disadvantage, they do not offer the large variety of combination and application possibilities as ionic liquids.

12.3.4 Enzymatic Reactions in Deep Eutectic Solutions

Possible applications of deep eutectic solutions in enzyme catalysis are their direct use as a nonvolatile solvent, taking into account the essential water concentration for the respective enzyme, or their use in two-phase systems (Paiva et al. 2014). In addition to the sometimes high viscosity, a disadvantage can be that, in contrast to organic solvents and ionic liquids, the addition or formation of water during the reaction can lead to a breakdown of the electrostatic interactions between the components of the deep eutectic solution. However, this effect can in turn be used specifically in the work-up of the reaction mixtures. In any

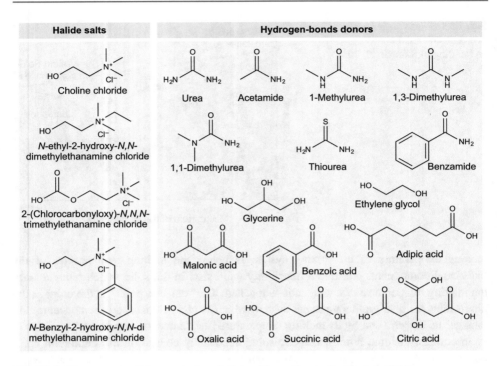

Fig. 12.4 Components of deep eutectic solutions. (According to Smith et al. 2014)

case, denaturing effects on the enzymes used are expected to be less than in the case of many organic solvents and ionic liquids.

Despite the inexpensive accessibility and easy producibility of deep eutectic solutions, there are so far only a manageable number of examples in the literature of enzyme reactions therein, mainly with lipases (Durand et al. 2013). One of the first published papers dates to 2008 and describes lipase-catalyzed transesterification using an ethyl fatty acid and 1-butanol as substrates. Other papers describe the use of these reaction systems in the lipase-catalyzed production of biodiesel from high-viscosity triacylglycerides and methanol, in *N-alkylations* of primary aromatic amines, and in epoxide hydrolysis.

Interest in deep eutectic solutions as new reaction media for enzymes is growing, even if the spectrum of possible applications is expected to be much smaller than for organic solvents and ionic liquids. Apart from their nonvolatility and nonflammability, their nontoxicity, good environmental compatibility, and the possibility of producing them on the basis of renewable raw materials are attractive for future industrial applications as "*green solvents*." Future challenges for enzyme catalysis are not only to exploit the existing potential but also to expand the application and combination possibilities of deep eutectic solutions by using further salts and hydrogen bridge donors in a similar way as has already been done for ionic liquids.

Questions
   7. What is the definition of an ionic liquid? What are ionic liquids composed of?
   8. What is an eutectic solution? What are "*natural eutectic sovents*" (NADES)?

12.3.5 Supercritical Fluids

Substances can occur in solid, liquid, and gaseous forms. This is influenced by temperature
and pressure (Fig. 12.5). At the so-called triple point, all three phases are in equilibrium.
Supercritical fluids are compounds whose properties lie between those of gases and liquids.
The so-called critical point is the state of a substance which is characterized by the
equalization of the densities of the liquid phase and the gas phase. The differences between
these two aggregate states cease to exist under these conditions. Above this point, a fluid is
said to be "supercritical" or in a "supercritical state." This occurs when a gas is subjected to
an ever-increasing pressure and the distances between the gas molecules decrease until, on
reaching the critical pressure, they become as large as those of molecules in the liquid
phase.
   The solvent properties of a supercritical fluid depend strongly on its respective density,
which can be adjusted over a relatively wide range as an advantage of these reaction
systems (Mesiano et al. 1999). In this context, a higher density generally also increases the
solubility of most substances, while a lower density lowers it. Supercritical fluids can thus
combine the solubility of liquids with the low viscosity of gases, which is a great advantage
because a substance dissolved in the supercritical fluid can be separated from the reaction
medium again very easily at normal pressure. In contrast to working with organic solvents,
ionic liquids, and deep eutectic solutions, however, carrying out reactions in supercritical
fluids requires in all cases significantly more complex plant and process technology since
work has to be carried out at alternating pressure.
   Carbon dioxide, ethane, propane, ethylene, and fluoroform are substances that are
commonly used as supercritical fluids (Hobbs and Thomas 2007). Supercritical carbon
dioxide, in particular, is a much-used reaction medium due to its cheap and easy availability, its nontoxicity when handled appropriately, and its ease of separation from substances
dissolved in it, and has long been used, for example, in food technology as a nontoxic
alternative for extraction purposes to decaffeinate coffee and tea. It is formed as soon as the
pressure and temperature exceed 304.13 K (30.980 °C) and the pressure exceeds
7.375 MPa (73.75 bar) and then exhibits significantly different properties than under
standard conditions.

12.3.6 Enzymatic Reactions in Supercritical Fluids

As is the case when working with enzymes in nonpolar organic solvents, it should be noted
when using enzymes in supercritical fluids that an essential water concentration on the

Fig. 12.5 Phase diagram for a supercritical fluid. (According to Hobbs and Thomas 2007)

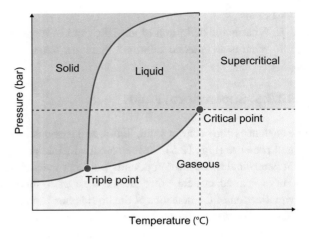

surface of the enzyme molecules is necessary for them to be active in the supercritical fluids. This can be achieved by water saturation, which is not a problem in the case of supercritical carbon dioxide (Hobbs and Thomas 2007).

As one of the first enzyme reactions in supercritical carbon dioxide again, the cleavage of the model compound *p-nitrophenyl* phosphate by alkaline phosphatase was described in 1985 (Matsuda 2013). Other enzymes that were subsequently used successfully in various supercritical fluids included subtilisin *Carlsberg* and various immobilized commercially available lipases when used for transesterification reactions. Among other things, it was shown that for kinetic resolutions therein, the enantiomeric rate can be affected by pressure and temperature. Furthermore, asymmetric reduction reactions with alcohol dehydrogenases as well as carboxylation reactions with various decarboxylases were successfully carried out in supercritical carbon dioxide (Matsuda 2013).

Despite the positive results and the advantages of the adjustability of the solvent properties by pressure and temperature as well as the possibility of simple product separation, the use of supercritical fluids in enzyme catalysis has not yet become widely established. The reasons for this may presumably be seen in the limited solvent spectrum and the necessary high equipment costs.

Questions
  9. What is a supercritical fluid? Which compounds are commonly used as supercritical fluids?

12.4   Enzymatic Reactions Under Complete Solvent-Free Conditions

As an alternative to the use of enzymes in aqueous reaction systems, with or in organic solvents, ionic liquids, deep eutectic solutions, or supercritical fluids, there is also the possibility of almost completely dispensing with the use of solvents (*solvent-free biocatalysis*), whereby substrates and products can be present either in the gas phase or in the liquid or solid phase. Even, if such an approach initially appears unusual, there are a number of examples in the literature of reactions with soluble and immobilized enzymes or whole cells that are used directly in the substrate without the addition of solvents (Hobbs and Thomas 2007).

The approach is relatively simple and involves bringing the biocatalyst directly into contact with the substrate or substrates, which are then also the reaction medium. In all cases, the water activity must again be taken into account.

When carrying out reactions in the gas phase, this can be done, for example, simply by passing the substrate over the respective biocatalyst. This can be living whole cells of microorganisms immobilized on a carrier. Subsequent separation of the substrate and product can then be accomplished by cold traps. Examples of this are cis-epoxidation reactions on gaseous alkenes.

When carrying out reactions in the liquid phase, in which the latter is both the substrate and solvent, the biocatalyst is added directly to the liquid substrate phase. Here, in addition to the essential water concentration for the enzyme, it should be noted that mixing can become the rate-determining step.

Solid-to-solid biocatalysis poses a particular challenge as the reaction only takes place at the interface between the enzyme and substrate. In some cases, minimal amounts of water or solvent must be added as an adjuvant. In this case, one also speaks of a "heterogeneous eutectic reaction." Examples of such reactions from the literature are various peptide syntheses with the immobilized proteases subtilisin or thermolysin, including the successful synthesis of the dipeptide sweetener precursor *Z-aspartame* in three-molar concentration or lipase-catalyzed syntheses of esters from sugar alcohols and fatty acids. With immobilized penicillin acylase, the successful synthesis of ampicillin from equimolar amounts of solid lyophylated 6-aminopenicillanic acid and D-phenylglycine-methylester was achieved, whereby the essentially required water was made available in the form of salt hydrates.

The scale-up of such solid-phase biocatalysis reactions poses a challenge in terms of equipment and process technology.

Answers
1. The main motivation for the use of enzymes in unusual reaction media was to significantly increase the solubility of substrates and products compared to aqueous reaction systems. In the meantime, the focus is on the use of hydrolytic enzymes for synthesis reactions in almost anhydrous reaction media in order to specifically exploit their properties of regioselectivity, stereoselectivity, and substrate specificity.
2. Besides classification into hydrophilic (water-miscible) and hydrophobic (non-water-miscible), polar and nonpolar solvents, or a classification according to their chemical composition, e.g., as chlorinated and nonchlorinated solvents, there are many other possibilities of classification, which are important for their safe handling, such as flammability, toxicity, water hazard, or biodegradability. Other important ways of

classifying solvents for practical applications are the eluotropic series and the log$P$ value.

3. The log$P$ value is a parameter used in physical chemistry that describes the solubility of a substance X at 25 °C in a mixture of water and n-octanol. It is defined as the logarithm of the partition coefficient of the respective substance X in a two-phase system of n-octanol and water. With the aid of this parameter, the solvents can be divided into four large groups: 1. completely water-miscible solvents (log$P$ -2.5 to 0), 2. partially water-miscible solvents (log$P$ 0 to 2), 3. low-water-miscible solvents (log$P$ 2 to 4), and 4. non-water-miscible solvents (log$P > 4$).

4. The water activity aW is defined as the ratio of the water vapor partial pressure in a product or protein (p) to the saturation vapor pressure of pure water (p0) at a certain temperature. The aW value is dimensionless and ranges between 0 (no water available) and 1 (formation of condensation water). In biochemical reactions, the aW value indicates the proportion of available water in the reaction system. Completely water-miscible solvents (log$P$ -2.5 to 0) can lead to a withdrawal of the hydrate shell since the affinity of these solvents for water is higher than that of proteins. In contrast, non-water-miscible solvents (log$P > 4$) can stabilize the hydrate shell of enzymes and thus prevent their denaturation.

5. Basic criteria for the selection of suitable solvents for enzyme reactions in two-phase systems are the capacity of the solvent for the reactant and product, their partition coefficients in the aqueous and organic phases, possible denaturing effects at the phase interface, and for large-scale industrial use, their flammability, toxicity, and possible reuse.

6. The essential water layer around the enzyme molecules should be maintained in order to preserve the enzyme activity. Before using the respective enzyme in an almost water-free solvent, it should be ensured that the "micro pH" corresponds to the pH optimum of the enzyme. Since the enzymes are no longer dissolved in the solvent, diffusion of the reactant to and of the product away from the enzyme surface should be ensured by stirring, shaking, or ultrasonication. Stabilization of the enzymes can be achieved either by addition of stabilizing agents such as nonactive proteins, sugars, polyalcohols, polymers, and salts or by immobilization.

7. Ionic liquids are salts that are liquid at temperatures below 100 °C without the addition of water. They are usually composed of organic cationic compounds such as imidazolium, pyridinium, pyrrolidinium, and guanidinium, which may be alkylated, and anionic compounds such as halides, tetrafluoroborates, trifluoroacetates, imides, or amides. When these compounds are combined with each other, charge delocalization and steric effects hinder the formation of a stable crystal lattice, the solid crystal structure in the individual compounds is broken down, and a so-called "ionic liquid" is formed.

8. A solution or alloy of different substances is called "eutectic," if its components are in such a ratio to each other that it becomes liquid or solid as a whole at a certain temperature. The corresponding point in the phase diagram is called a eutectic. The

melting temperature of the mixing system is below the melting temperatures of the individual components. Natural eutectic solvents (NADES) usually consist of an organic ammonium compound, namely choline or urea, in combination with organic compounds such as sugars, sugar alcohols, or carboxylic acids, which act as donors to form hydrogen bonds. All building blocks can be obtained from renewable raw materials and, as individual components, have melting points above 100 °C.

9. Substances can occur in solid, liquid, and gaseous forms. This is influenced by temperature and pressure. At the so-called triple point, all three phases are in equilibrium. Supercritical fluids are compounds whose properties lie between those of gases and liquids. The so-called critical point is the state of a substance, which is characterized by the equalization of the densities of the liquid phase and the gas phase. Above this point, a fluid is said to be "supercritical" or in a "supercritical state." Carbon dioxide, ethane, propane, ethylene, and fluoroform are substances that are commonly used as supercritical fluids. In particular, supercritical carbon dioxide is a much-used reaction medium due to its cheap and easy availability, its nontoxicity when handled appropriately, and its ease of separation from substances dissolved in it.

Take Home Message

Polar and nonpolar organic solvents, supercritical fluids, ionic liquids, and deep eutectic solutions can be used as unusual media for enzymatic reactions. Since many interesting target products of organic synthesis are much more soluble in unusual reaction media than in aqueous solutions and often have a much higher stability than in aqueous solutions, the motivations for the use of unusual reaction media are to significantly increase the solubility of substrates and products compared to aqueous reaction systems, to simplify the subsequent work-up and to avoid microbial contamination as well as side and degradation reactions. Additionally, many enzymes exhibit different properties in unusual reaction media and catalyze different reactions from those in aqueous systems, i.e., hydrolytic enzymes can be successfully used for synthesis reactions in almost anhydrous reaction media in order to specifically exploit their properties such as regioselectivity, stereoselectivity, and substrate specificity. An essential parameter to be considered with the use of unusual reaction media is water activity aW. 4. It is defined as the ratio of the water vapor partial pressure in a product or protein (p) to the saturation vapor pressure of pure water (p0) at a certain temperature. The aW value is dimensionless and ranges between 0 (= no water available) and 1 (= formation of condensation water). In biochemical reactions, the aW value indicates the proportion of available water in the reaction system. In the case of polar and nonpolar organic solvents, the $\log P$ value is another parameter which is used to classify the solvents into completely water-miscible solvents ($\log P$ −2.5 to 0), partially water-miscible solvents ($\log P$ 0 to 2),

(continued)

low-water-miscible solvents (log$P$ 2 to 4), and non-water-miscible solvents (log$P$ > 4). While completely water-miscible solvents (log$P$ -2.5 to 0) very often will lead to a withdrawal of the enzyme hydrate shell and loss of activity since the affinity of these solvents for water is higher than that of proteins, non-water-miscible solvents (log$P$ >4) can even stabilize the hydrate shell of enzymes and thus prevent their denaturation.

References

Castro GR, Knubovets T. Homogeneous biocatalysis in organic solvents and water-organic mixtures. Crit Rev Biotechnol. 2003;23:195–231.

Durand E, Lecomte J, Villeneuve P. Deep eutectic solvents: synthesis, application, and focus on lipase-catalyzed reactions. Eur J Lipid Sci Technol. 2013;155:379–85.

Eggers DK, Blanch HW. Enzymatic production of L-tryptophane in a reverse micelle reactor. Bioprocess Bioeng. 1988;3:83–91.

Faber K. Biotransformations in organic chemistry. 6th revised and corrected edition, Springer-Verlag; 2011.

Fukui S, Tanaka A. Enzymatic reactions in organic solvents. Endeavor New Ser. 1985;9:10–7.

Hobbs HR, Thomas NR. Biocatalysis in supercritical fluids, in fluorous solvents, and under solvent-free conditions. Chem Rev. 2007;107:2786–820.

Hudson EP, Eppler RK, Clark DS. Biocatalysis in semi-aqueous and nearly anhydrous conditions. Curr Opin Biotechnol. 2005;16:637–43.

Iyer PV, Ananthanarayan L. Enzyme stability and stabilization in aqueous and non-aqueous environment. Process Biochem. 2008;43:1019–32.

Laane C, Boeren S, Vos K, Veeger C. Rules for optimization of biocatalysis in organic solvents. Biotechnol Bioeng. 1987;30:81–7.

Matsuda T. Recent progress in biocatalysis using supercritical carbon dioxide. J Biosci Bioeng. 2013;115:233–41.

Mesiano AJ, Beckman EJ, Russell AJ. Supercritical biocatalysis. Chem Rev. 1999;99:623–33. Review

Moniruzzamana M, Nakashimab K, Kamiyaa N, Gotoa M. Recent advances of enzymatic reactions in ionic liquids. Biochem Eng J. 2010;48:295–314.

Mozhaev VV, Sergeeva MV, Belova AB, Khmelnitsky YL. Multipoint attachment to a support protects enzyme from inactivation by organic solvents: alpha-chymotrypsin in aqueous solutions of alcohols and diols. Biotechnol Bioeng. 1990;35:653–9.

Paiva A, Craivero R, Aroso I, Martins M, Ries RL, Duarte AR. Natural deep eutectic solvents—solvents for the 21st century. ACS Sustain Chem Eng. 2014;2:1063–71.

Potdar MK, Kelso GF, Schwarz L, Zhang C, Hearn MTW. Recent developments in chemical synthesis with biocatalysts in ionic liquids. Molecules. 2015;20:16788–816.

Rupley JA, Gratton E, Careri G. Water and globular proteins. Trends Biochem Sci. 1983;8:18–22.

Smith EL, Abbott AP, Ryder KS. Deep eutectic solvents (DES) and their applications. Chem Rev. 2014;114:11060–82.

Van Rantwijk F, Sheldon RA. Biocatalysis in ionic liquids. Chem Rev. 2007;107:2757–85.

Wei L, Zhang M, Zhang X, Xin H, Yang H. Pickering emulsion as an efficient platform for enzymatic reactions without stirring. ACS Sustain Chem Eng. 2016;2016:6838–43.

Zaks A, Klibanov AM. Enzymatic catalysis in organic media at 100°C. Science. 1984;224:1249–51.

Zaks A, Klibanov AM. The effect of water on enzyme activity in organic media. J Biol Chem. 1986;263:8017–23.

Zaks A, Russell J. Enzymes in organic solvents: properties and applications. J Biotechnol. 1988;8: 259–70.

Part III

Applications

Principles of Applied Biocatalysis

13

Selin Kara and Jan von Langermann

What You Will Learn from This Chapter

The use of enzymes represents a key technology for chemical synthesis, combining mild reaction conditions with high selectivities. After the enzyme itself, the reaction control of biocatalytic reactions is an essential part in the preparative synthesis of important products or intermediates. Thus, for the efficient performance of the planned biocatalytic processes, the understanding of the limiting factors and the use of appropriate strategies to overcome these challenges are essential. This chapter focuses primarily on the presentation of suitable concepts for the efficient feeding of reactants and the removal of the desired products from the reaction solution by integrating secondary techniques. This can be done by an integration separation process or the inclusion of secondary (bio)chemical reactions directly into the desired biocatalytic conversion. Advantages and disadvantages, as well as fundamental physicochemical limitations, are discussed and supported with application examples.

The application of enzymes to organic reactions is an emerging field, as biocatalysts enable exceptionally high selectivities under moderate reaction conditions. The previous chapters in this book focused primarily on the fundamental understanding of enzyme formulation

S. Kara (✉)
Institute of Technical Chemistry, Leibniz University Hannover, Hannover, Germany
e-mail: selin.kara@iftc.uni-hannover.de

J. von Langermann
Institute of Chemistry, Otto von Guericke University Magdeburg, Magdeburg, Germany
e-mail: jan.langermann@ovgu.de

(identification, characterization, and optimization). In contrast, this chapter focuses primarily on the principles of applied biocatalysis with emphasis on strategies for dealing with diverse process challenges such as regeneration of external cofactors, unfavorable reaction thermodynamics, substrate and product inhibition, and the low solubility of reagents in classical aqueous media.

Biocatalysis is generally regarded as "green chemistry" since the reactions usually occur in aqueous media and under environmentally friendly reaction conditions. In general, however, such reasoning is very simplified since wastewater produced must also be considered in the life cycle assessment. In contrast, many compounds of synthetic interest have limited solubility in such aqueous systems, and therefore high productivity cannot be achieved. For this reason, a large number of research groups have focused on the use of alternative reaction media for biocatalysis. Furthermore, the use of non-conventional reaction media, such as two-phase systems, is attractive for the reasons given above and is also a helpful approach in the selective extraction of the (co)products. Alternative adsorptive strategies are also pursued by using adsorber resins to selectively separate the products, which can also be used analogously for substrate dosing.

Moreover, cost-intensive cofactors must be recycled in oxidoreductase-catalyzed redox reactions to enable high conversions and productivity. Efficient cofactor regeneration approaches are introduced and discussed comparatively in this chapter. The chapter's end highlights future techniques for establishing efficient biotransformations.

13.1   Cofactor-Dependent Biotransformations

Cofactor-dependent enzymes catalyze a wide range of synthetically relevant reactions, and recycling the cofactor is essential in terms of economic and practical considerations. Due to their high price, direct use in stoichiometric quantities is practically impossible for cofactors, in addition to their low stability (Table 13.1; Paul et al. 2014).

Numerous scientific papers have been concerned with the development of reliable *in situ* regeneration methods for the most commonly used cofactors, especially nicotinamide cofactors (NAD(P)H and NAD(P)$^+$). The effectiveness of the regeneration process is measured by the number of cycles that can be achieved before the cofactor molecule is finally decomposed. This is referred to as the *total turnover number* (TTN), which is the total number of molar products per mole of cofactor formed over the lifetime of the cofactor. On a laboratory scale, TTN values of 1000–10,000 are sufficient for redox reactions, while TTN values of at least 100,000 are expected for engineering purposes. If the total lifetime of the cofactor has yet to be reached, the term *turnover number* (TON) is used to evaluate the system's productivity. Chemical systems, e.g., electrochemical and photochemical approaches, are equally applied for cofactor regeneration, however, the productivities are generally lower compared to biocatalytic approaches. Therefore, only biocatalytic cofactor regenerating systems such as substrate-coupled (Sect. 13.1.1), enzyme-coupled (Sect. 13.1.2), and self-sufficient cascades (Sect. 13.1.2) are discussed in the following sections (Sect. 13.1.3) (Kara et al. 2014).

Table 13.1 Current prices for nicotinamide cofactors (Paul et al. 2014)

Nicotinamide cofactor	Price ($€$ mol^{-1})
NAD$^+$	1410
NADH	2625
NADP$^+$	18,500
NADPH	70,835

13.1.1 Substrate-Coupled Systems

In a substrate-coupled system, the catalytically active form of the cofactor (e.g., NAD(P)H in reductions and NAD(P)$^+$ in oxidations) is regenerated by the same enzyme in a secondary redox reaction, which proceeds in the opposite direction with the consumption of a cosubstrate. This approach has some advantages but also disadvantages.

One of the main advantages is simplicity, as only one enzyme is required for the entire reaction. Secondly, the nicotinamide cofactor remains in the enzyme's active site, and thus, mass transfer limitations and stability problems can be mitigated. However, the enzyme activity is distributed between the primary and regeneration reactions (Fig. 13.1). Most importantly, this is a reversible system. When the substrates and products are chemically very similar (e.g., reducing a ketone to an alcohol), an inferior thermodynamic driving force is present. To alleviate this limitation, excess amounts of cosubstrates (e.g., isopropanol or ethanol for reductions and acetone or acetaldehyde for oxidations) are usually applied, e.g., 15–20 molar equivalents. Nevertheless, it is considered positive that in the case of a highly hydrophobic substrate (non-/poorly water soluble), using excess amounts of cosubstrates can be helpful, as they increase the solubility of the hydrophobic substrate compared to the purely aqueous medium. However, it should be mentioned that this approach can only be used as long as the enzyme is stable at these high cosubstrate concentrations. One of the significant challenges with this approach is the high amount of waste products generated due to the unreacted cosubstrate and the coproduct produced. Concerning the formation of by-products, in technical applications, the terms "E-factor" (*environmental factor,* i.e., (*mass of waste)/(mass of products*)) and atomic economy (i.e., (*molecular weight of the target product)/(sum of the molecular weights of all products*)) are employed.

To avoid the use of excess amounts of cosubstrates, the corresponding coproducts (e.g., acetone and acetaldehyde) can be removed from the reaction medium by stripping with nitrogen (or other inert gases) or pervaporation, as they are more volatile compared to the cosubstrates (e.g., isopropanol and ethanol). This way, the equilibrium of the reaction can be shifted in the desired direction. Furthermore, extraction can remove the product and/or by-product from the reaction medium.

Recently, two new concepts have been introduced that allow cosubstrates in stoichiometric amounts instead of high molar excesses. Both approaches are based on the formation of a thermodynamically stable by-product. For example, using activated ketones as a cosubstrate, only stoichiometric amounts are required, as the coproduct is

Fig. 13.1 Substrate-coupled approach for the regeneration of NAD(P)H to synthesize a chiral alcohol from a prochiral ketone. E1: Main enzyme and regenerating enzyme

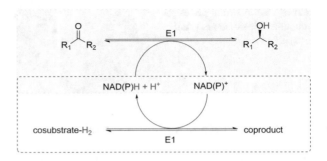

thermodynamically stabilized via intramolecular hydrogen bonds. In this study, the biotransformations were catalyzed by lyophilized cells of *Escherichia coli (E. coli)* overexpressing an alcohol dehydrogenase (ADH) from *Sphingobium yanoikuyae* (*SyADH*). Oxidation of several structurally different alcohols with high conversion (90% and higher) at 30 g L^{-1} concentration was achieved using only 1.1–1.5 molar equivalents of chloroacetone as the cosubstrate (Lavandera et al. 2008). A second approach was developed to regenerate reduced nicotinamide cofactors using 1,4-butanediol as a "smart cosubstrate" to promote NADH-dependent biotransformations (Kara et al. 2013). The thermodynamically stable coproduct γ-butyrolactone renders the regeneration reaction irreversible, and synthesis from 1,4-butanediol yields 2 molar equivalents of NADH. Consequently, the requirement for cosubstrate is drastically reduced (to 0.5 molar eq.).

13.1.2 Enzyme-Coupled Systems

In an enzyme-coupled system, the enzyme catalyzes the primary reaction, and a second enzyme recycles the cofactor in the regeneration reaction. Here, the enzymes should have a different substrate spectrum so that no cross-reaction can occur. Furthermore, the cosubstrate should be inert to the primary reaction so that the enzyme will not be deactivated or inhibited, and the cosubstrates should be inexpensive. So far, several enzyme-coupled approaches have been developed, and their distinct advantages and disadvantages are already well known. The following scheme (Fig. 13.2) shows the regeneration of NAD(P)H using a second enzyme.

Glucose dehydrogenase (GDH, EC 1.1.1.47) is applied to the oxidation of β-D-glucose to D-glucono-1,5-lactone to regenerate 1 molar equivalent of NAD(P)H. The lactone coproduct is spontaneously hydrolyzed to the corresponding acid, making the regeneration reaction practically irreversible. GDH from different organisms was used for cofactor regeneration. One of this system's most important advantages is that GDH applies to both NADH and NADPH. In addition, GDH is highly active and stable, cosubstrate glucose is very inexpensive, and the enzyme (from different organisms) is commercially available. However, the main disadvantage of this system is the pH change during the reaction due to the gluconic acid coproduct formed; therefore, pH control is needed. The

Fig. 13.2 Enzyme-coupled approach for the regeneration of NAD(P)H for the synthesis of a chiral alcohol from a prochiral ketone. E1: main enzyme, E2: regenerating enzyme

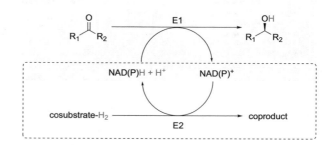

second disadvantage is that only 1 molar equivalent of NAD(P)H can be regenerated from such a large molecule (180 g mol^{-1}).

Formate dehydrogenase (FDH, EC 1.2.1.2) is another helpful enzyme for the regeneration of reduced nicotinamide cofactors. FDH catalyzes formate oxidation to CO_2 and reduces NAD(P)$^+$ to NAD(P)H. This process has several advantages, such as formate as a cosubstrate is inexpensive, both formate and CO_2 are generally harmless to the enzymes, and the FDH enzyme is commercially available. Most importantly, the coproduct CO_2 volatilizes rapidly from the reaction medium so that the chemical equilibrium can be shifted to the product side, and an additional workup step is not required. The main disadvantage of this system is that FDH is relatively expensive and has relatively low activity and stability (compared to GDH). However, protein engineering and immobilization can still mitigate these disadvantages, which are also described in this book. Usually, the formate/ FDH system is used for NADH regeneration. The most commonly used FDH from *Candida boidinii* is specific for NADH; however, its cofactor specificity can be modified. In addition, NADPH-dependent FDHs from various microorganisms are available.

The use of **hydrogenases** for the regeneration of the reduced nicotinamide cofactors by molecular hydrogen, H_2, represents a very clean process since no by-product is formed. Moreover, the cosubstrate H_2 is inexpensive, no coproduct is formed, and the workup of the target product is further simplified. However, the application of hydrogenases for cofactor regeneration is severely limited because the enzymes are generally less stable, mainly due to their high sensitivity to oxygen. On the other hand, high-pressure reactors are required due to the poor solubility of H_2 in aqueous media. For example, a hydrogenase was used to regenerate NADPH for the asymmetric reduction of acetophenone catalyzed by an ADH. The TON value of the cofactor was 100, and the product (*S*)-1-phenylethanol was synthesized with a very high enantiomeric excess (*ee*) (>99.5%) (Mertens et al. 2003).

The most commonly used enzyme for the regeneration of oxidized nicotinamide cofactors is **NADH oxidase** (NOx). The enzyme catalyzes the oxidation of NADH by simultaneous reduction of O_2 to either H_2O_2 or directly to H_2O (Weckbecker et al. 2010). The enzyme is isolated from organisms such as *Streptococcus mutans*, *Streptococcus faecalis*, *Archaeoglobus fulgidus*, *Lactobacillus brevis*, *Lactobacillus sanfranciscensis*, and *Borrelia burgdorferi*. Because of the deactivation of the enzyme by H_2O_2, water-forming NADH oxidases are of high interest. For example, the NADH oxidase from

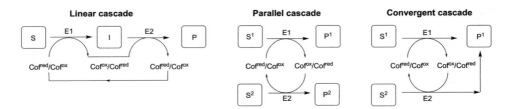

Fig. 13.3 Redox-neutral cascade reactions. S: Substrate, I: Intermediate, P: Product, E: Enzyme

Lactobacillus sanfranciscensis accepts NADH and NADPH. *L. sanfranciscensis* NOx and *L. brevis* NOx have been used for the deracemization of 1-phenylethanol to acetophenone and (*S*)-1-phenylethanol.

13.1.3 Self-Sufficient Cascades

The establishment of self-sufficient systems for redox cofactors requires the coupling of two synthetic reactions with opposite cofactor requirements (e.g., NAD(P)$^+$ and NAD(P)H) to form so-called "redox-neutral" cascades. Three redox-neutral cascade reactions are presented in the following sections: linear cascade, parallel cascade, and convergent cascade (Fig. 13.3; Kara et al. 2014).

In **linear cascades**, the product is formed starting from the substrate via the formation of an intermediate. In the present case, the intermediate is the product of the first enzyme reaction and, simultaneously, the substrate of the subsequent enzymatic reaction. These cascades exhibit high atomic efficiency because the maximum incorporation of the starting material into the target product can be achieved. A classic example of a linear cascade is the synthesis of ε-caprolactone by coupling an ADH-catalyzed oxidation of cyclohexanol with a Baeyer–Villiger monooxygenase (BVMO)-catalyzed oxidation of cyclohexanone (Fig. 13.4; Schmidt et al. 2015).

Parallel cascades couple the consumption of the two substrates and the synthesis of two or three products without forming an intermediate. An example is shown in Fig. 13.5 for the asymmetric reduction of a haloketone, coupled with the oxidation of racemic alcohol, for cofactor regeneration (corresponding to kinetic resolution). Here, the primary and regeneration reactions are catalyzed by the same enzyme ADH. Because of the thermodynamically stabilized product (Sect. 13.1.1), the entire reaction system is practically irreversible. In the end, two optically pure alcohols (*ee* > 99%) and an unreacted ketone as a by-product are obtained. High conversions (>80%) were achieved with a wide range of other substrates (Bisogno et al. 2009).

This parallel cascade system can also be used for the coupling of two different enzymes, e.g., Baeyer–Villiger monooxygenase (BVMO)-catalyzed asymmetric sulfoxidation in combination with the oxidation of racemic 2-octanol by an alcohol dehydrogenase. The main disadvantage of the parallel cascade systems is the separation of the different

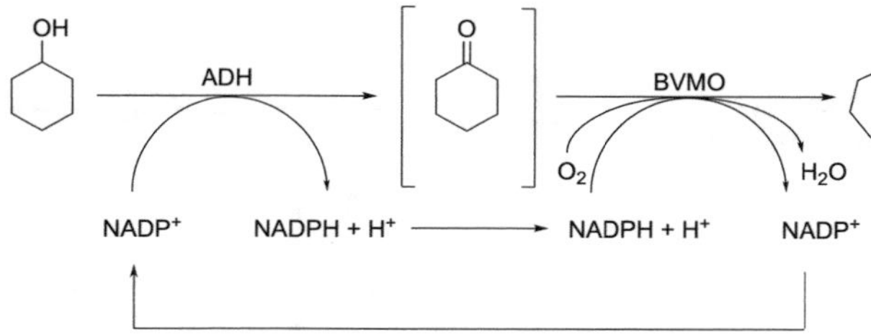

Fig. 13.4 Linear cascade for the synthesis of ε-caprolactone with alcohol dehydrogenase (ADH)-catalyzed oxidation of cyclohexanol and Baeyer–Villiger monooxygenase (BVMO)-catalyzed oxidation of the intermediate cyclohexanone

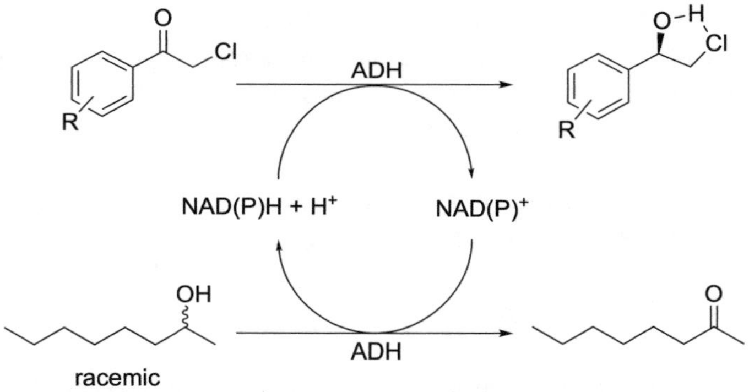

Fig. 13.5 Parallel cascade for the asymmetric reduction of α-haloketones coupled with 2-octanol for cofactor regeneration. ADH: alcohol dehydrogenase. R: substituent (adapted from Kara et al. 2013)

products, which can be complicated depending on the substance's physical-chemical properties.

Convergent cascades allow the maximum use of starting materials without forming one or more intermediates. An example of a convergent cascade is represented by the synthesis of ε-caprolactone, which couples a BVMO-catalyzed oxidation of cyclohexanone with an ADH-catalyzed oxidation of 1,6-hexanediol (for internal cofactor regeneration) coupling (Fig. 13.6; Bornadel et al. 2015). In the present case, 1,6-hexanediol represents a "double-smart cosubstrate" (Sect. 13.1.1).

On the one hand, using lactone-forming diol as a "smart cosubstrate" leads to a thermodynamically favored lactone by-product, which shifts the thermodynamic equilibrium towards the desired product. On the other hand, using 1,6-hexanediol as a "double-smart cosubstrate" allows the coupling of the BVMO-catalyzed oxidation of cyclohexanone (2 mol eq.) with the ADH-catalyzed oxidation of 1,6-hexanediol (1 mol eq.) to give the target compound ε-caprolactone (3 mol eq.). In this cascade, oxygen (2 mol eq.) is required as a cosubstrate, and only water (2 mol eq.) is formed as the by-product.

Fig. 13.6 Convergent cascade for the synthesis of ε-caprolactone. Combination of a BVMO-catalyzed oxidation of cyclohexanone and an ADH-catalyzed oxidation of 1,6-hexanediol (adapted from Bornadel et al. 2015)

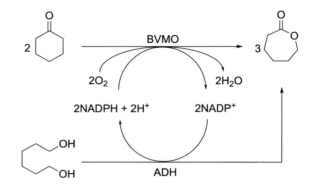

Questions

Q1: What are the basic forms of cofactor regeneration?

Q2: With the use of glucose dehydrogenase and format dehydrogenases comes an irreversible substep, which yields the equilibrium shift. Which are they?

Q3: What types of self-sufficient cascades exist.

13.2   Approaches to Substrate Dosing

13.2.1 Fed-Batch Processes

Biotransformations on a preparative scale are often carried out discontinuously in a *batch reactor* (Straathof and Adlercreutz 2000). For this purpose, the required substrates, cofactors, solvents, buffer salts, etc., are introduced into the reaction vessel at the beginning of the reaction. The desired reaction products are isolated from them after the reaction has taken place using a suitable method, e.g., by extraction. This procedure is not recommended for enzymatic reactions with potent substrate inhibition or toxicity or the occurrence of an undesired non-enzymatic side reaction of the substrate since this results in a significantly reduced productivity of the reaction system. Furthermore, the overall system yield decreases, and additional problems may occur during the workup.

Therefore, for such problems, a continuous substrate dosing via a so-called *fed-batch process* (Fig. 13.7) is often preferable. The substrate concentration controlled throughout the reaction minimizes undesirable effects, e.g., reduced reaction rates due to substrate inhibition/toxicity or side reactions.

The resulting increase in volume in the reaction vessel must also be considered in the feed process to prevent the reaction vessel from overflowing. The classic substrate dosing is usually carried out on a laboratory scale via pumps or valves, which, if necessary, adjust the feed rate computer-controlled according to the reaction control. In the simplest version, a dropping funnel or the incremental addition of the substrate is possible. Ideally, the reaction system can be started directly with an *in situ* product removal (ISPR) so that the desired target product can be continuously removed from the reaction medium (Sect. 13.3).

Fig. 13.7 Substrate dosing within a classical *fed-batch process*; PC-controlled dosing according to reaction control

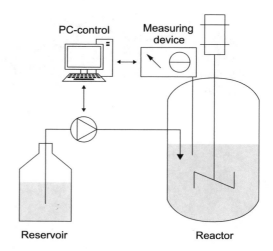

Similarly, the low solubility of a substrate in the reaction medium can be considered a continuous addition within a reaction solution. As an example, the d-hydantoinase-catalyzed hydrolysis of 5-(4-hydroxyphenyl)hydantoin (HPH) to N-carbamoyl-d-p-hydroxyphenylglycine (NCHPG) described by Lee and Kim (1998) can be given. The racemic substrate exists as a solid, a heterogeneous system, so to speak, during the reaction period in parallel with the d-hydantoinase reaction. Any consumed substrate is constantly replenished to the solution by constantly dissolving excess HPH to its maximum solubility. The product NCHPG possesses a significantly higher solubility. Eventually, it accumulates in the reaction mixture (Fig. 13.8). A conversion of 93% was achieved at 35 °C after a reaction time of 23 h (free enzyme) or 30 h (whole cell biotransformation). The reaction rate can be further increased with increasing temperature, mainly based on the increasing substrate solubility. However, at highly increased temperatures, a significant decrease in conversion was reported.

13.2.2 Use of Adsorber Materials

An alternative concept involves using adsorber materials that reversibly bind the required substrates and only release them into the reaction medium during the course of the reaction. The binding affinity and the capacity of the adsorber (corresponding to the adsorption isotherm) define the equilibrium between the adsorber and the reaction medium, which is established accordingly. In the ideal state, this behavior leads to a clear substrate excess in an adsorbed form on the solid phase and a significantly reduced substrate concentration in the reaction medium. The choice of a suitable adsorbent material is defined by the type of reactants, whereby silicates, zeolites, and inert ion exchange resins (e.g., Dowex, Amberlite, Duolite, etc.) are preferably used for enzymatic reactions with non-natural substrates. An example of the technical implementation of this process was presented by

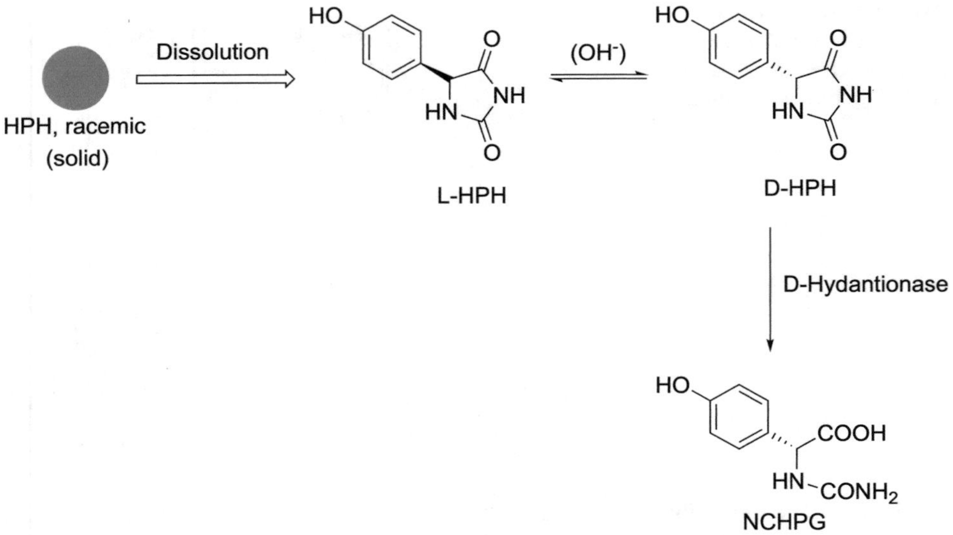

Fig. 13.8 Preparation of N-carbamoyl-d-p-hydroxyphenylglycine (NCHPG) with parallel post-dissolution of the substrate from the solid

Vicenzi et al. (1997) on a 300-L scale for the reduction of 3,4-methylene-dioxyphenylacetone (3,4-MDA) to (S)-3,4-methylene-dioxyphenylisopropanol (3,4-MDIPA) using *Zygosaccharomyces rouxii* (*Z. rouxii*).

In this reaction, both substrate and product have toxic effects on the microorganism used, so adsorption on the XAD-7 exchanger resin adjusts the concentration of both compounds in the reaction medium down to a maximum of 2 g L^{-1}. In total, 40 g L^{-1} was formally achieved in the heterogeneous reaction medium via adsorption on the exchanger resin. The new whole-cell biotransformation replaced a classical chemical synthesis and enabled to save 340 liters of solvents and avoided about 3 kg of chromium-containing waste per kilogram of the target product. The use of a classical two-phase system (Sects. 13.2.3 and 13.3.2) with a water-immiscible solvent, which in principle allows a similar distribution of the reaction compounds, was not successful due to the low extraction efficiency.

13.2.3 Multiphasic Systems

The use of a further solvent phase, analogous to the use of an adsorber, formally enables the supply of a substrate with a lower concentration into the reaction solution. As a rule, a buffer solution represents the reaction phase, and a water-immiscible organic solvent or an ionic liquid represents the extraction phase. The choice of organic solvent depends crucially on its compatibility with the biocatalyst, its toxicity, reactant capacity, and cost. Aliphatic alkanes, esters, ethers, and, less frequently, aromatics are primarily used. During the course of the reaction, the extraction phase contains a high concentration of the substrate, which is in equilibrium with the reaction phase according to the substrate's partition coefficient. After the reaction, the product is enriched similarly in the extraction phase, so that the desired product can be isolated from it after phase separation.

Of crucial relevance at this point are the activity and stability of the enzyme in the presence of the organic solvents, which must be determined accordingly. For whole-cell biotransformations, the effects of the organic solvents on the permeability of the cell walls must also be taken into account. Furthermore, within such a two-phase system, the substrate and product have different partition coefficients, which leads to changes in the concentration differences within the two phases. The corresponding differences can likewise be used for shifting the reaction equilibrium in the sense of an *in situ* product extraction.

13.2.4 Substrate Supply from the Gas Phase

Biotransformations can effectively be carried out in the gas phase as a solvent-free system. Using heterogeneous biocatalysts (e.g., immobilized enzymes) in the gas phase enables high turnovers and production rates, efficient transportation from the bulk phase, reduced diffusion limitations (due to low gas viscosity), and improved stability of enzymes. In addition, downstream processing is greatly simplified by the absence of a condensed solvent.

A related example is fermentation processes, which usually involve a direct continuous gassing of the reaction volume with oxygen, e.g., by air. The oxygen input here represents a continuous substrate input through the gas phase. This principle can also be transferred to other gaseous substrates present in the gaseous aggregate state under the reaction conditions, e.g., CO_2 and other substrates with a sufficiently low boiling point. During the reaction, the biocatalyst remains immobilized on a solid carrier, and the substrate flows over it (diluted in a carrier gas stream).

An example of this reaction control was presented by Ferloni et al. (2004) for the conversion of acetophenone to (R)-1-phenylethanol by alcohol dehydrogenase from *Lactobacillus brevis* (*LbADH*) on glass beads. The carrier gas nitrogen, which was enriched with the appropriate substrates, overflows the immobilized biocatalyst. The desired water activity (a_W) within the gas stream is adjusted here by the selective addition of (gaseous) water. Furthermore, the stability of the biocatalyst was significantly increased by co-immobilization with sucrose and space–time yields (STYs) of 107 g L^{-1} day^{-1}, and a TON value of about 700,000 was achieved. After the reaction is completed, the desired products can be condensed directly, which is an advantage over classical liquid reaction systems with downstream extraction and further isolation steps.

Questions

Q4: What are the forms of external substrate dosing?

Q5: What are the basic requirements for the secondary solvent in multiphase systems?

13.3   Approaches to Product Removal

For a successful *in situ* product separation directly from a reaction mixture, the product properties must be different from the properties of the other reaction partners. Differences in the physicochemical properties of the substances are particularly suitable for selective removal, e.g., the distillation of the product from the reaction mixture due to different boiling points. Unfortunately, such separation is only applicable to a few examples and predominantly involves the codistillation of other components, e.g., solvents and substrates. The resulting mixtures must then be separated in a further downstream-processing step and may need to be recycled.

13.3.1 Stripping of By-Products

A unique feature within the *in situ* product removal (ISPR) approaches is the selective removal of small, highly volatile coproducts, e.g., the removal of acetone from oxidoreductase-catalyzed reactions. The unfavorable thermodynamic driving force in substrate-coupled systems (Sect. 13.1.1) can be circumvented by removing the volatile component using stripping (Fig. 13.9).

Using the example of the enantioselective reduction of ketones by lyophilized *E. coli cells* overexpressing an ADH-A from *Rhodococcus ruber,* the productivity of the stripping procedure could be demonstrated. In this example, a continuous stream of air, saturated with water and isopropanol, is passed through the reaction mixture and continuously removes acetone from the reaction equilibrium (Fig. 13.9; Goldberg et al. 2006). In addition, the airflow causes improved mixing of the reaction components, eliminating the need for additional stirring/shaking. Applying the stripping process allowed conversions of up to >99% (Goldberg et al. 2006). As an improvement, elaborated microfabricated glass/silicon gas-liquid contactors, which use nitrogen as a purge gas to separate acetone, are suitable. Here, the gas-liquid interface is maintained by capillary action, and the separation efficiency increases with the increasing flow rate of the stripping gas.

13.3.2 Multiphasic Systems

Multiphasic systems can positively influence the concentration of the desired product. Here, the (partially) selective extraction of the product from the reaction medium is achieved using a different solvent system. In principle, the same solvent systems are

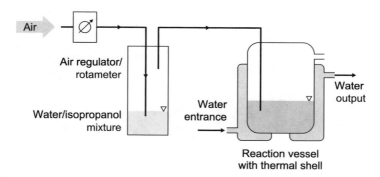

Fig. 13.9 Flow diagram of the stripping process. Compressed air is fed into the reaction solution after saturation with water and isopropanol and carries acetone out of the reaction vessel (adapted from Goldberg et al. 2006)

possible, which are used for the substrate dosing (Sect. 13.2.3). For this reason, in many applications, the continuous addition of the substrate and extraction of the product are combined in multiphasic systems. Overall, these two effects result in an improvement of the reaction yield and, thus, increased productivity. In addition, the toxicity of the substrate phase for the enzyme must be considered.

An unusual esterification in an aqueous-organic two-phase system was described by Duwensee et al. (2009) for the condensation of sebacic acid and 1,4-butanediol. The product (an oligomer) shows a high enrichment in the organic phase, which directly leads to a shift of the equilibrium towards the product. Finally, 40% higher product yields could be achieved compared to classical monophasic systems (Fig. 13.10). The *in situ* product extraction within the two-phase system also allows higher yields compared to the reactions in pure 1,4-butanediol since a potent substrate inhibition significantly reduces the productivity of the reaction system.

Another example of such a reaction system was investigated by von Langermann et al. (2007) for preparing acetophenone cyanohydrins, which have a very unfavorable equilibrium position in classical reaction systems. With an *in situ* product extraction, significantly higher reaction yields could be achieved, e.g., up to 36% for the substrate acetophenone. For the halogen-substituted acetophenone derivatives (2-fluoro-acetophenone), equilibrium conversions of up to 71% with an *ee* of >99% (*S*) could be achieved.

13.3.3 Direct Crystallization of the Product

Removing the desired product directly from a reaction medium utilizing crystallization is an alternative representing an exceptionally efficient variant of the ISPR strategy. The driving force of the separation is a significantly lower solubility of the target product compared to the other reactants, which leads to the product crystallizing spontaneously or after inoculating the reaction solution. This concept requires relatively similar process

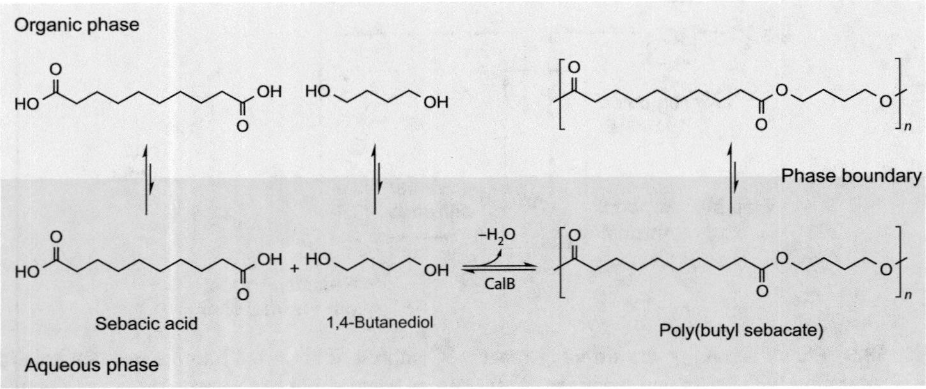

Fig. 13.10 Enzymatic polymerization with *in situ* extraction of the reaction product into an organic phase

conditions for both reaction and crystallization steps. Alternatively, the two sub-processes may be carried out spatially separately, provided that a simple change in environmental variables is possible. Likewise, another unit operation process, e.g., extraction, can be interposed. Nevertheless, only a few technical examples in the literature describe an efficient application of *in situ* product crystallization (ISPC). Possible applications were investigated by Buque-Taboada et al. (2005) using the synthesis of (6R)-dihydrooxoisophorone (DOIP) and discussed in terms of their applicability. The combination of a microbial reduction reaction and external crystallization proved to be fivefold more efficient and enabled yields of 85% with a selectivity of 98.7% (Fig. 13.11). This selective crystallization allows direct isolation of the target compound without additional downstream processing steps or chemicals. Overall, higher product purities could be achieved in addition to the improvement in catalyst stability.

The direct coupling of a biocatalytic reaction with an enantioselective crystallization allows the preparation of an enantiomerically pure compound from a racemic mixture. Using L-aspargine as an example, Würges et al. (2009) demonstrated dynamic kinetic resolution (DKR) via preferential crystallization and enzymatic racemization. The addition of enantiomerically pure seed crystals to a supersaturated solution results in the crystallization of the desired enantiomer on the offered surfaces with simultaneous depletion within the mother liquor before the undesired counter-enantiomer crystallizes out later as well. In contrast, adding amino acid racemase from *P. putida* KT2440 continuously resets the composition in the mother liquor to the racemic composition. This combination enabled high enantiomeric purities (≥92% *ee*), which can subsequently be enriched to enantiomerically pure l-aspargine (≥99.5% *ee*). Integrated racemization increased the yield significantly above the theoretical maximum yield of 50%. Unfortunately, the described method is only possible with conglomerate-forming compounds, which account for only 10% of all chiral systems.

Fig. 13.11 Coupling of a biocatalytic reaction with *in situ* product crystallization

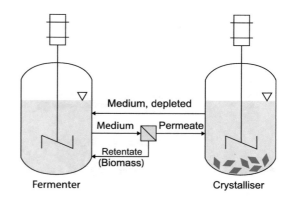

Medium, depleted

Medium Permeate

Retentate (Biomass)

Fermenter Crystalliser

13.3.4 Membrane Process

Enzyme Membrane Reactors (EMRs) are a commonly used reactor design for implementing efficient biocatalytic processes, including their use on larger scales. The continuous extraction of a product from an enzymatic reaction allows efficient use of the biocatalyst while maintaining high yields within a continuous process. In this concept, the enzyme is retained behind a membrane, whereas the products are selectively removed through the membrane. The enzyme can circulate here in dissolved or immobilized form or can be immobilized on the membrane itself. Membrane fouling and the unavoidable loss of enzyme activity over a more extended reaction time are potential limitations of this method. Therefore, efficient process control with optimized membrane materials, backflushing, etc., is significant for continuous product synthesis.

An example of a successful application of an EMR was shown by Stillger et al. (2006) using benzaldehyde lyase (BAL) from *Pseudomonas fluorescens.* The reaction studied resulted in the synthesis of 2-hydroxypropiophenone (2-HPP) from benzaldehyde and acetaldehyde (Stillger et al. 2006). Unfavorably, side reactions occur here, which have a negative effect on the desired product formation. In particular, the intermediate precipitation of benzoin is critical for the performance of the reaction. Within the study, the corresponding reaction kinetics were therefore investigated, and the benzoin synthesis was suppressed in favor of the synthesis of target product 2-HPP.

Within the EMR system, the overall reaction was optimized with respect to the differences in reaction kinetics starting from benzaldehyde, thus significantly reducing the benzoin concentration within the solution. Continuous deactivation of the biocatalyst slowly reduced the overall yield of the reaction, but STYs of up to 1120 g L^{-1} day^{-1} were still obtained with an *ee* of >99% (*R*). Therefore, the use of an EMR system here corresponds to a significant improvement over classical batch reactors.

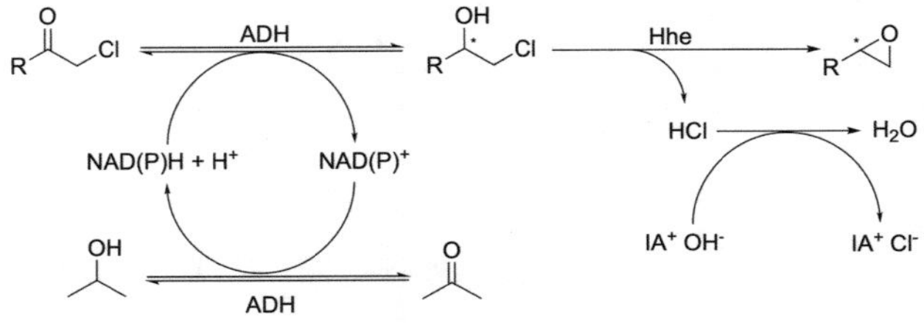

R = CH$_2$OPh, n-C$_6$H$_{13,}$ Ph, CH$_2$CO$_2$Me

Fig. 13.12 Use of ion exchangers (IA$^+$) in a cascade reaction for the preparation of epoxides

13.3.5 Use of Ion Exchange Resins

In addition to the above-listed methods to selectively remove the product (direct separation via the gas phase, secondary liquid phases, and direct crystallization of the product), the addition of solid exchange resins has become established for biocatalytic reactions. Here, the product is removed from the reaction mixture in an analogous manner, as mentioned above, whereby the equilibrium is also shifted to the product side. For example, Schrittwieser et al. (2009) reported the equilibrium shift in a biocatalytic cascade reaction by using an anion exchanger to improve the asymmetric bioreduction of a prochiral α-chloroketone to the corresponding β-chlorohydrin and, building on this, to the corresponding epoxide by halohydrin dehalogenase (Hhe) from *Mycobacterium* sp. (Fig. 13.12).

The unfavorable equilibrium position of the reversible epoxide ring closure was significantly increased by the removal of HCl to 93% conversion with respect to 1,2-epoxy-3-phenoxypropane after 24 h. The removal of HCl by the anion exchanger represents the crucial thermodynamically favored step here. In contrast, an alternative removal of chloride ions by precipitation with Ag$^+$ salts was not possible due to the simultaneous deactivation of halohydrin dehalogenase.

Questions
Q6: What are the advantages of removing a product directly from the reaction phase?
Q7: What is the basic requirement for the stripping of by-products?
Q8: What defines the amount of removed product in multiphase systems?

13.4   Approaches to Deracemization

Deracemization means the conversion of a racemic mixture (50% (*S*)- and 50% (*R*)-enantiomer) into an optically pure product (*ee* > 99.9%) with a maximum yield of theoretically 100%. Due to the high added value, deracemization approaches using

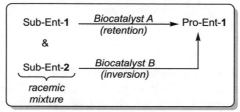

Fig. 13.13 Three basic deracemization methods. Sub: Substrate, Pro: Product, and Ent: Enantiomer

biocatalysts have attracted considerable attention, as very valuable enantiomerically pure products can be synthesized from readily available and inexpensive racemic substrates.

Basically, these methods can be divided into kinetic resolution, dynamic kinetic resolution, and enantioconvergent processes (Fig. 13.13).

Within a kinetic resolution, one enantiomer is preferably converted, which limits the maximum yield per enantiomer to 50%. This limitation can be overcome by a dynamic kinetic resolution using a racemization reaction that converts the remaining enantiomer back to the starting racemic mixture. In contrast, enantioconvergent processes require two biocatalysts, each of which produces an identical enantiomerically pure product.

This section describes established methods that have been developed for the deracemization of racemic compounds. Besides the classical kinetic resolution, often carried out with lipases, methods with a dynamic kinetic resolution (DKR) have been established. This includes, to a large extent, the synthesis of enantiomerically pure alcohols, carboxylic acids, amines, and amino acids (Turner 2010). For example, Haak et al. (2008) developed a DKR system for the synthesis of optically pure epoxides starting from a racemic halohydrin. In this, the racemization of the (*S*)-alcohol was catalyzed *in situ* by an iridium complex, and the subsequent epoxidation was catalyzed by a haloalcohol dehalogenase (HheC; Fig. 13.14, top) (Haak et al. 2008). A series of optically pure (*R*)-epoxides have been synthesized by using this chemo-enzymatic cascade in a single step.

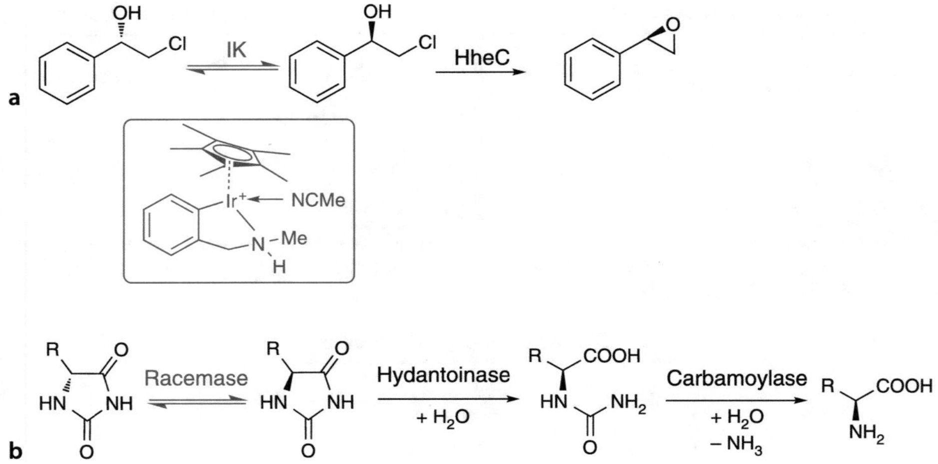

a

b

Fig. 13.14 (**a**) Dynamic kinetic racemate resolution (DKR) of a racemic halohydrin by haloalcohol dehydrogenase (HheC) in combination with an iridium catalyst (IK) for the racemization of the unwanted *(S)*-alcohol. (**b**) Synthesis of amino acids using racemase, hydantoinase, and carbamoylase

Evonik has also established a coupled cascade of racemase, hydantoinase, and carbamoylase for the synthesis of natural and non-natural L-amino acids (Fig. 13.14 below) on a technical scale.

Questions
Q9: What is the maximum yield of a simple kinetic resolution?
Q10: What is the difference between a kinetic resolution and a dynamic kinetic resolution?

13.5 Conclusion and Prospects

The approaches listed above describe various possible integrated process concepts for their use in biocatalytic reactions. The examples clearly describe that an intelligent interconnection of reactions and separation processes enables a significant increase in the productivity and selectivity of enzymatic synthesis routes. Future applications will also be more based on biocatalytic cascade reactions for preparing synthetically relevant complex chemical compounds. Here, whole-cell biotransformations and (partially) purified enzymes will be used for the respective cascade reactions. In addition to the classical coupling of two or more biocatalytic partial reactions ("bio-bio",) approaches involving chemical reactions ("bio-chemo") are also possible. The process conditions of the biocatalytic reaction must, of course, be compatible with the other reactions to enable effective coupling of individual reaction steps. Cascade reactions with opposing requirements on the reaction conditions can be combined by compartmentalization. Polymer-based inclusion immobilization, membrane processes, or cell-like process concepts can provide optimal conditions for the biocatalysts within the respective reaction environment.

The production of specific adsorber materials by 3D printing could enable significant advantages in the corresponding reaction concepts. In addition, using novel 3D printing techniques will enable the realization of reaction-specific reactors. This primarily includes microreactor concepts but can also be used for tailor-made classical reactors.

In addition, water as a reaction medium usually has a lower solubility for hydrophobic reactants, might lead to undesirable side reactions, may require complex processing steps, possibly triggers enzyme inhibition issues due to substrates and products dissolved in water, possibly leads to water-induced enzyme denaturation, and favors microbial contamination. For this reason, non-conventional media represent an essential alternative to aqueous reaction systems and are a current focus of research. Related examples of the use of non-conventional media are given in Chap. 12.

Take Home Message

The specific reaction control of biocatalytic reactions via integrated process concepts enables the efficient synthesis of the desired product on a preparative scale. This primarily involves overcoming process limitations that cannot be achieved by simply improving the biocatalyst itself, e.g., unfavorable equilibrium positions. For this purpose, the above-shown techniques are suitable: cofactor regeneration in oxidoreductases, efficient substrate dosing and product removal, and techniques in kinetic resolution of chiral compounds.

Answers

A1: Substrate-coupled and enzyme-coupled cofactor regeneration and additionally self-sufficient cascades are possible.

A2: Glucose dehydrogenase converts D-glucose to D-glucono-1,5-lactone, which spontaneously hydrolyzes with water to D-gluconic acid, removing the lactone from its original equilibrium. Formate dehydrogenase converts formate to CO_2, which volatilizes rapidly and thus is analogously removed from equilibrium.

A3: Linear, parallel, and convergent cascades exist for self-sufficient cascades.

A4: In the simplest form, substrate dosing is carried out via an external pump (liquid or gases). In an identical fashion, secondary phases (adsorbers, multiphase systems, etc.) can be used, which continuously deliver the substrate to the reaction phase.

A5: The additional solvent should be cheap, nontoxic, and compatible with the biocatalyst and should have a high capacity.

A6: An *in situ* product removal primarily enables the prevention of product inhibition and, in some circumstances, higher yields in equilibrium-limited reaction systems.

A7: The by-product should have a high vapor pressure/ low boiling point.

A8: The distribution between the reaction phase and secondary phase is determined by the partition coefficient. The coefficient itself is dependent on the chosen environmental parameters, e.g., temperature.

A9: In an ideal case, 50% yield for the substrate and 50% yield for the product.

A10: In both forms, one of the substrate enantiomers is preferentially converted, but within the dynamic kinetic resolution, the chiral starting material is continuously racemized. This enables product yields beyond 50%.

References

Bisogno FR, Lavandera I, Kroutil W, Gotor V. Tandem concurrent processes: one-pot single-catalyst biohydrogen transfer for the simultaneous preparation of enantiopure secondary alcohols. J Org Chem. 2009;74(4):1730–2. https://doi.org/10.1021/jo802350f.

Bornadel A, Hatti-Kaul R, Hollman F, Kara S. A bi-enzymatic convergent cascade for e-caprolactone synthesis employing 1,6-hexanediol as a 'double-smart cosubstrate'. ChemCatChem. 2015; https://doi.org/10.1002/cctc.201500511r1.

Buque-Taboada EM, Straathof AJJ, Heijnen JJ, Van Der Wielen LAM. Microbial reduction and in situ product crystallization coupled with biocatalyst cultivation during the synthesis of 6R-dihydrooxoisophorone. Adv Synth Catal. 2005;347(7–8):1147–54. https://doi.org/10.1002/adsc.200505024.

Duwensee J, Wenda S, Ruth W, Kragl U. Lipase-catalyzed polycondensation in water: a new approach for polyester synthesis. Org Process Res Dev. 2009;14(1):48–57.

Ferloni C, Heinemann M, Hummel W, Daussmann T, Buchs J. Optimization of enzymatic gas-phase reactions by increasing the long-term stability of the catalyst. Biotechnol Prog. 2004;20(3):975–8. https://doi.org/10.1021/bp0334334e.

Goldberg K, Edegger K, Kroutil W, Liese A. Overcoming the thermodynamic limitation in asymmetric hydrogen transfer reactions catalyzed by whole cells. Biotechnol Bioeng. 2006;95(1): 192–8. https://doi.org/10.1002/bit.21014.

Haak RM, Berthiol F, Jerphagnon T, Gayet AJA, Tarabiono C, Postema CP, Ritleng V, Pfeffer M, Janssen DB, Minnaard AJ, Feringa BL, de Vries JG. Dynamic kinetic resolution of racemic β-haloalcohols: direct access to enantioenriched epoxides. J Am Chem Soc. 2008;130(41): 13508–9. https://doi.org/10.1021/ja805128x.

Kara S, Schrittwieser JH, Hollmann F, Ansorge-Schumacher MB. Recent trends and novel concepts in cofactor-dependent biotransformations. Appl Microbiol Biotechnol. 2014;98(4):1517–29. https://doi.org/10.1007/s00253-013-5441-5.

Kara S, Spickermann D, Schrittwieser JH, Leggewie C, van Berkel WJH, Arends IWCE, Hollmann F. More efficient redox biocatalysis by utilising 1,4-butanediol as a 'smart cosubstrate'. Green Chem. 2013;15(2):330–5. https://doi.org/10.1039/c2gc36797a.

Lavandera IN, Kern A, Resch V, Ferreira-Silva B, Glieder A, Fabian WMF, de Wildeman S, Kroutil W. One-way biohydrogen transfer for oxidation ofsec-alcohols. Org Lett. 2008;10(11):2155–8. https://doi.org/10.1021/ol800549f.

Lee DC, Kim HS. Optimization of a heterogeneous reaction system for the production of optically active D-amino acids using thermostable D-hydantoinase. Biotechnol Bioeng. 1998;60(6): 729–38. https://doi.org/10.1002/(sici)1097-0290(19981220)60:6%3c729::aid-bit9%3e3.0.co;2-g.

Mertens R, Greiner L, van den Ban ECD, Haaker HBCM, Liese A. Practical applications of hydrogenase I from Pyrococcus furiosus for NADPH generation and regeneration. J Mol Catal B Enzyme. 2003;24–25:39–52. https://doi.org/10.1016/s1381-1177(03)00071-7.

Paul CE, Arends IWCE, Hollmann F. Is simpler better? Synthetic nicotinamide cofactor analogues for redox chemistry. ACS Catal. 2014;4(3):788–97. https://doi.org/10.1021/cs4011056.

Schmidt S, Scherkus C, Muschiol J, Menyes U, Winkler T, Hummel W, Gröger H, Liese A, Herz H-G, Bornscheuer UT. An enzyme cascade synthesis of ε-caprolactone and its oligomers. Angew Chem Int Ed. 2015;54(9):2784–7. https://doi.org/10.1002/anie.201410633.

Schrittwieser JH, Lavandera I, Seisser B, Mautner B, Spelberg JHL, Kroutil W. Shifting the equilibrium of a biocatalytic cascade synthesis to enantiopure epoxides using anion exchangers. Tetrahedron Asymmetry. 2009;20(4):483–8. https://doi.org/10.1016/j.tetasy.2009.02.035.

Stillger T, Pohl M, Wandrey C, Liese A. Reaction engineering of benzaldehyde lyase from Pseudomonas fluorescens catalyzing enantioselective C-C bond formation. Org Process Res Dev. 2006;10(6):1172–7. https://doi.org/10.1021/op0601316.

Straathof AJJ, Adlercreutz P. Applied biocatalysis. Boca Raton: CRC Press; 2000.

Turner NJ. Deracemisation methods. Curr Opin Chem Biol. 2010;14(2):115–21. https://doi.org/10.1016/j.cbpa.2009.11.027.

Vicenzi JT, Zmijewski MJ, Reinhard MR, Landen BE, Muth WL, Marler PG. Large-scale stereoselective enzymatic ketone reduction with in situ product removal via polymeric adsorbent resins. Enzym Microb Technol. 1997;20(7):494–9. https://doi.org/10.1016/s0141-0229(96)00177-9.

von Langermann J, Mell A, Paetzold E, Daußmann T, Kragl U. Hydroxynitrile lyase in organic solvent-free systems to overcome thermodynamic limitations. Adv Synth Catal. 2007;349(8–9):1418–24. https://doi.org/10.1002/adsc.200700016.

Weckbecker A, Gröger H, Hummel W. Regeneration of nicotinamide coenzymes: principles and applications for the synthesis of chiral compounds. Biosyst Eng I. 2010;120:195–242. https://doi.org/10.1007/10_2009_55.

Würges K, Petrusevska-Seebach K, Elsner MP, Lütz S. Enzyme-assisted physicochemical enantioseparation processes—part III: overcoming yield limitations by dynamic kinetic resolution of asparagine via preferential crystallization and enzymatic racemization. Biotechnol Bioeng. 2009;104(6):1235–9.

Enzymes in the Chemical and Pharmaceutical Industry

14

Jenny Schwarz, Jan Volmer, and Stephan Lütz

What You Will Learn in This Chapter

Enzymes are used today in the chemical and pharmaceutical industry to produce a wide range of products. In this chapter, you will be introduced to product categories like bulk chemicals, fine chemicals, or active pharmaceutical ingredients (APIs) and some areas in which enzymes can offer benefits over existing production processes, like reduction of waste or higher product purity. In addition to these production processes, in which the enzyme acts as a catalyst for a particular reaction, you will learn that in the pharmaceutical industry enzymes are studied as well for their biological functions. These functions may range from being a target of a drug, i.e., an enzyme that should be inhibited, to being involved in the metabolism of a drug to being a drug itself. Overall, you will get to know the historical and present-day applications of enzymes in these two industries, as well as some recent developments like synthetic biology.

The application of enzymes for the synthesis and production of chemicals has a long history. In this field, discoveries and applications on a laboratory scale and the implementation in industrial processes have cross-fertilized each other and contributed overall to the development of the field of biocatalysis. Nowadays, enzymes are not only established tools in organic synthesis but are also used as efficient catalysts in industrial processes. The

J. Schwarz · J. Volmer · S. Lütz (✉)

Bioprocess Engineering, Bio- and Chemical Engineering, TU Dortmund, Dortmund, Germany
e-mail: jenny.schwarz@tu-dortmund.de; jan.volmer@tu-dortmund.de;
stephan.luetz@tu-dortmund.de

© The Author(s), under exclusive license to Springer Nature Switzerland AG 2024
K.-E. Jaeger et al. (eds.), *Introduction to Enzyme Technology*, Learning Materials in Biosciences,
https://doi.org/10.1007/978-3-031-42999-6_14

driving force behind this development of biocatalysis has always been the development of products and processes that had economic advantages over existing chemical processes, e.g., through shorter synthesis routes or better product properties that saved purification steps.

A whole range of enzymatic processes have thus become state of the art in both the chemical and pharmaceutical industries. Enzymes are now used to produce a wide variety of substances, ranging from basic chemicals with high annual production and compara- tively low price (Sect. 14.2.3.1) to chiral building blocks (Sect. 14.2.1.1) and active pharmaceutical ingredient (API) syntheses (Sect. 14.3.2.1) with low tonnage but high added value. It speaks for enzymes as catalysts that they can be used over the entire range of the different requirements of these processes. For example, the production of basic chemicals requires extremely high yields and high cost efficiency, while pharmaceutical products involve a highly documented and often multi-step manufacturing process that is monitored by the relevant authorities and must, of course, be equally economical. The number of excipients used and the total amount of waste produced in the process can be calculated, for example, with the E-factor (amount of waste produced per amount of product) (Sheldon 2008). The amount of waste varies significantly depending on the product group (Table 14.1).

Against the background of the United Nation's sustainable development goals (SDGs), the conservation of resources, and the biologization of industries, enzymatic processes are seen as having a high potential for achieving industrial processes with lower and less problematic waste streams. This chapter will outline how, based on historical knowledge and modern methods of reaction engineering and molecular biology (Chap. 8), enzymes are used in the chemical and pharmaceutical industries. Depending on the process control and the complexity of the reaction, enzymes can be used either in (partly) isolated form (enzymatic biotransformation), intracellularly in living, resting, or dead cells (whole-cell biotransformation) or as an enzyme network for multi-step reactions in growing cells (fermentation). Here, the actual value-added reaction is linked to the growth and energy metabolism of the microorganism. It should be noted that fermentation is also necessary for the first and second cases to produce the enzymes and the cells, respectively.

Applications of Enzymes in the Chemical and Pharmaceutical Industry
1. Single enzyme—from crude cell extract to purified, isolated enzyme preparation
2. Whole-cell biotransformation—individual reaction steps by enzymes in living, resting, or dead cells
3. Fermentation—multi-step reaction of enzyme networks in growing cells

Q1: What is the difference between an enzymatic biotransformation and a whole-cell biotransformation?

Table 14.1 Classification of chemical and pharmaceutical products and their E-factors (Sheldon 2008)

Product group	Production volume (t a^{-1})	E-Factor (kg$_{Waste}$ kg$_{Product}$$^{-1}$)
Petrochemicals	10^6–10^8	<0.1
Basic chemicals	10^4–10^6	<1–5
Fine chemicals	10^2–10^4	5–>50
Pharmaceutical products	10–10^3	25–>100

14.1 Origins of Enzyme Use in the Chemical and Pharmaceutical Industry

At the beginning of the nineteenth century, the first microbiological processes were introduced in the chemical and pharmaceutical industries. Due to inexpensive agricultural raw materials, the first biotechnological processes were mainly used for the production of basic chemicals such as lactic, butyric, and acetic acid, as well as ethanol (a. o. Weizmann Process; Weizmann 1919). After the collapse of agricultural production in the wake of World War I, biotechnology was increasingly used to produce complex specialty products, such as optically active drugs or high-molecular-weight compounds, whose synthesis was chemically impossible (Marschall 2000). During the World Wars, some normally uneconomical processes, such as the protol process, were used due to shortages of raw materials. This process is based on the shift of the equilibrium of the alcoholic fermentation by *Saccharomyces cerevisiae* in the direction of glycerol, which is normally formed only in small proportions. By adding sodium sulfite, acetaldehyde, as a direct precursor of ethanol, is captured so that it can no longer act as a hydrogen acceptor for the reduced NADH$_2$. This subsequently transfers the hydrogen to dihydroxyacetone phosphate, a precursor of glycerol. This fermentatively produced glycerol was primarily used for dynamite production.

One of the first and still operational industrial applications of enzymes for chemical synthesis is the decarboxylation of pyruvate to acetaldehyde, followed by condensation on benzaldehyde by the enzyme pyruvate decarboxylase from the baker's yeast *Saccharomyces cerevisiae* (Fig. 14.1a). The reaction was described by Neuberg and Hirsch in 1921 and commercialized by Knoll A.G. (later BASF AG, as of 2015 Siegfried Holding) from 1930 onwards. The optically active product *(R)*-phenylacetylcarbinol is then chemically converted to L-(−)-ephedrine. The process is remarkable because although Buchner had already shown in 1897 that yeast cells ground up with sand are still able to carry out alcoholic fermentation, and thus the knowledge was available that resting cells or even cell extracts can be used for chemical conversions, fundamental microbial metabolic pathways such as glycolysis and the citrate cycle were not fully elucidated until the 1930s.

Almost at the same time, in 1923, the bacterium *Acetobacter suboxydans* was isolated. Because of its ability to carry out incomplete oxidations, it was used from 1934 in the Reichstein-Grüssner synthesis for the production of vitamin C (L-ascorbic acid)

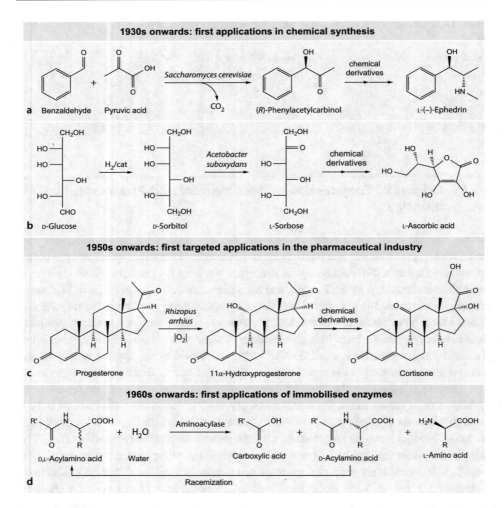

Fig. 14.1 Important development steps in the implementation of biotechnological processes in industrial chemical synthesis. All processes are still used industrially, sometimes in modified form. (**a**) Condensation of benzaldehyde and pyruvate to *(R)*-phenylacetylcarbinol, an ephedrine precursor, with the aid of pyruvate decarboxylase from *S. cerevisiae*. (**b**) Reichstein-Grüssner synthesis, oxidation of D-sorbitol to L-sorbose, a vitamin C precursor, with the aid of *Acetobacter suboxydans*. (**c**) Hydroxylation of progesterone to 11α-hydroxyprogesterone, a precursor of the hormone cortisone, using *Rhizopus arrhius*. (**d**) Hydrolytic cleavage of a racemic mixture of *N-acetylated amino acids* to yield L-amino acids using aminoacylases

(Fig. 14.1b). In this process, D-glucose is first chemocatalytically reduced to D-sorbitol, which serves as a substrate for biocatalytic oxidation to L-sorbose. L-Sorbose is chemically converted to L-ascorbic acid via diacetone-L-sorbose, diacetone-2-keto-L-gulonic acid, and 2-keto-L-gulonic acid (Reichstein and Grüssner 1934). In a slightly modified form, large-scale vitamin C production is still based on this original process. In the meantime, it is also possible to produce 2-keto-L-gulonic acid fermentatively on the basis of glucose (Genencor-Eastman process). This process combines the four necessary enzymes in recombinant *Escherichia coli cells* (Bommarius and Riebel 2004).

An important process in the pharmaceutical industry is based on the discovery by Peterson et al. in the 1950s that *Rhizopus arrhius* is able to selectively oxidize steroids. Among other things, this ability could be used to hydroxylate progesterone specifically at the C11 atom of the steroid skeleton to 11α-hydroxyprogesterone, a precursor of cortisone (Fig. 14.1c; Peterson et al. 1952). The previously used chemical synthesis, which was extremely uneconomical due to more than 30 synthetic steps and a substrate input of 615 kg of deoxycholic acid per kilogram of cortisone (cf. E factor, Table 14.1), could thus be replaced by an economical, combined process consisting of a total of 11 chemical and biocatalyzed steps.

Due to a lack of fundamental knowledge about enzymes, the first biotechnological processes in the chemical and pharmaceutical industries were mainly based on the use of whole cells. In 1954, the first technical application of an isolated enzyme was carried out on an industrial scale in Japan at Tanabe Seiyaku for the production of L-amino acids using an aminoacylase. In this process, L-amino acids are obtained by hydrolytic cleavage from a racemic mixture of N-acetylated amino acids (Fig. 14.1d). The remaining N-acetylated D-amino acids are separated via ion exchangers or crystallization and, after racemization, are fed to a new hydrolysis. For cost reasons, the aminoacylase from *Aspergillus oryzae* was immobilized on DEAE-Sephadex beads in fixed-bed reactors from 1969 onwards. This was the first industrial application of immobilized enzymes. From 1982, membrane reactors were used at Degussa for enzyme retention (Liese et al. 2006).

With the development and introduction of genetic engineering methods in the 1980s, the use of biocatalysts increasingly developed into a competitive technology. This development was supported by the discovery that enzyme catalysis could in part also be carried out under non-physiological conditions, e.g., at high temperatures or in organic solvents, which led to increased acceptance in organic chemistry. In addition, the FDA, among other things in the context of the thalidomide scandal, tightened the approval guidelines for the use of racemic mixtures to the effect that both enantiomers must be tested separately with regard to their pharmacological activity. Therefore, biocatalysis is now often the method of choice due to its enantioselectivity, especially in the stereoselective synthesis of compounds that are chemically difficult to access. Moreover, not only substrates structurally similar to the product but also inexpensive carbon sources can be used to make products accessible by fermentation. The steady advancement of molecular and genetic engineering methods, as well as the introduction of bioinformatics, now make it possible not only to produce almost any known enzyme recombinantly in easily cultivated microorganisms but also to adapt it to the appropriate process conditions by rational and evolutionary design. The steadily growing DNA, protein, and metabolome databases also contribute to this, making it possible, in combination with high-throughput screening methods and increasingly inexpensive artificial gene syntheses, to also apply tailor-made multi-enzyme complexes (Syldatk et al. 2001).

Q2: Name three microorganisms that were used in whole-cell biotransformations from 1930 onwards!

Q3: Which unphysiological conditions might be encountered by enzymes in chemical production processes?

14.2   Enzymes in the Chemical Industry

The chemical industry has been using enzymes in its processes for several decades (Sect. 14.1). The use of enzymes is particularly attractive when the enzymatic reaction has advantages over the chemical alternative. For example, due to their properties as highly selective biocatalysts with respect to substrate specificity as well as diastereo-, regio-, and enantioselectivity, products of higher purity can be obtained by the use of enzymes. In addition, enzymes often offer ecological and economic advantages, which are becoming increasingly important in the course of "green chemistry." For example, enzymes have a high environmental compatibility because they are obtained from renewable raw materials and are biocompatible, biodegradable, and non-toxic. In addition, the requirements for the reaction conditions are generally moderate (neutral pH values, moderate temperatures, and pressures), which reduces the use of organic solvents and metal catalysts and thus the amount of waste (see E-factor, Table 14.1), as well as the energy requirement accordingly. Since enzymes are difficult to recover from aqueous solutions, e.g., by ultrafiltration, they must either be produced very cheaply for single use or made accessible for reuse via retention devices such as membranes (Chap. 10) or immobilization (Chap. 11). This often makes production and recovery expensive and laborious. Lack of stability and inhibitory effects at high substrate and product concentrations, sometimes low reaction rates and high substrate specificity, therefore often require time-consuming and costly biocatalyst development. Especially in rapid process developments, chemical catalysis is therefore often still superior to biocatalysis.

The total market for industrially applied enzymes (US$4.6 billion in 2016) is not yet very large compared to the biotherapeutics market (US$192.2 billion in 2016). However, relative to the total catalyst market (US$17.1 billion, 2014), biocatalysts already account for more than a quarter of sales, while the corresponding share of biopharmaceuticals is around 20% of the global pharmaceutical market. What both markets have in common is that they are forecast to grow strongly in the coming years.

Although an enzyme catalyzing the corresponding reaction is now known for almost every reaction in organic chemistry, hydrolases play the most important role among industrially applied enzymes, ahead of lyases and transferases. This preferential role is due to the easy handling of hydrolases. They do not require cofactors, have high stability in organic solvents and in lyophilized form, and possess a broad substrate spectrum.

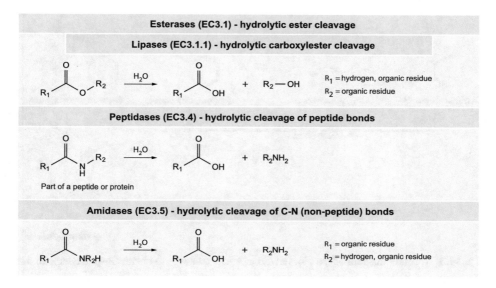

Fig. 14.2 Selection of industrially important hydrolytic enzyme families and their reactions

Depending on the type of hydrolytically cleaved bond, the hydrolases (EC 3) are divided into further enzyme families (Fig. 14.2).

In 2002, the number of biotechnological processes was already well over 100, with a strong upward trend. Therefore, only a few selected biotransformations of different enzyme classes can be presented in the following, which are of particular importance, e.g., due to their production scale or other remarkable properties. For a detailed overview of industrially applied biotransformations, the textbook "Industrial Biotransformations" by Liese et al. (2006) is recommended.

14.2.1 Lipase (EC 3.1.1.3)

14.2.1.1 Preparation of *(R)*-Phenylethylamine

The preparation of enantiomerically pure *(R)*-phenylethylamine is based on the catalytic promiscuity of a lipase from *Burkholderia plantarii*. This process is worth noting because lipases normally catalyze the hydrolysis of carboxyl esters (Fig. 14.2), but the triacylglycerol lipase from *Burkholderia plantarii* is promiscuously capable of enantioselectively acylating racemic amines by having the amine, rather than water, which acts as a nucleophile (Grunwald 2015). This capability has been applied on a scale >1000 t a^{-1} since 1993 at BASF AG for the resolution of racemic amines (Liese et al. 2006). Here, the lipase immobilized on polyacrylate is used for the acetylation of racemic 1-phenylethylamine with ethyl methoxyacetate (Fig. 14.3).

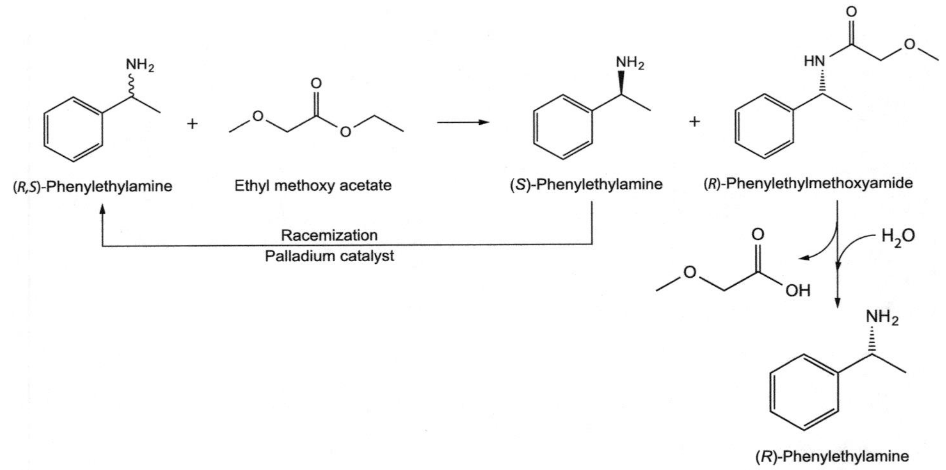

Fig. 14.3 Reaction diagram of the biocatalytic racemate cleavage to obtain *(R)*-phenylethylamine

The reaction is carried out in *tert*-methyl butyl ether (MTBE) in a continuous plug-flow reactor (PFR), which allows high substrate concentrations. Deactivation of the lipase can be prevented by freeze-drying in the presence of fatty acids.

A common problem with the resolution of racemates is that the yield is limited to 50% since by definition a racemate consists of equal parts of the two optical antipodes. However, unreacted *(S)*-phenylethylamine can be reracemized on palladium catalysts, after distillative separation, and recycled to the reaction so that a complete conversion of the racemic mixture can be achieved. The *(R)*-phenylethylmethoxyamide formed is converted to *(R)*-phenylethylamine in a simple hydrolysis. Chiral amines are important building blocks in the production of pharmaceuticals (Grunwald 2015).

Q4: Which change in the reagents allows the lipase to perform an acylation reaction rather than a hydrolysis reaction in the production of enantiopure *(R)*-phenylethylamine?

14.2.2 D-Hydantoinase (EC 3.5.2.2)

14.2.2.1 Preparation of D-*p*-Hydroxyphenylglycine

Another example of overcoming the yield threshold in racemate cleavage reactions is the production of D-*p*-hydroxyphenylglycine with the aid of 5,6-dihydropyridine amidohydrolase in immobilized *Bacillus brevis* cells. From a racemic mixture of D,L-5-(*p*-hydroxyphenyl)-hydantoin obtained by Mannich condensation of phenol, glyoxylic acid, and urea, the D-enantiomer is converted by D-hydantoinase to D-N-carbamoyl-*p*-hydroxyphenylglycine (Fig. 14.4; Liese et al. 2006).

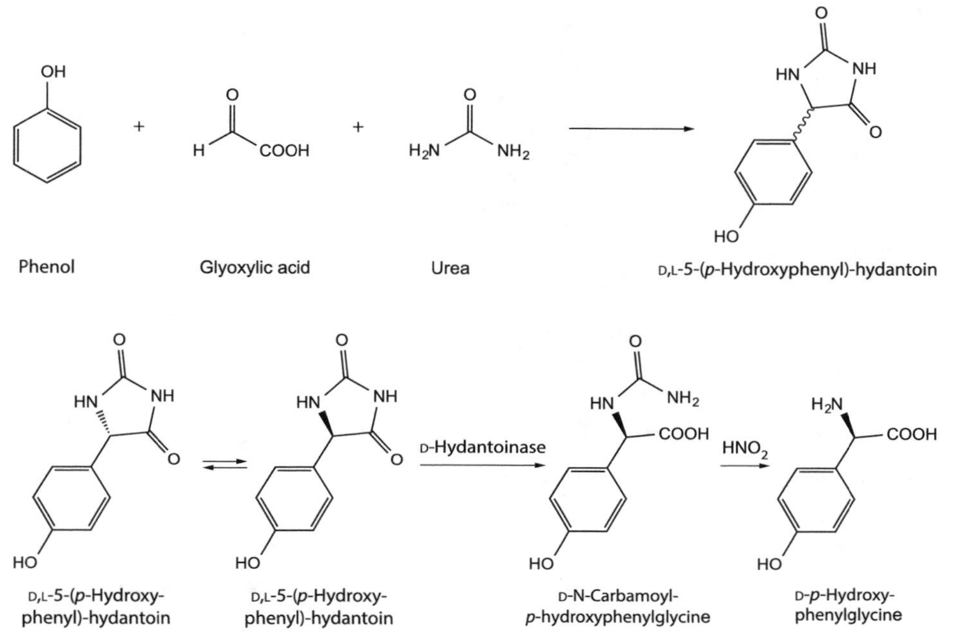

Fig. 14.4 Reaction scheme of the enzymatic racemate cleavage for the stereospecific recovery of D-p-hydroxyphenylglycine

Under the aqueous conditions of the enzymatic hydrolysis, a continuous racemization of the L-5-(p-hydroxyphenyl)-hydantoin takes place so that a quantitative conversion is also achieved here. Cleavage of the carbamoyl group, either enzymatically using a carbamoylase or chemically with sodium nitrite, yields the D-p-hydroxyphenylglycine, which can be used as a precursor for the side chains of semisynthetic β-lactam antibiotics. The described process, which is applied on a scale of several hundred tons at Kanegafuchi Chemical Industries Co., Ltd., among others, is also suitable for the stereospecific preparation of other D-amino acids.

Q5: How does the theoretical yield limitation of 50% for resolution reactions overcome in the case of D-p-hydroxyphenylglycine production with hydatoinase?

14.2.3 Nitrile Hydratase (EC 4.2.1.84)

14.2.3.1 Production of Acrylamide
By far the largest industrial enzymatic process outside the food industry (Sect. 16) is the biocatalytic production of acrylamide from acrylonitrile with the aid of nitrile hydratase (Fig. 14.5).

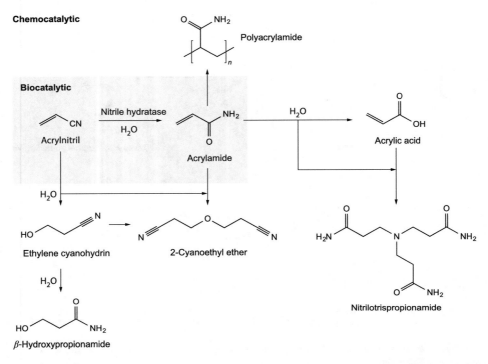

Fig. 14.5 Reaction schemes of the biocatalytic and the chemical acrylamide production. The indicated by-products only occur during chemical synthesis

Acrylamide is mainly used in the production of flocculants for wastewater treatment and paper manufacture, as well as in adhesives, paints, and tertiary oil recovery. The biocatalytic process was first introduced in 1991 at Nitto Chemical Industry Co., Ltd. (Japan) and has been continuously improved since then. With a production scale of approx. 100,000 tons per year as early as 2001, the process can clearly be classified as a bulk or base chemical and is thus the first example that biotechnological processes are also capable of replacing petrochemical processes on this scale. New plants for acrylamide synthesis are based exclusively on the biocatalytic process because, as Fig. 14.5 shows, they offer clear advantages over the chemical production process. For example, yield, conversion, and selectivity are all above 99.99% and no by-products are produced. In 2014 and 2016, BASF opened new world-scale bio-acrylamide plants in the US and UK, and in China in 2017. The capacity of the plant at the Nanjing site alone is already over 50,000 metric tons per year, so that the total biocatalytic production scale is now well over 100,000 metric tons per year. The process is based on the immobilization of whole *Rhodococcus rhodochrous J1 cells* (Fig. 14.6).

After cultivation of the cells and induction of nitrile hydratase by addition of urea, the cells are immobilized in a polyacrylamide gel. At temperatures of 0–15 °C to stabilize the acrylamide and prevent polymerization, acrylonitrile is continuously added to a batch

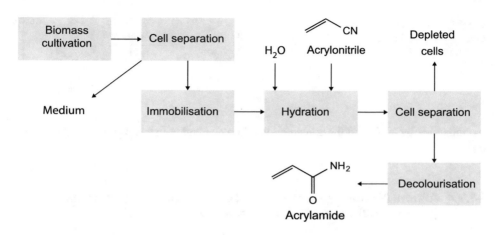

Fig. 14.6 Biotechnological acrylamide process

reactor in aqueous solution. After about 24 h, a final acrylamide concentration of 50% is achieved (Ashina et al. 2010). This is followed by separation of the biocatalyst and decolorization of the acrylamide to meet customer requirements. Major advantages of the biocatalytic process, besides the low temperatures and atmospheric pressure, are mainly due to the complete conversion of acrylonitrile. Thus, neither excess acrylonitrile nor the catalyst needs to be separated, and the extremely high selectivity eliminates the need to separate by-products that are generated in the copper-catalyzed chemical process due to the high reaction temperature of 100 °C (Fig. 14.5). As a result, the biocatalytic process is not only more advantageous in terms of energy consumption and CO_2 production, but also considerably cheaper.

Q6: What are some advantages the enzymatic production process for acrylamide has over the classical chemical approach?

14.2.3.2 Preparation of Nicotinamide
In addition to the production of acrylamide, nitrile hydratase is also capable of hydrating many other nitriles with a yield of 100%. Particularly noteworthy here are the extremely high substrate and product concentrations. For example, in the production of nicotinamide from 3-cyanopyridine using nitrile hydratase from *Rhodococcus rhodochrous* J1, up to 12 M (1353 g L^{-1}) of 3-cyanopyridine can be used as a substrate, from which 1464 g L^{-1} of nicotinamide is produced at 100% conversion (Fig. 14.7). At these concentrations, both the substrate at the beginning of the reaction and the product after complete hydration are in solid form, whereas during the course of the reaction both components are in solution. The advantage of this process, compared to chemical alkaline hydrolysis, is the absence of by-products such as nicotinic acid. In 2010, Lonza AG expanded its total capacity for

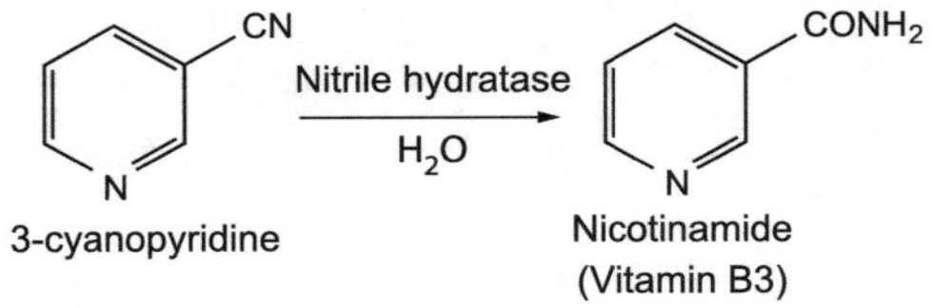

Fig. 14.7 Reaction scheme of biocatalytic nicotinamide production

3-cyanopyridine → (Nitrile hydratase, H_2O) → Nicotinamide (Vitamin B3)

nicotinamide by 40% with the construction of a new plant with a capacity of 15,000 metric tons a^{-1}.

14.2.4 Alkane Monooxygenase (EC 1.14.15.3) and ω-Transaminase (EC 2.6.1.62)-Based Multienzyme Process

14.2.4.1 Preparation of ω-Amino Lauric Acid

One example of the coupling of several enzymes is the production of ω-amino lauric acid from dodecanoic acid methyl ester based on palm kernel oil. ω-amino lauric acid can be used as an alternative to petroleum-based lauric lactam for the production of polyamide 12 and has been produced biotechnologically by Evonik Industries AG in a pilot plant in Slovakia since 2013. The process is based on whole-cell biotransformation with recombinant *Escherichia coli* cells. Dodecanoic acid methyl ester is terminally hydroxylated by the alkane monoxygenase AlkBGT from *Pseudomonas putida* GPo1 and further oxidized to 12-oxododecanoic acid methyl ester (Ladkau et al. 2016). This serves as the initial substrate for terminal amination to ω-amino lauric acid by ω-transaminase CV2025 from *Chromobacterium violaceum*. Further oxidation of the 12-oxododecanoic acid methyl ester produces the by-product dodecanoic acid monomethyl ester (Fig. 14.8).

By co-expressing the alcohol dehydrogenase AlkJ from *P. putida* GPo1 and increasing the intracellular L-alanine concentration, either by the alanine dehydrogenase AlaD from *Bacillus subtilis* or an external L-alanine addition, product formation can be directed towards ω-amino lauric acid. In this process, AlkJ increases the intracellular concentration of 12-oxododecanoic acid methyl ester so that transamination proceeds preferentially due to the enzyme kinetics, while a high L-alanine concentration has a positive effect on reversible transamination (Ladkau et al. 2016).

14.3 Enzymes in the Pharmaceutical Industry

The pharmaceutical industry differs in many aspects from the chemical industry. The research and development times for new products are much longer. As a rule, it takes ten years or more for a new compound to reach the market. In addition, many aspects of product approval and production are much more regulated by the relevant authorities, e.g., the Food and Drug Administration, FDA, in the United States of America and the European

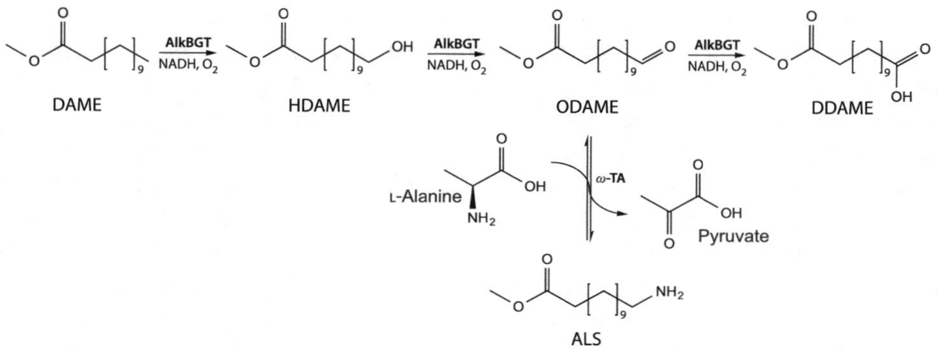

Fig. 14.8 Reaction scheme of terminal oxy- and amino-functionalization of dodecanoic acid methyl ester (DAME) to produce ω-amino lauric acid (ALS). HDAME, 12-hydroxydodecanoic acid methyl ester; ODAME, 12-oxododecanoic acid methyl ester; DDAME, dodecanoic acid monomethyl ester; AlkBGT, alkane monooxygenase from *P. putida* GPo1; ω-TA, ω-transaminase CV2025 from *C. violaceum*. (Adapted from Ladkau et al. 2016)

Medicines Agency, EMEA, respectively. It is estimated that over 80% of APIs are already chiral, and the proportion is expected to increase further.

In the pharmaceutical industry, enzymes are used in a wide variety of applications because of their properties mentioned in Sect. 14.2, e.g., substrate specificity and enantioselectivity:

- Synthesis of precursors and building blocks
- API synthesis
- Studies on drug metabolism
- Synthesis of natural products
- As drugs and as targets of drugs

These areas are explained in more detail and described with examples in the following sections.

14.3.1 Enzymes for the Synthesis of Precursors and Building Blocks

In addition to biocatalysis by lipases, ketoreductases and transaminases have also proven themselves to be useful for the production of chiral building blocks. Exemplary processes are presented in the following.

14.3.1.1 Preparation of Hydroxynitrile

Lipitor® is a cholesterol-lowering drug containing the API atorvastatin calcium. This is an HMG-CoA reductase inhibitor that blocks cholesterol biosynthesis in the liver. An important chiral building block for atorvastatin is ethyl *(R)*-4-cyano-3-hydroxybutyrate, also called hydroxynitrile. Chemically, this building block is produced by the reaction of halohydrin with cyanide in an alkaline environment at increased temperatures. However, both the substrate and the product are sensitive to bases, resulting in much by-product that must be removed. The key step in the synthesis of hydroxynitrile is cyanide attachment to the epoxide. Accordingly, a biocatalyst capable of performing this reaction under mild conditions and neutral pH was sought to minimize by-product formation. Halohydrin dehalogenase is an enzyme that catalyzes the reaction of halohydrins via elimination to the corresponding epoxides. In combination with a ketoreductase and a glucose dehydrogenase, the desired hydroxynitrile is thus synthesized from ethyl 4-chloroacetoacetate, as shown in Fig. 14.9. These three enzymes were adapted to the previously determined process parameters via directed evolution, here using DNA shuffling and bioinformatic evaluation. For example, both the activity and stability of the three enzymes have to be improved while maintaining their enantioselectivity. In addition, since the original halohydrin dehalogenase is inhibited by the product, it had to be made less sensitive towards hydroxynitrile.

By exploiting the high selectivity of the enzymes, avoiding alkaline by-products, and recycling the solvent used for extraction, a lot of waste can be avoided. Moreover, the glucose required to obtain the cofactors is a renewable raw material, and the gluconate produced is biodegradable. Overall, the three-step enzymatic synthesis of hydroxynitrile has enabled a greener process to be developed compared to the original processes.

Q7: What function does the enzyme HMG-CoA-reductase have in the human body?
Q8: Which enzyme properties of the halohydrin dehalogenase have to be improved by directed evolution for its application in the hydroxynitrile production process?

14.3.1.2 Preparation of Cipargamine

The active ingredient cipargamine (NITD609, see Fig. 14.10) belongs to the group of spiroindolones, which exhibit antiplasmoidal activity and can thus be a potential drug against the malaria pathogen *Plasmodium falciparum.* Especially with regard to emerging resistance of *Plasmodium falciparum,* new drugs are of interest. The essential chiral building block for this compound is a tryptamine. This can be produced biotechnologically by a transaminase transferring an amino group from a donor molecule, in this case isopropylamine, to the substrate. Acetone is formed as a by-product from isopropylamine. Through a series of further steps, the resulting tryptamine is converted to the active compound NITD609. As in the case of hydroxynitrile, a molecularly improved enzyme variant is also used in this reaction.

14.3.1.3 Production of Antibiotics

β-lactam antibiotics, e.g., penicillins and cephalosporins, already accounted for 65% of the world market for antibiotics at the turn of the millennium and continue to be important in today's medicine.

Fig. 14.9 Two-step conversion of ethyl 4-chloroacetoacetate to hydroxynitrile with three enzymes

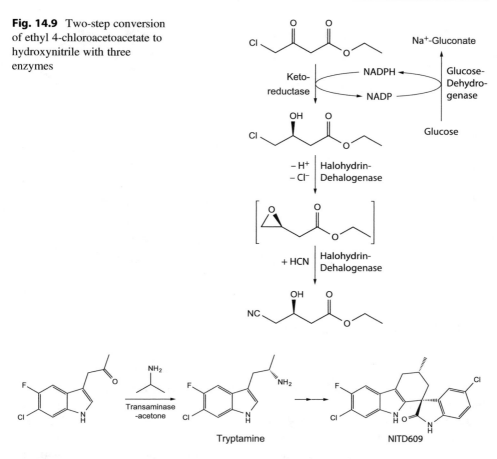

Fig. 14.10 Enzymatic synthesis of the tryptamine building block of NITD609

Most of the representatives of this class of drugs today are semisynthetic, i.e., compounds derived and modified from natural compounds. The syntheses of these derivatives are based on the building blocks 6-aminopenicillanic acid (6-APA) and 7-aminocephalosporanic acid (7-ACA). Accordingly, large quantities of these precursor molecules are required and are currently produced by enzymatic cleavage of natural products obtained by fermentation. 6-APA is the starting molecule for the industrial production of many semisynthetic penicillins such as amoxicillin and ampicillin. Figure 14.11 shows the enzymatic production of 6-APA by cleavage of penicillin G compared with the less advantageous chemical synthesis.

The enzyme amidohydrolase, also known as acylase, breaks down phenylacetic acid with the addition of water to form 6-APA.

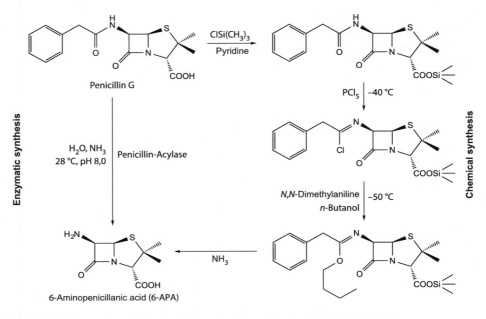

Fig. 14.11 Comparison of the chemical and enzymatic hydrolysis of penicillin G

Analogous to this synthesis, there are also enzymatic pathways for the synthesis of 7-ACA that, unlike chemical synthesis, can proceed at room temperature, as shown in Fig. 14.12.

In the standard pathway, two enzymes, a D-amino acid oxidase and a glutaryl amidase, are involved in the conversion. The former oxidizes the amino acid side chain to the α-keto acid, whereupon a spontaneous chemical cleavage of CO_2 takes place. Subsequently, the glutaryl amidase still cleaves the now shortened side chain as glutaric acid, and the desired product 7-ACA is formed (Buchholz et al. 2012). The literature even reports a one-step enzymatic industrial process from the Sandoz company. This is based on a modified glutaryl amidase, which is capable of hydrolyzing the complete cephalosporin C molecule to 7-ACA without prior oxidation (Boniello et al. 2010).

14.3.2 Enzymes in the Synthesis of APIs

14.3.2.1 Preparation of Sitagliptin Phosphate

Sitagliptin belongs to the class of dipeptidyl peptidase IV inhibitors and is used to treat type II diabetes. In this case, the patient has a permanently elevated blood glucose level as a result of insulin resistance. During eating, incretin hormones such as glucagon-like peptide 1 (GLP-1) are released in the gastrointestinal tract, stimulated by glucose uptake. These in turn stimulate the release of insulin from the pancreas. A disadvantage is the rapid

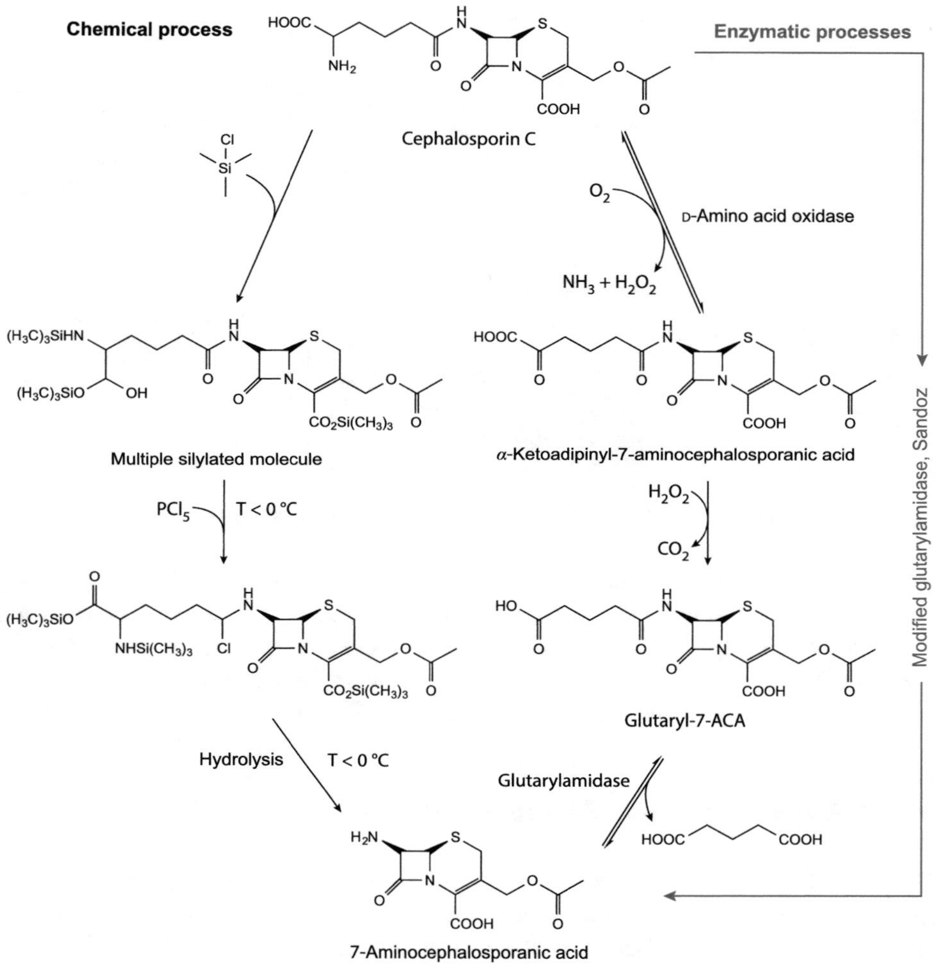

Fig. 14.12 Comparison of the chemical and enzymatic hydrolysis of cephalosporin C

degradation of GLP-1 by the enzyme dipeptidyl peptidase IV (DPP-IV). The drug sitagliptin is a competitive inhibitor of DPP-IV that reversibly binds to the enzyme, resulting in increased GLP-1 concentrations. Figure 14.13 shows a comparison of the chemical and enzymatic synthesis routes for sitagliptin phosphate.

The enzymatic route has several advantages over the chemical route. The chemical synthesis uses a rhodium catalyst and operates at high pressure of 17 bar to convert the precursor prositagliptin ketone. This produces a mixture with an optical purity of only 97% *ee* (enantiomeric excess), which is also contaminated with rhodium residues and requires further work-up. Using a transaminase and the cofactor pyridoxal phosphate (PLP), the prositagliptin ketone can be reductively aminated directly, and due to the enantioselectivity

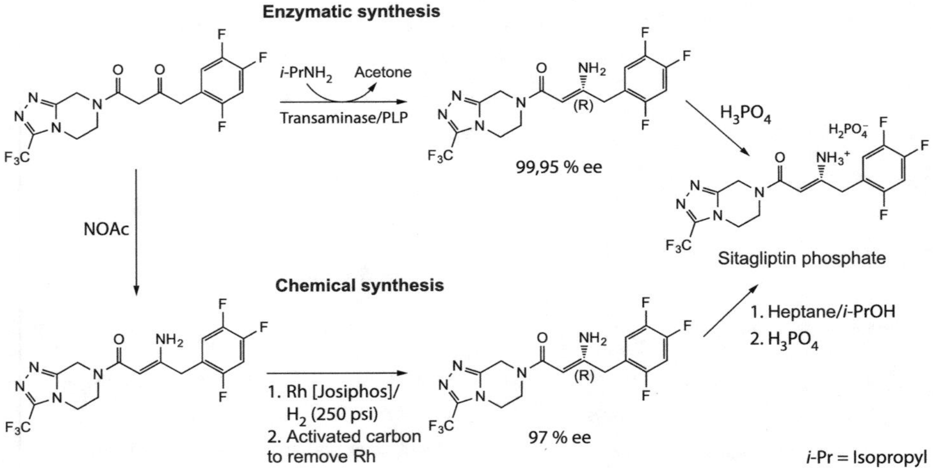

Fig. 14.13 Comparison of the chemical and enzymatic synthesis of sitagliptin

of the enzyme, an enantiomerically pure product with 99.95% *ee* is obtained. This only needs to be converted to its phosphate salt at the end (Savile et al. 2010). The processes considered previously (Sect. 14.3.1) each use enzymes to produce building blocks for drug molecules. With the sitagliptin process, however, it was shown that biocatalysis can also be successfully used in manufacturing processes at the API stage.

14.3.3 Enzymes for Drug Metabolism Studies

During the development of a new drug, extensive studies have to be carried out on the candidate molecules. An important field for these studies is called ADME and includes studies on absorption, distribution, metabolism, and excretion. This involves investigating how the drug enters the body and how it is distributed in organs and tissues, as well as how it is converted in the body and finally transported out of the body again. In this section, the role of isoenzymes, inhibition of metabolizing enzymes, metabolite identification, and prodrugs are considered in more detail.

Enzymes are particularly important in the study of metabolism and are used in various forms in the pharmaceutical industry. The reactions of drug metabolism are divided into two phases. The enzymes of the first phase facilitate the functionalization of the substrate molecule, mainly via oxidation reactions, while the enzymes of the second phase facilitate the formation of soluble compounds, which are better secreted, by conjugation reactions. Figure 14.14 shows an overview of the enzymes involved and their localization (Schroer et al. 2010).

Probably the most important enzymes of the two phases are the cytochrome P450 monooxygenases (CYPs), since 75% of the drug metabolism is CYP-mediated. The largest

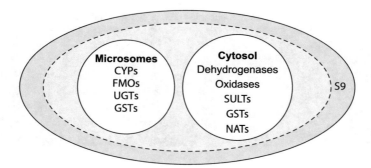

Fig. 14.14 Distribution of metabolizing liver enzymes. CYP = cytochrome P450 monooxygenase, FMO = flavin-dependent monooxygenase, GST = glutathione S-transferase, NAT = *N-acetyltransferase*, SULT = sulfotransferase, and UGT = UDP-glucuronosyl transferase. (According to Schroer et al. 2010)

CYP concentration is found in the liver, more specifically in the endoplasmic reticulum of hepatocytes (Schroer et al. 2010).

Q9: What are the two phases of drug metabolism and which group of enzymes is particularly important in the first phase?

14.3.3.1 Isoenzymes

Isoenzymes are by definition enzymes that differ in their primary structure but perform the same or a similar function within a species. Due to differences in charge distribution on the surface, there may also be different localizations of the molecules within the cell. Some isozymes also exhibit tissue or cell specificity. Different isoenzymes dominate in different tissue types, and over a lifetime their composition also changes. It can be assumed that each isoenzyme has a specific role in cell metabolism, as they allow fine adaptation to, for example, environmental changes. They often also differ in their properties, such as the optimum substrate concentration, electrophoretic mobility, or substrate specificity. Genetically determined differences in the type and concentration of isoenzymes also explain abnormalities in metabolism perceived as hereditary diseases and may also be responsible for the sensitivity of some individuals to diverse drugs. Therefore, a crucial step in drug metabolism studies is to find out which isoenzyme is responsible for the degradation or conversion of the drug. Table 14.2 lists the six most important isoenzymes of the CYPs that are mainly responsible for the metabolism of drugs, together with a selection of their substrates.

Some active substances, e.g., caffeine or the antidepressant amitriptyline, can be metabolized by several isoenzymes (Table 14.2). This may be a reason for the side effects as well as toxic and carcinogenic effects of xenobiotics. In addition, it is important to know by which isoenzyme or isoenzymes a drug is metabolized in order to detect possible interactions with other drugs.

Table 14.2 The most important six cytochrome P450 isozymes with a selection of their substrates

Enzyme	Substrate
CYP1A2	Amitriptyline, caffeine, (R)-warfarin, propranolol
CYP2C9	Amitriptyline, diclofenac, (S)-warfarin, naproxen
CYP2C19	Amitriptyline, diazepam, omeprazole, citalopram
CYP2D6	Amitriptyline, codeine, methadone, propranolol
CYP2E1	Ethanol, caffeine, acetaminophen
CYP3A4/5	Amitriptyline, cyclosporine, caffeine, codeine

Isoenzymes can also be helpful in the identification of cancer cells. New patterns in isoenzyme distribution occur primarily when fully differentiated cells undergo malignant transformation into tumors. Accordingly, isoenzymes can be used to determine whether a tumor is benign or malignant. A specific example of this is the use of the pyruvate kinase isoenzyme pyruvate kinase M2 as a tumor metabolism marker in stool. Normal pyruvate kinase (PK) is a glycolytic enzyme that converts phosphoenolpyruvate to pyruvate with the release of energy-rich ATP and GTP. This stimulates cell proliferation. In normal cells, there are the following tissue-specific isoenzymes of PK: M1-PK in muscle and brain tissue, L-PK in liver and kidney tissue, and R-PK in erythrocytes. The tumor-specific variant is called M2-PK and is synthesized in tumors to a much greater extent than the native isoenzymes of PK. Enzymatic tests of stool samples for their M2-PK concentration can therefore be used to diagnose intestinal tumors at an early stage due to their tumor-specificity.

14.3.3.2 Inhibition

Inhibition of drug-metabolizing enzymes leads to increased drug concentrations in the blood, which in turn can lead to severe toxic side effects. Due to this, inhibition of these enzymes is of great clinical interest. Inhibitor studies can therefore be used to exclude certain isoforms as responsible enzymes in the metabolism. On the other hand, preclinical studies are used to determine whether a drug candidate inhibits important metabolizing enzymes. In this way, negative consequences of comedications can be prevented. As already explained in Chap. 4, there are different types of enzyme inhibition. Competitive inhibition occurs primarily in cytochrome P450 monooxygenases that can metabolize several substrates and in which diverse cosubstrates also compete with each other. In this case, inhibition decreases with increasing substrate concentration. A known P450 inhibitor, and thus also the inhibitor of drug metabolism, is cimetidine. It interacts with varying affinity with different P450 isoenzymes. Assays also exist to determine IC_{50}, the concentration that reduces enzyme activity to 50%, relative to cytochrome P450 for various compounds. Furthermore, it can be determined whether the inhibition by the tested substance is reversible or irreversible.

14.3.3.3 Metabolite Identification

The synthesis of drug metabolites is on the one hand of interest in order to use them for the characterization of drug candidates or as reference compounds in the elucidation of the metabolism of the drug candidate. On the other hand, drug metabolites are used to investigate toxicity, biological activity, and drug-drug interactions. Metabolites derived from the drug candidate may also have new or modified biological properties and thus be the starting point for drug development. The chemical synthesis of drug metabolites often consists of many steps, including the addition and removal·of protecting groups. In contrast, biocatalytic synthesis often has advantages. Recombinant expression of human drug metabolizing enzymes in microorganisms and their use as whole-cell biocatalysts is an elegant method for the production of larger amounts of drug metabolites, especially for phase I metabolites.

14.3.3.4 Prodrugs

A prodrug is a pharmacologically inactive, reversible derivative of a drug, which can be converted in vivo either enzymatically or chemically into the drug molecule. This principle can be used to temporarily alter especially the undesirable physicochemical properties of the drug. These problematic properties can be, for example, low oral bioavailability, poor water or fat solubility, chemical instability, or toxicity, and also the lack of site specificity or poor patient acceptance due to poor odor or taste of the active ingredient. Many of these problems can be circumvented with the help of prodrugs.

An illustrative example of the enzymatic conversion of a prodrug into its active form is antihistamine fexofenadine, which can only slightly cross the blood-brain barrier. Its derivative terfenadine, the precursor, on the other hand, can cross it rapidly. In the body, terfenadine is metabolized to hydroxyl-terfenadine and converted by CYP3A4 to the pharmaceutically active form fexofenadine.

14.3.4 Enzymes in the Synthesis of Natural Products

Natural products are defined as low molecular weight chemical compounds synthesized by biological organisms (Breinbauer et al. 2002). Over time, they have emerged as a very good source of disease-regulating agents and still play important roles in medicinal chemistry and pharmaceutical drug development. However, their distinct biological activity is outweighed by the fact that they interact with proteins as substrates and targets both during their biosynthesis and during their biological task.

Evans et al. from Merck defined the term privileged structure at the end of the 1980s. They used it to describe substance classes that can bind to diverse protein receptor surfaces. Substance classes derived or inspired from these can be considered biologically relevant and are valuable starting points for medicinal chemistry. In order to optimize both the binding affinity and selectivity of these natural products, modifications to the basic structure must be made, and derivatives of the natural product must be developed

(Breinbauer et al. 2002). Enzymes are suitable tools for these modifications of the complex basic structure.

14.3.4.1 Preparation of Artemisinin

Despite enormous progress in organic chemistry, not all natural products can be produced economically via total synthesis. The extraction of natural products from conventional plants is usually not productive enough either. An alternative is offered by synthetic biology, i.e., the introduction of completely new biological synthesis pathways into microorganisms, fungi, plants, or even animals. One example of the application of synthetic biology is artemisinin, which is a sesquiterpene and a potent drug precursor for an antimalarial drug. In nature, the substance is produced by the plant *Artemisia annua,* the annual mugwort, and has been known in traditional Chinese medicine for centuries. The purely herbal production of the natural product is associated with an unsteady supply due to weather influences and crop failures. Therefore, a semisynthetic production of artemisinin is an option, in which the precursor molecule artemisinic acid is biotechnologically produced and subsequently chemically converted into artemisinin. Industrially, this process, which exploits the selectivity and regio- and enantioselectivity of enzymes from different organisms, is used with *Saccharomyces cerevisiae* at Sanofi. Figure 14.15 shows the enzymatic production of artemisinic acid. The most important step here is the cyclization of farnesyl diphosphate by amorpha-4,11-diene synthase to amorpha-4,11-diene. Subsequently, this is oxidized to artemisinic acid by CYP71AV1, a cytochrome C-P450 monooxygenase. All steps up to the intermediate farnesyl diphosphate are carried out by enzymes of the yeast, while the subsequent steps up to artemisinic acid are catalyzed by plant enzymes from *Artemisia annua,* which have been introduced into the yeast.

After biosynthesis, the intermediate is chemically converted to artemisinin in a few steps. There are also other processes that use *Escherichia coli* as producers and, in addition to their own genes, also use genes for enzymes from *Saccharomyces cerevisiae, Staphylococcus aureus*, and *Artemisia annua. In* addition to artemisinin, there are also approaches to produce other complex molecules such as taxol via synthetic biology.

14.3.4.2 Enzymes for the Modification of Complex Structures

In late-stage functionalization (LSF), the relatively inert C-H bonds are regarded as functional groups, which are considered as starting points for potential diversification to generate new analogs of existing lead structures without having to resort to de novo synthesis. This approach is particularly advantageous as it additionally provides rapid access to putative drug metabolites for, for example, ADME studies.

Modifying enzymes, so-called tailoring enzymes, already increase the diversity of natural products by introducing functional groups into natural product scaffolds. Often the functional groups are responsible for the occurring biological activity of the final compound. Examples of subsequent enzymatic modifications are chemo- and regioselective cyclizations, redox reactions, halogenations, glycosylations, and alkylations and acylations. For industrial use, the natural enzymes still have to be optimized, which is

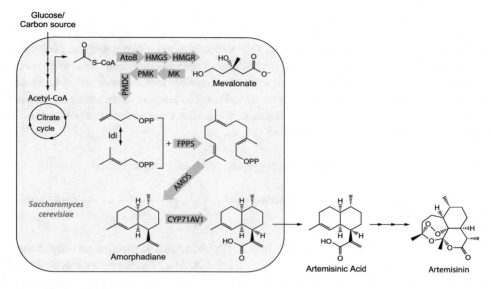

Fig. 14.15 Production of artemisinic acid with *Saccharomyces cerevisiae*

done by protein engineering. In particular, the catalytic activity is improved and the substrate specificity is extended to enable the conversion of unnatural substrates.

Examples include terminal thioesterases that perform regioselective macrocyclizations of linear non-ribosomal peptides (NRPs). Such regioselective macrocyclizations are synthetically difficult, but necessary for the bioactivity of NRPs.

Another example of enzymatic LSF is the diastereoselective hydroxylation of milbemycin A_4 by the actinomycete *Streptomyces violascens*. The epoxide of the compound is formed as a by-product. Furthermore, the fungus *Cunninghamella elegans* can hydroxylate dianilinophthalimides once and also twice, making metabolites accessible for biological and pharmacological studies (Schmid and Urlacher 2007).

14.3.5 Enzymes as Drugs

However, enzymes themselves can also be the active ingredient of a drug. First of all, the pineapple enzyme bromelain should be mentioned as an example. It belongs to a group of proteolytic enzymes that have very low toxicity and are used to treat inflammation and dissolve blood clots, e.g., as additives in radiotherapy or surgery to prevent edema and improve wound healing.

Other examples are lactase capsules containing β-galactosidase and urate oxidase. The former is used to break down milk sugar into glucose and galactose and is prescribed to lactose-intolerant patients. The latter breaks down uric acid in the body and is used in cases of hyperuricemia, i.e., gout-like side effects of chemotherapy.

Further therapies with enzymes as active ingredients are already in development. Phenylketonuria is an inherited metabolic disorder in which toxic concentrations of phenylalanine lead to mental retardation. One treatment option is the enzyme phenylalanine ammonia lyase (PAL), which metabolizes phenylalanine into non-toxic *trans*-cinnamic acid and ammonia. The small amounts of ammonia can be converted to urea in the body. Preclinical studies using mouse models of this enzyme as an active agent were promising, showing reduced phenylalanine concentrations and reduced manifestation of symptoms in both brain tissue and blood vessels.

Q10: Which type of diseases are currently treated with enzymes as drugs?

14.4 Conclusion

Today, enzymes are used in many ways in both the chemical and pharmaceutical industry, from the synthesis of chiral specialty chemicals and building blocks to the production of bulk chemicals or APIs. In addition, they are versatile tools for drug discovery and development. Due to advancements in protein optimization and the trend towards raw material change in the chemical industry, among other factors, it can be assumed that the importance of enzymes will continue to increase in both industries.

Take Home Message
- Enzymes have become established catalysts in the chemical and pharmaceutical industry.
- The largest application of an enzymatic biotransformation in the chemical industry is the production of acrylamide with nitrile hydratase.
- APIs or their building blocks can be produced with enzymes.

Answers

A1: In an enzymatic biotransformation, the protein catalyst is used in (partly) isolated form, whereas a whole-cell biotransformation uses the enzymes as catalysts intracellularly in living, resting, or dead cells.

A2: *Saccharomyces cerevisiae*, *Acetobacter suboxydans*, and *Rhizopus arrhius*

A3: Among other influences, higher temperatures or the presence of organic solvents might be encountered as unphysiological conditions for enzyme catalysts in chemical processes.

A4: The use of an amine as a nucleophile instead of water allows the acylation of said amine to form an amide. The absence of water prevents the hydrolysis reaction.

A5: In an aqueous solution, a continuous racemization of the L-5-(*p*-hydroxyphenyl)-hydantoin substrates takes place so that a quantitative conversion of the substrate is possible. This so-called dynamic resolution is a general way to overcome the 50% yield limitation in resolution reactions.

A6: The enzymatic process is run at temperatures of 0–15 °C. This stabilizes the acrylamide and prevents polymerization. In addition, the conversion is higher and more selective compared to the chemical process; hence no residual substrates and no by-products need to be removed. Overall, the process is also more economical.

A7: HMG-CoA-Reductase is involved in the biosynthesis of cholesterol in the human body.

A8: The halohydrin dehalogenase enzyme has to be improved by directed evolution in terms of its activity (i.e., overall rate of catalysis), its stability (i.e., life-time in the process), and reduced inhibition in the presence of hydroxynitrile.

A9: In metabolism, phase I reactions are called functionalization reactions, and phase II reactions are called conjugation reactions. The most important enzymes in phase I are cytochrome P450 monooxygenases (CYPs).

A10: Bromelain from pineapple is used to treat inflammations. Lactase is used as a food additive for lactose intolerance, and urate oxidase is used to treat hyperuricemia. Other enzyme therapies are in development.

References

Ashina Y, Suto M, Endo T. Nitrile hydratase. Encyclopedia of industrial biotechnology. Hoboken: Wiley; 2010.

Bommarius AS, Riebel B. Biocatalysis. Weinheim: Wiley-VCH Verlag GmbH; 2004.

Boniello C, Mayr T, Klimant I, Koenig B, Riethorst W, Nidetzky B. Biotechnol Bioeng. 2010;106(4): 528–40.

Breinbauer R, Vetter IR, Waldmann H. Angew Chem Int Ed. 2002;41(16):2878.

Buchholz K, Kasche V, Bornscheuer UT. Biocatalysts and enzyme technology. 2nd ed. Weinheim: Wiley-VCH Verlag GmbH; 2012.

Buchner E. Ber Dtsch Chem Ges. 1897;30:117–24.

Grunwald P, editor. Industrial biocatalysis. Singapore: Pan Stanford Publishing Pte. Ltd.; 2015.

Ladkau N, Assmann M, Schrewe M, Julsing MK, Schmid A, Bühler B. Metab Eng. 2016;36:1–9.

Liese A, Seelbach K, Wandrey C, editors. Industrial biotransformations. 2nd ed. Weinheim: Wiley-VCH Verlag GmbH; 2006.

Marschall L. Im Schatten der chemischen Synthese. Industrial biotechnology in Germany (1900–1907). Frankfurt a. M: Campus Verlag; 2000.

Neuberg C, Hirsch J. Biochem Z. 1921;115:282–310.

Peterson DH, Murray HC, Eppstein SH, Reineke LM, Weintraub A, Meister PD, Leigh HM. J Am Chem Soc. 1952;74:5933–396.

Reichstein T, Grüssner A. Helv Chem Acta. 1934;17:311–28.

Savile CK, Janey JM, Mundorff EC, Moore JC, Tam S, Jarvis WR, Colbeck JC, Krebber A, Fleitz FJ, Brands J, Devine PN, Huisman GW, Hughes GJ. Science. 2010;329(5989):305–9.

Schmid RD, Urlacher VB. Modern biooxidation: enzymes, reactions and applications. Weinheim: Wiley-VCH; 2007.

Schroer K, Kittelmann M, Lütz S. Biotechnol Bioeng. 2010;106(5):699.

Sheldon RA. Chem Commun. 2008:3352–65.
Syldatk C, Hauer B, May O. BIOspektrum. 2001;2(01):145–7.
Weizmann C. 1919; US Patent 1,315,585.

Enzymes for the Degradation of Biomass

15

Christin Cürten and Antje C. Spieß

What You Will Learn in This Chapter

Enzymatic hydrolysis in a biorefinery process plays a key role in the conversion of lignocellulosic biomass into sugars, which in turn are used as substrates for fermentation into chemicals or fuels. This chapter focuses on the wide range of lignocellulose degrading enzymes like cellulases, hemicellulases, pectinases, and ligninases. These enzymes vary in their substrate as well as their mode of action, which will be shown exemplarily for cellulases. The role these enzymes play in the biorefinery process, also necessary background information about lignocellulosic biomass, and possible pretreatment methods are given. Section 15.1 describes the composition of lignocellulose from cellulose, hemicellulose, and lignin and its influence on hydrolysis. Cellulose is degraded by fungal secreted cellulases into the monomer glucose (Sect. 15.2), while hemicellulose and lignin are degraded by hemicellulases and ligninases (Sect. 15.3). The choice of a suitable enzyme cocktail depends on the composition of the biomass, which in turn depends on the pretreatment processes (Sect. 15.4). Finally, special challenges are highlighted using the example of an ethanol biorefinery (Sect. 15.5).

C. Cürten
Chair of Biochemical Engineering, RWTH Aachen University, Aachen, Germany

A. C. Spieß (✉)
Institute for Biochemical Engineering, TU Braunschweig, Braunschweig, Germany
e-mail: a.spiess@tu-braunschweig.de

Introduction

In view of dwindling fossil carbon sources and increasing demand for fuels and chemicals, the use of lignocellulose, i.e., woody biomass, is becoming more important. In contrast to the utilization of biomass containing starch or oil, the utilization of lignocellulose, e.g., wood, straw, and grass, is not in direct competition with food production. Nevertheless, it can be available as a raw material in addition to or after being used as bedding in stables, as animal feed, or as erosion control in fields. In such a cascade use, raw materials are only recycled after they have been used in the stable or on the field. In addition to the agricultural waste straw and grass, lignocellulose also includes hard and soft woods as well as paper waste.

From this biomass, its sugar components are split off by means of enzymatic hydrolysis, with the aim of achieving the highest possible sugar concentrations in the hydrolysate possible. The hydrolysates can then be fermented into ethanol or other basic chemicals. Lignocellulose is thus the basis for second-generation biofuels as well as a starting material for the production of chemicals and is presented below in its structural composition as a substrate (Van Dyk and Pletschke 2012).

15.1   Composition of Biomass

Lignocellulose is found in the cell walls of the plant cell and consists of macrofibrils with a diameter of 1025 nm, which in turn contain cellulose microfibrils (Fig. 15.1). Lignin and hemicellulose are found in the space between the microfibrils.

The microfibrils consist of cellulose fibers and make up the largest mass fraction of the biomass with about 50%. This is followed by lignin with 25% and hemicellulose with about 20%. The remaining 5% are pectin and mineral components. However, the composition of biomass varies to some extent depending on the plant species, function of the plant part, cultivation conditions, and region and fertilizer use (Van Dyk and Pletschke 2012). The structure and function of biomass-derived polymers are shown below.

15.1.1 Cellulose

Cellulose is a long-chain, unbranched polysaccharide consisting of glucose units linked by a glycosidic β-1,4 bond. Approximately 24 cellulose strands each assemble into tightly packed microfibrils. Hydrogen bonds between the individual cellulose strands lead to a mostly crystalline structure. However, cellulose can also be less ordered in areas, i.e., amorphous. The degree of crystallinity can be described with the crystallinity index (CrI). This is determined with the aid of X-ray diffraction, nuclear magnetic resonance (NMR), or differential scanning calorimetry. The length of cellulose fibers is determined by the number of glucose monomers. This number is called the degree of polymerization and

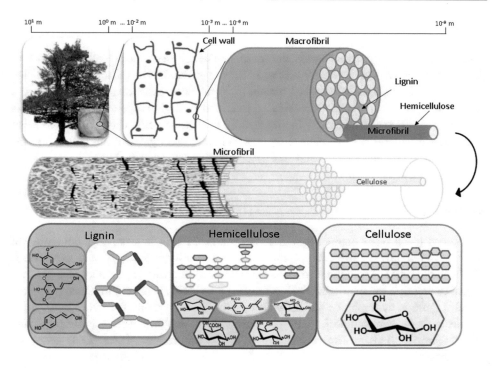

Fig. 15.1 Occurrence and structure of lignocellulosic biomass as well as structure of the three main components cellulose, consisting of glucose, hemicellulose, consisting of xylose, arabinose, glucuronic acid, galactose and ferulic acid, and lignin, consisting of coniferyl alcohol, p-coumaryl alcohol, and sinapyl alcohol

varies between 10,000 and 15,000 depending on the biomass (Agbor et al. 2011; Payne et al. 2015).

15.1.2 Hemicellulose

Hemicellulose is a polysaccharide like cellulose, but hemicellulose is branched and has a much lower degree of polymerization. The basic building blocks of hemicellulose are pentoses (xylose and arabinose) and hexoses (galactose, glucose, and mannose). These building blocks are the basis for the nomenclature, e.g., hemicellulose with a high xylose content is called xylan. If the two building blocks xylose and mannan predominate, this leads to the name xylomannan. The composition of hemicellulose varies depending on the plant species. Glucuronoarabinoxylan, for example, occurs mainly in grasses, galactoglucomannan in softwoods, and 4-O-methylglucuronoxylan mainly in hardwoods. The hemicelluloses rearrange the cellulose, as shown in Fig. 15.1 (Himmel et al. 2007).

15.1.3 Pectin

Pectin is a branched polysaccharide similar to hemicellulose and accounts for 0.5–4% of the fresh plant weight. Building blocks of the main chain are galacturonic acid and rhamnose. Side chains of various combinations of arabinose, galactose, xylose, and fucose extend from this main chain. Occasionally, esterified methyl groups and acetylations are also present. Pectin occurs mainly in tough, unlignified plant parts. Citrus fruits and sugar beets are particularly rich in pectin. Like hemicellulose and lignin, pectin is found in the matrix between the microfibrils and acts as an adhesive (Himmel et al. 2007).

15.1.4 Lignin

In contrast to all other components of lignocellulose, lignin is a polymer composed of aromatic components. The main components are coniferyl, synapyl, and *p-coumaryl alcohol* (Fig. 15.1). These three subunits form a widely branched hydrophobic network.

This network is the matrix in which cellulose microfibrils, hemicellulose, and pectin are found. Lignin is hydrophobic and difficult to degrade. As a result, the rather hydrophilic cellulose and hemicellulose are protected from environmental influences and microorganisms. This protective mechanism makes pretreatment (Sect. 15.4) necessary for biomass recovery. However, utilization of the resulting lignin is challenging. Due to the aromatic functionality of lignin, recycling is not exclusively thermal and attractive (Roth and Spiess 2015).

The enzymes for cellulose, hemicellulose, and potential lignin degradation are discussed below.

Questions:
1. What are the main components of lignocellulose?
2. Which component is a polymer of glucose?
3. Which plants contain pectin?
4. What is the difference between lignin and the other components of lignocellulose?

15.2   Cellulases

Cellulases are enzymes that contribute to the degradation of cellulose. The reaction mechanisms of fungal cellulases were originally studied by the US Army during World War II, as cotton-based equipment was affected by fungal infestation. As a result of the oil price crisis, research attention since the 1970s has focused on their potential for biomass degradation (Montenecourt 1983). Since biomass degradation mostly occurs by hydrolysis, cellulases are classified into enzyme class EC 3.2. Two reaction mechanisms are distinguished: inverting and retaining hydrolysis (Fig. 15.2).

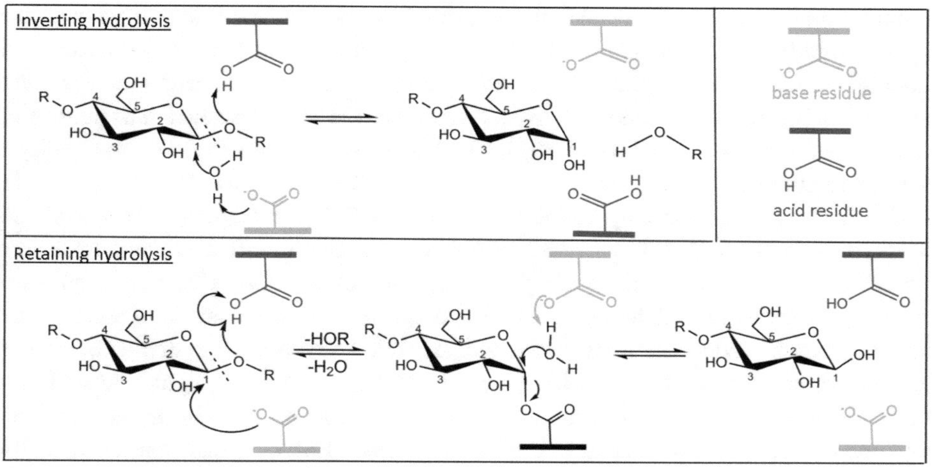

Fig. 15.2 Reaction mechanisms of inverting and retaining hydrolysis (modified after Payne et al. 2015)

In the inverting hydrolysis, the bond between two glucose molecules is cleaved in one step by a water molecule, which has lost a proton due to the base residue of an amino acid (AS 1) of the enzyme, nucleophilically attacking the C1 atom. This attachment of the generated hydroxyl anion weakens the bond between the C1 atom and the β-O atom of cellulose. In addition, an acid group (AS 2) of the enzyme provides another proton to finally break the bond. After such a reaction, the position of the base and the acid of the enzyme are swapped, i.e., inverted, due to the proton transfer (Payne et al. 2015).

In the case of retaining hydrolysis, the functional group of the catalytic amino acids in the active site of the enzyme does not change before and after the reaction; however, the reaction requires two steps. In the first step, the base (AS 1) of the enzyme attacks the C1 atom, while the acid residue provides a proton (AS 2). This leads to a break in the bond to the rest of the cellulose strand, while a glycosyl-enzyme complex is formed. In the second step, the entry of a water molecule into the active site breaks the complex by nucleophilically attacking the C1 atom and adding a proton to the base residue (AS 2) (Payne et al. 2015).

Since cellulose hydrolysis is a heterogeneous surface reaction, the binding equilibrium of the enzymes to the cellulose influences the reaction rate. The reaction mechanism for cellulose hydrolysis can be divided into five steps:

1. Adsorption of cellulase to the cellulose strand
2. Breaking of the crystalline structure of the microfibril of the attacked cellulose strand
3. Hydrolysis of the cellulose chain
4. Release of cellobiose by cellulase
5. Detachment and desorption of cellulase from the cellulose strand

To prevent the active site from detaching from the substrate each time, most cellulases have a carbohydrate binding module (CBM). This is connected to the actual catalytic module by a connecting peptide. It binds to the cellulose strand and enables the enzyme to continue on the cellulose strand after hydrolysis and release of the product, thus performing several cleavages in succession. Thus, a CBM increases processivity, i.e., the ability to perform multiple cleavages in succession, and thus cellulase activity (Payne et al. 2015).

Cellulose is hydrolyzed by cellobiohydrolases (CBHs), endoglucanases (EGs), and β-glucosidases (BG). In this process, cellobiohydrolase splits off cellobiose units from the ends of a cellulose strand. Endoglucanases can cut cellulose strands in amorphous regions, where cellulose is less densely packed and thus more amenable to attack. Thus, endoglucanases produce more available ends or cellobiohydrolases to attack. Since both endoglucanases and cellobiohydrolases are inhibited by cellobiose, β-glucosidase breaks down the resulting cellobiose units into glucose that can be used by microorganisms. Another cellulase is polysaccharide monooxygenase (PMO), which causes oxidative cleavage of crystalline cellulose into two cellulose strands, one of which has a terminal carbonyl group.

Microorganisms that specialize in cellulose as a substrate, such as the fungi *Trichoderma reesei*, *Aspergillus niger*, *Ustilago maydis*, and *Neurospora crassa*, usually secrete a set of different cellulases. The best studied are the enzymes of the filamentous fungus *T. reesei*. This secretes high concentrations of 100 g L^{-1} enzymes, with the mixture consisting of two cellobiohydrolases, at least five endoglucanases, one β-glucosidase, and two PMOs that jointly degrade cellulose (Fig. 15.3; Lombard et al. 2014; Wilson 2009).

Like all other glycosidic enzymes, cellulases are assigned to different GH *(glycoside hydrolase)* families according to their amino acid sequence. Currently, 340,000 enzymes from more than 330 families are known and listed in the CAZy database (*carbohydrate-active enzymes*, http://www.cazy.org/; Lombard et al. 2014). The individual GH families share a similar structure as well as the same reaction mechanism. The cellulolytic *Trichoderma reesei* enzymes, which are generally used as the main source of cellulose cocktails, are presented in more detail below.

15.2.1 Cellobiohydrolases (CBHs)

Cellobiohydrolases are also referred to as exoglucanases, as they usually attack cellulose strands from the end *(exo)*. Two types are distinguished: CBH I (EC 3.2.1.91), which attacks from the reducing end, and CBH II (EC 3.2.1.176), which attacks from the nonreducing end of the cellulose strand. The ability to cleave cellulose in the *endo* or *exo position* is due to the structure of the active site. In the case of CBHs from *T. reesei*, the active site forms a tunnel through which the cellulose strand is passed (Fig. 15.3). This tunnel consists of four amino acid loops in CBH I and two in CBH II. According to recent findings, at least some CBHs can also perform *endo-cleavages*. The ability to perform

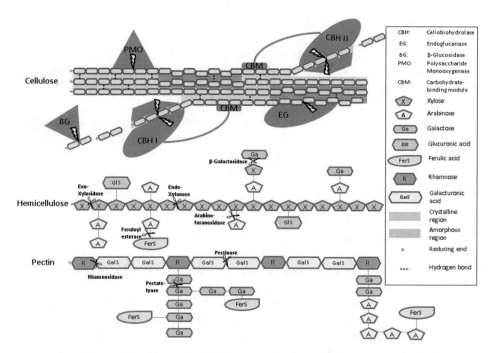

Fig. 15.3 Attack positions of cellulases on crystalline and amorphous cellulose and interfaces of hemicellulases and pectinases on hemicellulose and pectin, respectively (modified after Montenecourt 1983; Neufeld and Pietruszka 2012)

endo-cleavages is made possible by the movable loops and the associated opening of the tunnel (Payne et al. 2015).

The main product of CBH is cellobiose, the β-1,4-linked disaccharide of glucose.

15.2.2 Endoglucanases (EGs)

Like most cellulose-cleaving fungi, *T. reesei* secretes several endoglucanases (EC 3.2.1.4). The enzymes EG I and EG II, which were the first to be discovered and are the best studied, have a similar structure to CBH II, but the active site is more open and thus cleaved. Therefore, endoglucanases can adsorb anywhere on cellulose but prefer amorphous regions of cellulose, where the likelihood of proper positioning of the active site is greater. The activity of endoglucanases on crystalline cellulose is very low. Since endoglucanases randomly adsorb on and cleave the substrate to be hydrolyzed, they produce oligosaccharides of different chain lengths in addition to glucose and cellobiose (Payne et al. 2015).

15.2.3 β-Glucosidases (BG)

In addition to cellobiohydrolases and endoglucanases, a β-glucosidase (EC 3.2.1.21) is also part of the enzyme set of *T. reesei*. However, the β-glucosidase activity in the commercial enzyme mixtures from the secreted proteins from *T. reesei* is low, as they remain bound to the mycelium.

β-Glucosidases cleave cellobiose and short-chain cellooligomers, which inhibit mainly cellobiohydrolases and endoglucanases. Since the degradation of these cellooligomers has a positive effect on the activity of the enzyme mixture, β-glucosidases are usually added to commercial enzyme mixtures, e.g., high activity β-glucosidases from *Aspergillus niger* (Payne et al. 2015).

15.2.4 Polysaccharide Monooxygenases (PMOs)

Polysaccharide monooxygenases were initially misclassified as GH 61 enzymes and thus as hydrolases, but are copper-dependent oxidases. PMOs have since been grouped into the AA9 and AA10 families. AA *(auxiliary activities)* families include CBMs and PMOs as well as other enzymes that contribute to the degradation of lignocellulose.

With regard to the cleavage mechanism, there are two PMO types. Type I inserts an oxygen molecule at position C1, and type II at position C4 of the glucose building blocks. This leads to destabilization and thus to the breaking of the ether bond. This reaction step is energetically favored, since the cellulose fiber does not have to be lifted from the crystalline cellulose beforehand, as is the case with CBHs (Fig. 15.4).

Determining the activity of PMOs is fraught with difficulties: on the one hand, the products of cleavage may still adhere to the crystalline substrate; on the other hand, PMOs require an electron donor, which in the natural environment is probably produced by cellobiose dehydrogenases (CDH). Therefore, little PMO activity is observed in the degradation of pure crystalline cellulose. However, the activity of other cellulases is enhanced by PMOs, especially in the enzymatic hydrolysis of lignin-containing biomass, as components of lignin can act as electron donors (Neufeld and Pietruszka 2012).

Questions:
5. Which are the main steps of cellulose hydrolysis?
6. Which enzymes are called cellulases?
7. What is the difference between a cellobiohydrolase and an endoglucanase?
8. Why are beta-glucosidases often added to commercial cellulase mixtures to increase activity?

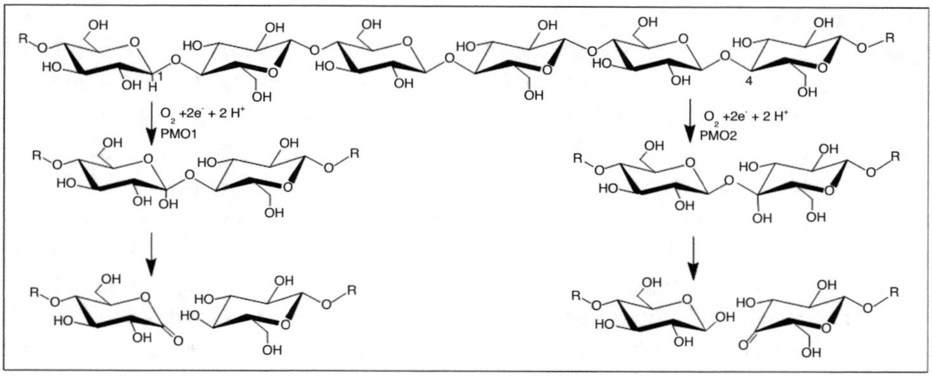

Fig. 15.4 Reaction scheme of PMO 1 and PMO 2 with cellulose (adapted from Neufeld and Pietruszka 2012)

15.3 Additional Enzymes for Biomass Degradation

The degradation of hemicellulose and pectin primarily leads to an increase in the pentose content of the hydrolysate. Pentoses can be used by some microorganisms as a C source for fermentation. In addition to the utilization of various by-products, the degradation of hemicellulose increases the surface area of cellulose accessible to cellulases. At the same time, the proportions of hemicellulose and pectin decrease and thus the diffusion resistance for the cellulases. This increases the probability of a hydrolysis reaction. Since lignin present increases the proportion of inactive cellulases, lignin-modifying and lignin-degrading enzymes have a positive effect on cellulase activity. In summary, additive enzymes can therefore increase the activity of cellulases and thus the yield of glucose (Roth and Spiess 2015).

Due to the different bonds within the hemicellulose, pectin, and lignin, a high number of enzymes are also required here.

15.3.1 Hemicellulases

The term hemicellulases is a collective term for all enzymes that contribute to the cleavage of the bonds in hemicellulose. In the following, a small selection of enzymes is presented, and their respective functions are described. With the exception of feruloyl esterase, all the enzymes mentioned here cleave a glycosidic bond in exactly the same way as cellulases. Therefore, the mechanism is similar to the reaction pathway described in Sect. 15.2.

Endo-1,4-β-Xylanases (EC 3.2.1.8)

Xylanases perform *endo-cleavages* of the β-1,4-bonds at any position of the xylan chain in hemicellulose. Xylanases are mainly grouped in the GH 10 and GH 11 families; however, enzymes with xylanase activity can also be found in GH families 5, 8, and 43. Thus, some xylanases have multiple active sites or binding modules for xylan or cellulose. These binding modules are connected to the catalytic unit by short-linking peptides. Usually, a microorganism produces several xylanases with different properties, structures, and activities; for example, the fungi *A. niger* and *T. reesei* produce 13 and 15 different extracellular xylanases, respectively. A small amount of extracellular xylanases is constitutively expressed, producing xylose and xylooligomers in the presence of a substrate. The soluble product xylose is taken up by the microorganisms and induces the expression of further xylanases (Collins et al. 2005).

β-Xylosidase (EC 3.2.1.37)

Xylosidases attack xylan chains at the nonreducing end and cleave xylose. Represented in GH families 1, 3, and 43, they increase the activity of the enzyme mixture in conjunction with other lignocellulose-degrading enzymes, as they degrade the inhibiting short-chain xylans to xylose (Collins et al. 2005).

Arabinofuranosidase (EC 3.2.1.55)

Arabinofuranosidases hydrolyze α-l-arabinofuranoside residues at the nonreducing end of arabinan and arabinoxylan to arabinose and short-chain arabinans. Most representatives of these enzymes are found in GH families 43 and 51.

α-l-Rhamnosidase (EC 3.2.1.40)

Rhamnosidases catalyze the hydrolysis of terminal, nonreducing α-l-rhamnose residues linked to other sugar polymers. Most rhamnosidases belong to the GH family 78.

β-Mannanase (EC 3.2.1.78)

Mannanases catalyze the random *endo-cleavage* of 1,4-β-d bonds in mannan to form mannooligomers and mannose. They are classified in the GH families 26 and 113.

β-Mannosidase (EC 3.2.1.25)

Mannosidases cleave mannose units from the nonreducing end of mannan. Representatives of this enzyme class are found in the GH families 1, 2, and 5.

β-Galactosidase (EC 3.2.1.23)

Galactosidases catalyze the cleavage of galactose from the nonreducing end of a galactoside chain. They are divided into the GH families 2, 35, 42, and 59.

Feruloyl Esterase (EC 3.1.1.73)

In contrast to the enzymes mentioned so far, feruloyl esterase (FAE) hydrolyzes ester bonds. Depending on substrate affinity, a distinction is made between type A and type B FAEs, both of which cleave feruloylic acid from arabinose residues. Type A FAEs prefer substrates in which the C3 or C5 position of the phenol ring is methoxylated. Type B FAEs prefer substrates in which the phenol ring is occupied by hydroxyl groups.

15.3.2 Pectinases

Pectinases are produced by fungi, bacteria, and yeasts, as well as by plants. The fungus *A. niger* is frequently used for industrial production. Pectinases can generally be divided into three groups: Protopectinases (PPases), depolymerases (polygalacturonases, pectate lyases), and pectin esterases (Jayani et al. 2005).

Protopectinases

Protopectin is water-insoluble pectin. This is transformed by protopectinases with the inclusion of water into water-soluble, highly polymerized pectin. A distinction is made between type A and type B PPases. Type A PPases catalyze reactions on the main chains of galacturonic acid, while type B PPases attack the side chains consisting of sugars (Jayani et al. 2005).

Polygalacturonases

Polygalacturonases (PGAs) are the best-studied pectinases. They hydrolytically cleave the α-1,4-bonds between two polygalacturonic acid units in the pectin main chain. Two PGAses can be distinguished depending on the substrate binding site, the endo-PGAse (EC 3.2.1.15), which cuts the main chain at a random position, and the exo-PGAse (EC 3.2.1.67), which attacks sequentially from the end of the chain. Endo-PGAses are common in fungi, bacteria, and yeasts, as well as in more highly developed plants. In contrast, exo-PGAses occur less frequently. A distinction is made between bacterial exo-PGAses, whose main product is the dimer of galacturonic acid, and fungal exo-PGAses, which mainly produce galacturonic acid as a monomer (Jayani et al. 2005).

Pectate Lyases

Under the collective term pectate lyases, two enzyme types can be distinguished according to their substrate: the polygalacturonate lyases (EC 4.2.2.2, EC 4.2.2.9) and the polymethylgalacturonate lyases (EC 4.2.2.10). For both enzyme types, there are representatives that attack the respective substrate by an *endo cleavage* at a random position of the chain or by an *exo cleavage* at the chain end. The pectate lyases are also called *transeliminases* because they perform a *trans-eliminative* cleavage that forms a double bond between the C4 and C5 positions of one of the products (Jayani et al. 2005).

Pectin Esterases

Pectin esterases (EC 3.1.1.11) catalyze the cleavage of methoxy groups from the pectin main chain. Polygalacturonic acid and methanol are formed as products. They occur mainly in fungi and plants. Fungal pectin esterases usually carry out cleavages at random positions, while the plant pectin esterases attack the methoxyl groups at the ends of the pectin chain (Jayani et al. 2005).

Other Enzymes for Pectin Degradation

In addition to the enzymes listed so far, there are other pectin-degrading enzymes: exo-polygalacturonisidases (EC 3.2.1.82), rhamnogalacturonases (EC 3.2.1.-), pectin acetylesterases (EC 3.1.1.-), and rhamnogalacturonan acetylesterases (EC 3.1.1.-) (Jayani et al. 2005).

15.3.3 Ligninases

The search for enzymes for the degradation of lignin usually focuses on lignolytic enzymes from white-rot fungi, the so-called *basidiomycota*. The enzyme sets of these fungi have so far been considered the most efficient for lignin degradation. As a rule, white-rot fungi attack cellulose, hemicellulose, and lignin simultaneously. However, some subspecies, such as *Ceriporiopsis subvermispora*, specialize in lignin-only degradation and thus offer high potential for applications in which the sugar-containing structures of the biomass are to be preserved and lignin selectively removed. Since ligninolytic enzymes cannot yet be produced recombinantly in large quantities, their use has been limited to date.

Lignin-degrading enzymes are also found in the CAZy database as enzymes supporting carbohydrate degradation (Lombard et al. 2014) and fall into three categories: phenol oxidases, heme peroxidases, and supportive enzymes. They are outlined below.

Phenol Oxidases (EC. 1.10.3.2)

Phenol oxidases, also called laccases, are glycosylated oxidoreductases and contain four copper ions as the only cofactor in the active site. The reaction sequence of lignin degradation with laccases can be explained with the aid of the laccase mediator system. In this process, the reduced laccase is oxidized by molecular oxygen. This then binds a mediator molecule and oxidizes it to form a free radical. The free radical then oxidizes possible lignin substrates by a redox reaction. The mediator, which is reduced again in this step, can then be re-oxidized by an oxidized laccase (Roth and Spiess 2015).

Phenol oxidases can carry out oxidations on phenols, but arylamines, anilines, or thiols are also possible substrates. An example reaction is the oxidation of benzenediol. With regard to lignin degradation, reactions involving the laccase-mediator system can contribute to both depolymerization and polymerization, as well as to the chemical modification of lignin (Dashtban et al. 2010).

Heme Peroxidases

Heme peroxidases are glycosylated proteins with a heme group in the active site, which is connected to the surface by a tunnel. Heme peroxidases require hydrogen peroxide (H_2O_2) as a cosubstrate for the oxidation of substrates. Among the heme peroxidases, lignin peroxidases (LiP), manganese peroxidases (MnP), and versatile peroxidases are distinguished with regard to their substrate spectrum and possible reaction pathways (Dashtban et al. 2010).

Lignin Peroxidases (EC.1.11.1.14)

Although lignin peroxidases are able to depolymerize the nonphenolic portion of lignin, which accounts for about 80–90%, they are not produced by all fungi specialized in the degradation of lignin. In addition to nonphenolic lignin, the substrate spectrum includes phenolic molecules with small molar masses or redox potentials above 1.4 V. The reaction occurs via a mediator that is formed in the active site, detaches from it, and then oxidizes the substrate molecule upon contact with it. In this way, cleavage of side chains, polymerization, and depolymerization can occur (Dashtban et al. 2010).

Manganese Peroxidases (EC. 1.11.1.13)

The substrate spectrum of manganese peroxidases mainly includes phenolic components found in lignin. Reactions with nonphenolic components require an additional reaction step and therefore occur less frequently. The reaction sequence is similar to that of lignin peroxidase. The mediator is manganese, which is oxidized in the first step in the active site of Mn(II) to Mn(III). It then leaves the active site bound in a chelate complex, e.g., with oxalates. Outside the enzyme, this reactive Mn(III) complex can then react with phenolic lignin. For oxidation with nonphenolic lignin, the Mn(III) complex first requires another mediator, e.g., organic acids. This leads to the formation of reactive carbon radicals, such as acetic acid radicals, which can oxidize nonphenolic lignin (Dashtban et al. 2010).

Versatile Peroxidases (EC: 1.11.1.16)

Versatile peroxidases have the greatest potential of the various peroxidases, as they have a broad substrate and reaction spectrum due to their hybrid molecular structure. They can oxidize both phenolic and nonphenolic substances and are more efficient than lignin or manganese peroxidases. Versatile peroxidases have a hybrid structure with multiple binding sites. Here, the active site is located inside the enzyme but is connected to the surface by two tunnels. The first tunnel is similar to that of LiP, while the oxidation of manganese takes place in the second tunnel, as in MnP (Dashtban et al. 2010).

Supporting Enzymes

In addition to the laccases and heme peroxidases already presented, there are other enzymes that support lignin degradation. Oxidases support the activity of heme peroxidases by producing their cosubstrate H_2O_2. These enzymes include aryl alcohol oxidase (EC 1.1.3.7) and glyoxal oxidase (EC 1.1.3.-). In addition, aryl alcohol dehydrogenases, quinone reductases, and cellobiose dehydrogenase (CDH) contribute to the further degradation of lignin. CDH plays a major role in this process as it degrades cellulose and hemicellulose in addition to lignin. It oxidizes disaccharides and oligomers of glucose and other sugars with a β-1,4 bond. The reaction with lignin occurs via free hydroxyl radicals, which can be formed in the presence of hydrogen peroxide (Dashtban et al. 2010).

Following this detailed presentation of the relevant enzyme activities for biomass degradation, the following sections place enzymatic biomass hydrolysis in the context of biomass pretreatment and an ethanol biorefinery.

Questions:
9. What is the difference between cellulases and hemicellulases?
10. Why is it complicated to get pure ligninases?

15.4 Pretreatment of Biomass

Due to the strongly cross-linked structure of lignocellulose, access to cellulose and hemicellulose for cellulases and hemicellulases is limited. In addition, parts of the lignin inhibit enzymatic hydrolysis. Therefore, biomass is pretreated prior to hydrolysis to reduce particle size, remove lignin and/or hemicellulose, and reduce the crystallinity of cellulose. A distinction is made between physical, chemical, and biological pretreatment, each of which has different effects on the biomass. There are also mixed forms of these methods, which are not discussed in more detail here.

15.4.1 Physical Pretreatment

Physical pretreatment methods include mechanical and thermal processes. Mechanical particle size reduction is usually done by cutting or grinding to a particle size in the millimeter to centimeter range and precedes virtually all other pretreatment methods. Thermal pretreatment methods use water at increased temperatures to break up the lignocellulosic structure. A distinction is made between the *liquid hot water* method and the *steam explosion*. Both methods process biomass with a particle size in the millimeter to centimeter range.

Liquid Hot Water Pretreatment
In hot water hydrolysis, the hemicellulose fraction of the shredded biomass is hydrolysed with hot water at increased pressure and at temperatures between 190 °C and 230 °C in a reactor. After a certain residence time, the material can be separated into a liquid phase containing hemicellulose and a solid residue consisting of cellulose and lignin (Agbor et al. 2011).

Steam Explosion (Steam Explosion)
Steam explosion uses the energy of the expansion of water during sudden evaporation. For this purpose, biomass soaked in water is placed in a reactor and heated to a temperature of 160–260 °C at high pressure. After a dwell time under these conditions, the pressure is abruptly reduced to ambient pressure. This leads to spontaneous evaporation of the water within the biomass. The associated expansion ruptures the cellular structure of the biomass and thus increases its surface area. After the reactor has cooled, a liquid phase is present, as in hot water hydrolysis, which contains most of the hemicellulose. The remaining swollen solid contains cellulose and lignin. Compared to the hot water method, steam explosion more effectively detaches the hemicellulose from the solid. In addition, the reduction in particle size makes the biomass more usable (Agbor et al. 2011).

15.4.2 Chemical Pretreatment

Chemical pretreatment processes include methods of pretreating biomass for enzymatic hydrolysis using various chemicals, such as acids, bases, organic solvents, and ionic liquids.

Acid Pretreatment
Acid pretreatment can use both concentrated and dilute sulphuric, hydrochloric, and nitric acids. The main effect of acid-based pretreatment is hydrolysis and thus removal of the hemicellulose. In particular, pretreatment with concentrated acids also dissolves and precipitates some of the lignin. However, the corrosive effect of concentrated acids causes high equipment costs and hinders enzymatic hydrolysis due to the low pH, so dilute acids are preferred for pretreatment (Agbor et al. 2011).

Alkaline Pretreatment
For alkaline pretreatment, sodium hydroxide, ammonia (e.g., for *ammonia fiber expansion*, AFEX), and calcium hydroxide are used, which effectively remove lignin from the biomass. However, the structure of the lignin is altered in the process, making further processing difficult. Alkaline pretreatment also leads to an increase in the surface area of the cellulose due to swelling and depolymerization and dissolves part of the hemicellulose (Agbor et al. 2011).

Organosolv Pretreatment
In the organosolv process, an organic solvent, usually ethanol, and also other short-chain alcohols, ketones, or organic acids, is used in combination with water to pretreat the biomass. The solvents remove lignin and part of hemicellulose from the biomass. The effect can be enhanced by adding a catalyst namely oxalic acid. At the end of this so called organocat process, the liquid consists of an organic phase rich in lignin and an aqueous phase rich in hemicellulose. Disadvantages of solvent pretreatments are the relatively high

solvent costs, which require separation and recycling of the solvent for the profitability of the process (Agbor et al. 2011).

Ionic Liquids

Ionic liquids (IL) are organic salts that exist in liquid form at room temperature. For example, 1-ethyl-3-methylimidazolium acetate (EMIM Ac) or 1,3-dimethylimidazolium dimethyl phosphate (MMIM DMP) is used in the pretreatment of biomass. The treatment of biomass with ionic liquids leads to the dissolution of hemicellulose and lignin in the liquid phase. Pure cellulose, whose crystallinity has been greatly reduced, remains as the solid phase, which facilitates subsequent hydrolysis. However, ionic liquids are very expensive, despite the fact that they should be recycled. In addition, the cellulose obtained must be washed to avoid inactivation of the celluloses by traces of the ionic liquid (Engel et al. 2010).

15.4.3 Biological Pretreatment Methods

Biological pretreatment methods of biomass mean the use of microorganisms, usually fungi, to degrade lignin. Various white, brown, or soft rot fungi are used for this purpose. Since most fungi that produce enzymes for the degradation of lignin also degrade cellulose and hemicellulose, such strains that strongly prefer lignin as a substrate are targeted for reduction of cellulase activity. Since the biological pretreatment does not require any further chemicals, it is interesting in terms of energy and safety. However, the effort required for reaction control and, above all, the long reaction times still make these biological processes uneconomical.

Questions:
11. Which processes can reduce the lignin content of the substrate?
12. Which process mainly dissolves hemicellulose?
13. Which process leads to pure cellulose?

15.5   Process Overview

The energetic or material use of lignocellulose is subject to a large number of regional as well as general influencing factors. The regional factors include:

- Which biomass is available as a raw material at which time of year?
- What influence does the utilization of this biomass have on the agricultural economy and soil quality in the region?
- Does the infrastructure exist to profitably utilize by-products such as waste heat or biogas?

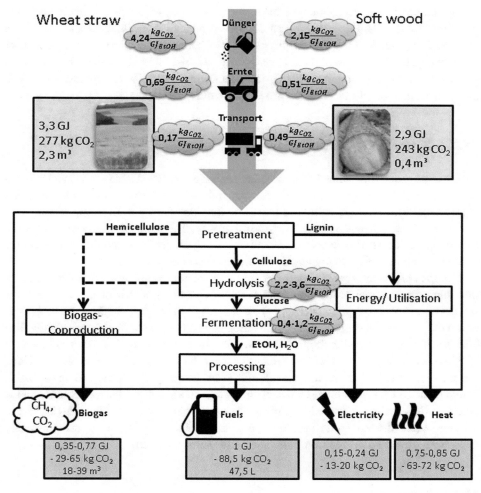

Fig. 15.5 Two example processes for the production of the target quantity 1 GJ ethanol, starting from wheat straw (left) and softwood (right) by means of hydrolysis and fermentation with parallel biogas co-production. Applied energies are given in kg CO_2 equivalent per GJ ethanol produced (according to Börjesson et al. 2013)

This results in the general factors:

- Which process management is optimal under the given circumstances?
- What is the greenhouse gas saving potential of the process?

Part of this complexity is exemplified by a biorefinery which, in addition to the main product ethanol, also produces biogas, electricity, and heat. Two exemplary process routes start with the raw materials straw and softwood (Fig. 15.5).

In addition to the production of fuels, the processing of biomass aims to save the greenhouse gas CO_2. However, the production of ethanol from biomass also leads to a release of CO_2 due to the use of fertilizer as well as harvesting and transport. On the other hand, the reduction in crop waste leads to reduced N_2O release, which also acts as a greenhouse gas (Börjesson et al. 2013). The balance in Fig. 15.5 is not completely closed, as the data situation of this complex process control is incomplete.

15.5.1 Biomass Cultivation

On the one hand, lignocellulosic biomass, such as straw and wood residues, is available as a waste material to the agricultural and forestry industries. On the other hand, wood and highly cellulosic plants are cultivated for the production of ethanol. Both approaches have direct or indirect impacts on land use and soil quality. The removal of residual and waste materials such as straw reduces the amount remaining on the field, and more nutrients are removed from the soil, making additional nitrogen fertilization necessary.

If crops are planted specifically for biomass utilization, competition with food production, e.g., through fuel production from biomass containing starch or oil, has been the subject of controversial public debate in recent years *(food vs. fuel conflict)*. Although this does not apply to lignocellulose, the cultivation of this biomass indirectly means that less land is available for growing food. If new areas for the agricultural and forestry industry are designated not only in the rainforest regions of the tropics but also in Central Europe *(land use change, LUC)*, this leads to a reduction in plant diversity and to the release of carbon bound in the soil. Therefore, the choice of cultivation area is included in a calculation of the potential CO_2 savings. Currently, plants are being developed that provide a high yield of cellulose and at the same time have low demands on soil quality in order to use previously unusable barren land for biomass cultivation (Börjesson et al. 2013).

15.5.2 Transport

Biomass feedstocks for biofuel production are usually distributed over large areas. Therefore, the biomass has to be transported from its decentralized production sites to the biorefinery. In relation to the energy content of the biomass moved, the transport costs vary. In particular, for feedstocks with low energy density, transport is only profitable over limited distances. On the one hand, small decentralized biorefineries can reduce the transport distance, and on the other hand, the energy density of the raw materials can be increased decentrally by pretreatment (Börjesson et al. 2013).

15.5.3 Pretreatment

Different pretreatment processes lead to different biomass compositions (Sect. 15.4). In a biorefinery, separation of lignin is usually desirable in order not to inhibit enzymatic hydrolysis. In addition, lignin can be burnt to generate the necessary process energy, e.g., in the Inbicon process (Larsen et al. 2012). However, the material use of lignin is more desirable as it significantly increases the profitability of the overall process. However, so far lignin has only been utilized as a material on a small scale.

15.5.4 Enzymatic Hydrolysis

Enzyme production costs have a major impact on the profitability of a biorefinery. Therefore, research is being conducted worldwide on enzyme mixtures that achieve a high turnover in a short time and that can be produced at a low cost. By increasing the enzyme activity, the required enzyme loading for sufficient biomass conversion can be reduced from the current 19–26 mg of enzyme per gram of biomass to 10 mg g^{-1}. Since enzyme activity decreases over time and a portion cannot be recycled, enzymes must be constantly replenished. In this regard, research by enzyme manufacturers Novozymes, DuPont, and Verenium has led to a reduction in enzyme costs to about $0.13 per liter of ethanol produced if the enzymes are used immediately after they are produced. Further potential exists in genetic modifications of the enzymes that lead to higher specific activity (Wilson 2009).

Since a high glucose yield is the goal of enzymatic hydrolysis, a high biomass concentration is required at the beginning of hydrolysis. However, this is limited due to the high viscosity, poor mixing, low water activity, and product inhibition. Process design calculations for a biorefinery currently assume a solid loading of 20 wt% to achieve a sugar concentration of 100 g L^{-1}. However, higher loadings are being targeted (Larsen et al. 2012). The hydrolysate is filtered to retain biomass residues and enzymes and is then fed to the fermentation.

15.5.5 Ethanol Production

The nonaromatic components of the biomass, cellulose, and hemicellulose can be used separately or simultaneously, depending on the type of process. In separate or simultaneous hydrolysis and fermentation (SHF, SSF), only the cellulose is used. The advantage of SHF is that hydrolysis and fermentation can be carried out in separate reactors under optimum conditions. If both processes take place in one reactor, as in the case of SSF, a compromise has to be found with regard to temperature and pH value, but the equipment required is considerably reduced. In addition to glucose, cofermentation also utilizes the sugars of hemicellulose, especially xylose. Xylose and other pentoses are poorly or not at all utilized by most ethanol-producing microorganisms. Therefore, strains are being developed that also efficiently convert pentoses to ethanol. Alternatively, xylose is utilized in biogas coproduction. After fermentation, ethanol is present in concentrations of about 10% by volume. Therefore, purification and concentration are necessary. This is usually done by rectification, but vacuum stripping can be carried out beforehand. In vacuum stripping, enzymes are not denatured and can therefore be recycled. Separation by rectification only achieves an ethanol concentration of 96 vol% under normal conditions. To obtain pure ethanol, further purification steps, such as the use of molecular sieves, are necessary (Larsen et al. 2012).

Questions:
14. What is the main problem with the transport of lignocellulose?
15. What are the advantages and disadvantages of separate and simultaneous hydrolysis and fermentation?

Take Home Message
- There are many different lignocellulosic structures depending on the plant and the function of the plant part.
- For each component of lignocellulose, there are specialized enzymes produced by fungi, which use it as a substrate.
- Cellulose is a polymer of glucose, hemicellulose is a polymer of pentoses and hexoses, while lignin is a polymer of aromatic components.
- The fungus *Trichoderma reesei* is widely used to produce cellulases.
- Cellobiohydrolases, endoglucanases, and beta-hydrolases are the most used cellulases.

Answers:
1. The main components are cellulose, hemicellulose, and lignin. Some plants also contain pectin.
2. Cellulose.
3. For example, citrus fruits and sugar beets.
4. Lignin contains aromatic monomers which form a complex structure. These aromatic components make lignin hydrophobic.
5. Adsorption, breaking of the crystalline structure, hydrolysis, release of products, and desorption.
6. Cellobiohydrolases (CBHs), endoglucanases (EGs), beta-glucosidases, and polysaccharide monooxygenases.
7. Cellobiohydrolases attack the cellulose strands from the ends and produce cellobiose while endoglucanase cuts cellulose in amorphous regions inside the strand and thus produces new ends and oligomers of different lengths.

8. During the production of celluloses from the fungus *T. reesei*, which is widely used, the beta-glucosidases remain bound to the mycelium and thus lack in the produced enzyme cocktail.
9. The main difference is the substrate for which they are specialized.
10. Most fungi that produce ligninases also produce cellulases, the activity of which is enhanced by ligninases.
11. Alkaline pretreatment, organosolv, and ionic liquids.
12. Liquid hot water, steam explosion, and acid pretreatment.
13. Pretreatment with ionic liquids.
14. The energy content of substrates like straw is slow compared to the volume and effort needed for transport.
15. Performing hydrolysis and fermentation separately has the advantage that both processes can run at optimum temperatures and pH, while there has to be found a compromise during the simultaneous process. However, the equipment cost is higher since two vessels are needed for the process.

References

Agbor VB, Cicek N, Sparling R, et al. Biomass pretreatment: fundamentals toward application. Biotechnol Adv. 2011;29:675–85. https://doi.org/10.1016/j.biotechadv.2011.05.005.

Börjesson P, Ahlgren S, Barta Z, et al. Sustainable performance of lignocellulose-based ethanol and biogas co-produced in innovative biorefinery systems. Department of Environmental and Energy Systems Studies; 2013.

Collins T, Gerday C, Feller G. Xylanases, xylanase families and extremophilic xylanases. FEMS Microbiol Rev. 2005;29:3–23. https://doi.org/10.1016/j.femsre.2004.06.005.

Dashtban M, Schraft H, Syed TA, Qin W. Fungal biodegradation and enzymatic modification of lignin. Int J Biochem Mol Biol. 2010;1:36–50.

Engel P, Mladenov R, Wulfhorst H, et al. Point by point analysis: how ionic liquid affects the enzymatic hydrolysis of native and modified cellulose. Green Chem. 2010;12:1959–66.

Himmel ME, Ding S, Johnson DK, et al. Biomass recalcitrance: engineering plants and enzymes for biofuel production. Science. 2007;315:804–7.

Jayani RS, Saxena S, Gupta R. Microbial pectinolytic enzymes: a review. Process Biochem. 2005;40: 2931–44. https://doi.org/10.1016/j.procbio.2005.03.026.

Larsen J, Haven MØ, Thirup L. Inbicon makes lignocellulosic ethanol a commercial reality. Biomass Bioenergy. 2012;46:36–45. https://doi.org/10.1016/j.biombioe.2012.03.033.

Lombard V, Golaconda Ramulu H, Drula E, et al. The carbohydrate-active enzymes database (CAZy) in 2013. Nucleic Acids Res. 2014;42:490–5. https://doi.org/10.1093/nar/gkt1178.

Montenecourt BS. *Trichoderma reesei* cellulases. Trends Biotechnol. 1983;1:156–61. https://doi.org/10.1016/0167-7799(83)90007-0.

Neufeld K, Pietruszka J. Understanding nature—towards the enzymatic degradation of cellulose with monooxygenases. ChemCatChem. 2012;4:1239–40.

Payne CM, Knott BC, Mayes HB, et al. Fungal cellulases. Chem Rev. 2015;115:1308–448. https://doi.org/10.1021/cr500351c.

Roth S, Spiess AC. Laccases for biorefinery applications: a critical review on challenges and perspectives. Bioprocess Biosyst Eng. 2015; https://doi.org/10.1007/s00449-015-1475-7.

Van Dyk JS, Pletschke BI. A review of lignocellulose bioconversion using enzymatic hydrolysis and synergistic cooperation between enzymes-Factors affecting enzymes, conversion and synergy. Biotechnol Adv. 2012;30:1458–80. https://doi.org/10.1016/j.biotechadv.2012.03.002.

Wilson DB. Cellulases and biofuels. Curr Opin Biotechnol. 2009;20:295–9.

Enzymes in Food Production

16

Lutz Fischer

What You Will Learn in This Chapter

The use of enzymes in the food industry is very different compared to their use in the chemical or pharmaceutical industries. Firstly, the use of separately added enzymes is legally restricted for safety reasons and the EU regulation describes an authorization requirement for such *food enzymes*. Secondly, the food enzymes have to be cheap since the foods are cheap and traded in weight and not in performance as it is the case with high-priced pharmaceuticals or chemicals. Thus, the production of new food enzymes is most often done recombinantly using suitable host microorganisms. The food enzymes have to be active and stable in presence of highly complex food compositions during food processing but, on the other hand, they have to be easily inactivated by technical means. The conditions of the particular food manufacturing determine the conditions for the enzymatic reaction and cannot be adjusted to the enzyme's requirements due to the food law. So, enzymes with a broad tolerance in pH, temperature, and ionic strength of the settings are beneficial. In this chapter important examples of application are presented for main enzyme classes EC 1–6 according to the scientific classification of the International Union of Biochemistry and Molecular Biology (IUBMB). For example, lipases are used in the production of

(continued)

L. Fischer (✉)
Department of Biotechnology and Enzyme Science, Institute of Food Science and Biotechnology, University of Hohenheim, Stuttgart, Germany
e-mail: sekretariat-bt@uni-hohenheim.de

© The Author(s), under exclusive license to Springer Nature Switzerland AG 2024
K.-E. Jaeger et al. (eds.), *Introduction to Enzyme Technology*, Learning Materials in Biosciences,
https://doi.org/10.1007/978-3-031-42999-6_16

certain cheese products and varieties or margarines, and glycosidases (e.g., *β*-galactosidases = lactases) are used in the production of lactose-free milk or prebiotic galactooligosaccharides (GOS). Peptidases (e.g., chymosin = rennin) are required, for example, for milk thickening in cheese production, and various microbial peptidases are required for protein hydrolysates of specialty foods (e.g., baby foods, sports foods) and food seasonings. The best-known *isomerase* converts D-glucose intramolecularly to D-fructose up to equilibrium, is used immobilized and is used for the production of sugar syrup for soft drinks and other foods on a million-ton scale. New isomerases (cellobiose 2-epimerases) can directly convert lactose found in milk to lactulose and epilactose, two potentially prebiotic sugars. All examples make it clear that the targeted use of enzymes can positively change the properties of food and thus achieve higher quality and added value of the products in terms of their shelf life and texture, taste, and physiological functionality.

16.1   Determining Factors of Enzymes in the Food Industry

Foods serve, according to their definition, the nourishment of humans and the drug-free preservation of humans life. Foods consist of edible plants and animals or parts thereof in an unprocessed or—mostly—processed state. Foods that have not yet been further processed are typically referred to as *raw foods* or *raw materials*, respectively. By their very nature, these consist of cells and possess all intracellular components such as DNA, RNA, lipids, sugars, proteins, and hence enzymes, vitamins, low molecular weight metabolites, minerals, and water. The "Food and Feed Code" (LFBG) replaced the previous "Foodstuffs and Consumer Goods Act" (LMBG) in 2005 and legally regulates all production and processing stages of plant and animal raw materials along the *food value chain* for the production of foodstuffs, consumer goods, animal feed, and cosmetics. Foodstuffs must be absolutely safe in terms of health and must meet consumers' sensory (taste, smell, texture) and visual expectations. Furthermore, emotional, cultural, and religious aspects also come into play. The use of enzymes in the food industry is therefore completely different than in chemical or pharmaceutical industries and subject to various important peculiarities, which are addressed here first.

16.1.1 Endogenous and Exogenous Enzymes in Foodstuffs

The raw foods, as mentioned above, consist of cells and contain their intrinsic enzymes, which are called *endogenous enzymes*. The type and amount of endogenous enzymes in a raw commodity is organism-specific and depends on the physiological state of the cells at the time of harvest or slaughter. These endogenous enzymes are naturally active and cause changes in the raw foods during transport, storage and processing, depending on water content, pH, temperature, and ionic strength. This may have to be taken into account in the subsequent processing of the foodstuffs. In addition, raw foods are not obtained, transported, and stored under aseptic conditions and may naturally be associated or contaminated with microorganisms, respectively. The unavoidable microbial intra- and extracellular enzymes from a raw food product are also considered as *endogenous enzymes*, can vary a lot in quantity dependent on the treatment of the raw foods, and play a crucial role especially in the fermentation of food products. If the natural contamination of particularly sensitive raw material, such as milk, is unacceptable for further use, a techno-logical inactivation step to avoid the growth of the microorganisms and the activity of endogenous enzymes is carried out as soon as possible (for example, for raw milk: a heating or high-pressure process or addition of H_2O_2).

Exogenous enzymes are all enzymes added separately to the raw foods or raw materials, respectively. Exogenous enzymes are added in highly concentrated, mostly liquid form or, less frequently, as powder or granules. This depends on the consistency of the food materials. Exogenous enzymes play a decisive role in the processing and production of most foodstuffs for the product-specific properties, quality, and/or product yield. As an example of product quality, the selective casein cleavage through the use of chymosin, a specific peptidase/protease from a calf stomach (bovine, camel), fungus *(Rhizomucor miehei)* or recombinant from a microorganism *(Aspergillus niger, Kluyveromyces lactis)*, is applied in the production of certain types of cheese. For example, microbial pectinase preparations consisting of a mixture of different pectin-degrading enzymes are used in fruit juice production to significantly increase the product yield. The latter also improve the fruit juice quality, as the enzymatic pectin degradation simultaneously reduces the viscosity of the juice.

The exogenous enzymes in the food industry often bear trivial names such as "CakeZyme®" (lipase from DSM, Heerlen, The Netherlands), "Yieldmax®" (phospholi-pase A1, Christian Hansen, Hoersholm, Denmark), or "FoodPro® 51FP" (endo-/exopeptidase, DuPont Industrial Bioscience, Brabrand, Denmark), which are used by the enzyme manufacturers for marketing reasons. Thus, the correct enzyme names determined by the International Union of Biochemistry and Molekular Biology (IUBMB) should be ascertained when discussing science (see enzyme classes EC below). It should also be noted that in the vast majority of cases, the exogenous enzymes represent a mixture of different enzymes and therewith different activities. Still, the enzyme activity indicated by the manufacturer corresponds only to the main activity of the enzyme compound. It is therefore better for scientific reasons to speak of enzyme preparations as this makes it clear that the enzyme compounds are not pure enzymes and, that secondary activities are present. A scientifically well studied example for this is the enzyme preparation "Flavourzyme®" (exopeptidase with endopeptidase side activities, Novozyme, Bagsværd, Denmark). Flavourzyme is used, as the trivial name should express, e.g., for flavor formation in plant protein hydrolysates. It is marketed as an aminopeptidase preparation and is standardized to so-called leucine aminopeptidase units (LAPU). In fact, Flavouryzme is much more complex: it is a mixture of at least two aminopeptidases, two dipeptidyl peptidases, three proteases (endopeptidases) and one α-amylase (Merz et al. 2015). This mixture of peptidases shows synergistic proteolytic activities and so, each of the peptidases contributes to the final hydrolysis result in its own manner.

Questions
1. What belongs to endogenous and to exogenous enzymes in the field of food manufacturing?
2. What is special about the names and the activity of exogenous enzymes used in the food industry?

16.1.2 Requirements for the Use of Exogenous Enzymes in Food Processing

The addition of exogenous enzymes in industrial food production enables the selective and gentle influence on the functional, i.e., the nutritional, sensory, and technological properties of foods. For these reasons, enzymes have been used in many manufacturing processes for decades. The trend is still increasing, as the catalytic selectivity and efficiency of enzymes, while maintaining the sensory and nutritional value of the food, cannot be replaced by exclusively process engineering measures. The combination of both aspects, on the other hand, enables the development and production of new, often even higher quality foods. In principle, however, the addition of exogenous enzymes must be economically justifiable. This means that the use of enzymes must make the production of the food more cost-effective (yield increase, energy saving) and/or improve the functional quality.

Enzymes are known to belong to the substance class of proteins and are basically edible. Nevertheless, the use of exogenous enzymes in the food industry is legally restricted and prohibited by the above mentioned LFGB in § 6 (2) with reference to Regulation (EC) No. 1332/2008 on food enzymes of the European Parliament and of the Council, because the use of enzymes in food is to be regulated uniformly throughout Europe in the future. The EU regulation describes an authorization requirement for food enzymes. Accordingly, an application for approval must be submitted to the EU, which will be evaluated by a commission of experts (European Food Safety Authority; EFSA). Only if the EU Council then decides in favor of the enzyme will it be included on a positive list and can be used in food in the future. However, until the compilation of the positive list for food enzymes which are already on the market before the regulation is completed, the national regulations remain in force. In Germany, a distinction is currently made between so-called food enzymes which are added as *processing aids* and which are not declared and do not have to be declared, and those that are used as *food additives* and have to be and must be approved. The vast majority of enzymes that were already on the market before the EU regulation of 2008 fulfil the requirements of processing aids and can still be used in food production in Germany. The criteria for a processing aid are:

- It is not consumed as a food ingredient itself but is used in the processing of raw materials, foodstuffs, or their ingredients for technological reasons.
- Its presence in food is unintentional, technically unavoidable and harmless to health.
- It has no technological effect on the variability of the final product.

The latter means that the endogenous food enzymes used as processing aids must be inactivated or denatured by means of process technology after they have carried out their catalytic work.

Questions
   3. Why are exogenous enzymes used in food manufacturing?
   4. What has to be considered when exogenous enzymes shall be used for food production?

16.1.3 Recombinant Production of Enzymes for the Food Industry

The food industry processes very large quantities of substances, which, however, have to be produced at extremely low costs. The possibility of using exogenous enzymes thus depends largely on the price. Microorganisms are the most biotechnologically effective enzyme producers and realize the highest possible productivity in terms of product quantity per volume and time. However, enzyme formation in microorganisms living in nature (wild strains) is very well regulated at the gene level according to their homeostasis and evolutionary pressure in the field. This means that there is normally no overproduction of one or more enzymes in a microorganism for cellular energetic reasons.

Before the 1980s, enzyme overproduction in microorganisms was achieved by the time-consuming and resource-intensive screening of chemically or X-ray mutagenized cells (classical mutagenesis methods). In the microorganisms obtained in this way, the desired enzyme overproduction had been accomplished by random mutations in the various gene elements responsible for enzyme expression (promoter, operon, regulatory proteins, etc.).

Nowadays, through the rational and targeted use of genetic engineering methods, the natural enzyme regulation of a microorganism is disabled in such a way that enormous enzyme overproduction is achieved using known efficient expression systems instead. This so-called *recombinant expression* can be done either *homologous* in the same species where the enzyme originates from or, *heterologous* in another species which is specialized in overproduction of proteins. Whenever possible, a simultaneous secretion of the recombinant enzyme into the medium is tried. The success of the latter is dependent on several intrinsic factors of an enzyme, such as gene and/or amino acid sequence, conformation, folding, and the kind of suitable secretion pathway. The secretion of an enzyme would be desired as it enables an additional cost reduction, since the microorganism does not have to be disrupted, but the desired enzyme can be obtained by processing the cell-free supernatant.

The production of food enzymes with a genetically modified microorganism (GMO) should follow the *food-grade concept*. This means that the host organism (species) in which the genetic modification is carried out must be a safe microorganism of the lowest risk class 1. Depending on the region of use, it must also have GRAS status (*generally recognized as safe*) for the USA or QPS status (*qualified presumption of safety*) for Europe. The typical selection markers based on antibiotic resistance, which are required for the genetic engineering methods during GMO design, must be exchanged in the final production strain. Antibiotic resitances are prohibited in such recombinant production strains. The genetically engineered modifications must be described in all details and reduced to a minimum. Whenever possible, the use of heterologous enzyme production should be limited to closely related species compared to the wild-type species that also occur naturally in food. It must be warranted that any formation of harmful substances in the host organism is excluded. Due to the classical or modern methods of mutagenesis of microorganisms, the costs for food enzymes are generally between 30 and 100 € per kilogram of enzyme preparation.

The following sections describe in more detail the first six main classes (EC 1–6) of enzymes used in food processing. The new seventh main class (EC 7), translocases, is irrelevant for the food production. Each enzyme is categorized conforming its catalyzed reaction and has in total four digits, starting with the number of the main class and three further subclasses and sub-subclasses (for more details see www.enzyme-database.org).

Questions
  5. Why are food enzymes recombinantly produced nowadays in most cases?
  6. What means food-grade concept for the recombinant production of enzymes?

16.2   Oxidoreductases

16.2.1 Importance to the Food Industry

Oxidoreductases constitute the first class of enzymes according to the International Union of Biochemistry and Molecular Biology (IUBMB; http://www.sbcs.qmul.ac.uk/iubmb/) and have the number EC 1.x.x.x within the four-digit EC classification system. Oxidoreductases catalyze redox reactions, i.e., for each reduction, an oxidation is also carried out simultaneously. The subclasses (EC 1.1.x.x to 1.23.x.x and 1.97.x.x to 1.99.x.x) of oxidoreductases are classified according to the functional group that acts as a donor for electrons. Of particular relevance to the food industry are certain oxidoreductases of classes EC 1.1.x.x (CH-OH groups as donors), EC 1.10.-.- (diphenols and their relatives as donors), and EC 1.11.x.x (peroxides as acceptors). In detail, these are the *glucose oxidase* (EC 1.1.3.4), *laccase* (EC 1.10.3.2), and the *catalase* (EC 1.11.1.6). Oxygen or hydrogen transfer to a substrate takes place via prosthetic groups linked to the enzyme (e.g., FAD) or with the aid of so-called transport metabolites (e.g., NAD^+).

16.2.2 Applications in the Food Industry

The glucose oxidase (GOD; EC 1.1.3.4) catalyzes the conversion of glucose in the presence of oxygen to glucono-δ-lactone and H_2O_2 (Fig. 16.1) and is produced primarily from *Aspergillus* or *Penicillium* species (Bauer et al. 2022). GOD is selective for β-D-glucose and can barely oxidize α-D-glucose, hexoses, and xyloses. The chemical formation of gluconic acid happens in presence of water and causes a pH drop in the food, so that on the one hand a preservative effect occurs because of the pH shift and, at the same time Maillard reactions between amino groups and the reducing ends from glucose are prevented since glucose is converted. GOD is mainly added to lemonades. It is often used in combination with catalase to remove H_2O_2 (see below). Glucono-δ-lactone is also used in baking powders, dairy products, and in raw sausage seasoning. In wheat doughs, the H_2O_2 acts as a strong oxidizing agent and solidifies the gluten structure by cross-linking free sulfhydryl groups to form disulfide bridges. Another advantage of using GOD in baked goods is that ascorbic acid can be excluded, as ascorbic acid is not allowed to be used for traditional breads in countries such as France.

The laccases (EC 1.10.3.2) are also called "blue copper proteins" or "multicopper oxidases." This is due to the four copper atoms in their active site, which are responsible for the blue color of the laccases. Laccases are mainly extracellular enzymes and catalyze the oxidation of one electron of a wide range of phenolic substrates, such as *o*- and *p-benzenediols*, polyphenols, aminophenols, polyamines, lignins, aryldiamines, and various inorganic ions. It requires molecular oxygen as a co-substrate and it yields water as the sole by-product. They are used in the textile and paper industry, in the purification of waste water, and as catalysts for chemoenzymatic syntheses. In the food industry, and here primarily in the bakery and beverage industry, they are mainly used to remove undesirable phenolic compounds and free oxygen (Mayolo-Deloisa et al. 2020). Their use can also influence sensory and functional properties of foods. In beer production, laccases not only improve stability but also extend the shelf life of beers. By using laccases after extraction of the original wort, undesirable polyphenols are eliminated. Polyphenol complexes are formed which are removed by filtration. Thus, there is no turbidity of the beer. The prolonged shelf life can be explained by the conversion of free oxygen in a water molecule. Removing oxygen from the beer can also ensure that no off-flavor (a flavor that is perceived as unsuitable for a particular food) is chemically formed, as otherwise precursors of off-flavor components are formed through the oxidation of amino acids, proteins, fatty acids, and alcohols. However, in Germany, the use of laccases in beer is prohibited due to the German purity law. Another application of laccases is the stabilization of wines. Here, too, polyphenols are oxidized by laccases. The oxidized polyphenols subsequently polymerize and can be removed from the wine by clarification. Laccases can also be used to reduce the cloudiness and color changes of fruit juices. In food research, laccases are described to create structures in yoghurt. Here, casein is cross-linked with the help of so-called mediators (e.g., vanillic acid).

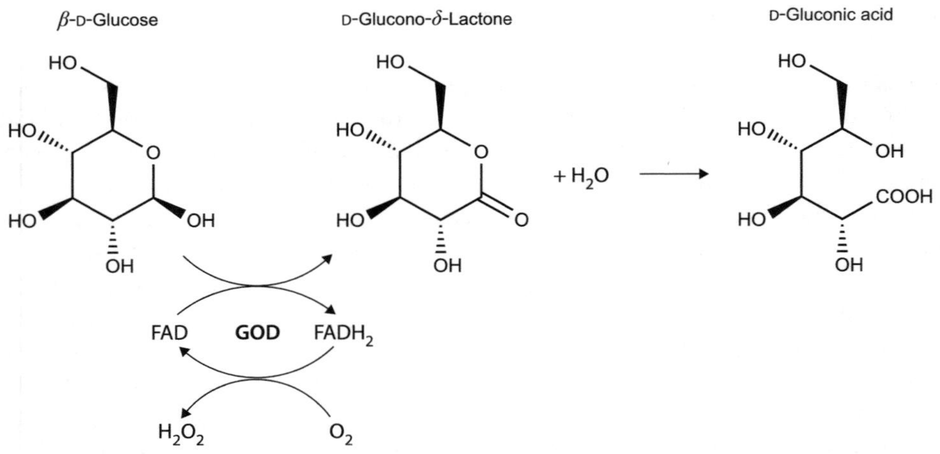

β-D-Glucose   D-Glucono-δ-Lactone   D-Gluconic acid

Fig. 16.1 Schematic representation of the reaction catalyzed by glucose oxidase (GOD)

Catalase (EC 1.11.1.6) catalyzes the degradation of hydrogen peroxide to oxygen and water. It is found in nearly all living organisms and has one of the highest turn-over numbers of all enzymes as it has the capacity to decompose more than one million molecules of H_2O_2 per molecule of enzyme per second. It acts as an antioxidants and protects the cell against oxidative stress. In food, it is therefore used to remove H_2O_2 that was either produced by other enzymes (see GOD) or added for disinfection reasons (Kaushal et al. 2018). For example, H_2O_2 is used for so-called cold pasteurization of milk, mainly in developing countries, and of some special cheeses. Similarly, several food packages are disinfected with hydrogen peroxide and then treated with catalase.

Question
  7. What are the reasons for using a glucose oxidase in the food industry?

16.3   Transferases

16.3.1 Importance to the Food Industry

Transferases belong to main class 2 (EC 2.x.x.x) and catalyze the transfer of a group X (e.g., glycosyl or methyl group) from one compound, generally regarded as donor A (A-X), to another compound, generally regarded as acceptor B (B-X). In many cases, the donor molecule is a cofactor (coenzyme) carrying the group to be transferred. Within main class 2, transferases are subdivided into ten subclasses, depending on the group they transfer (EC 2.1.x.x to EC 2.10.x.x). In the food sector, transferases from subclass EC 2.3.x.x (acyltransferases) and subclass EC 2.4.x.x (glycosyltransferases) are of particular relevance. Acyltransferases transfer acyl groups and form either esters or amides, and glycosyltransferases transfer glycosyl groups. Sub-subclasses are based on the acyl group that is transferred: acyl groups other than amino-acyl groups (EC 2.3.1.x), aminoacyltransferases (EC 2.3.2.x), and acyl groups that are converted into alkyl groups on transfer (EC 2.3.3.x). Some glycosyltransferases also catalyze hydrolysis, which can be regarded as the transfer of a glycosyl residue from the donor to water. Also, inorganic phosphate can act as acceptor in the case of phosphorylases; phosphorolysis of glycogen is regarded as transfer of one sugar residue from glycogen to phosphate. However, the more general case is the transfer of a sugar from an oligosaccharide or a high-energy compound to another carbohydrate molecule that acts as the acceptor. The glycosyltransferases can be further sub-subdivided depending on the nature of the sugar residue being transferred. There are hexosyltransferases (EC 2.4.1.x), pentosyltransferases (EC 2.4.2.x) and those that transfer other glycosyl groups (EC 2.4.99.x).

16.3.2 Applications in the Food Industry

The enzyme transglutaminase (EC 2.3.2.13) catalyzes the reaction between a γ-carboxyamide group of a peptide-bound glutaminyl residue act as donor and, an ε-amino group of a peptide-bound lysine residue act as acceptor to give intra- and inter-molecular N^6-(5-glutamyl)-lysine crosslinks (Fig. 16.2a). In addition to lysine residues, other primary amines can also serve as acceptors (Fig. 16.2b). If no primary amines are present, water reacts as a nucleophile, leading to deamidation of the glutaminyl residue (Fig. 16.2c).

All three reactions catalyzed by transglutaminase can be used in the food industry to modify the functional properties of proteins in foods. Their use lead to an increase in the strength of the protein matrix and its water-holding capacity, to an increase in the viscosity of protein solutions and to an improvement in the formation ability and thermal stability of gels. Thus, the texture of foods is modified. Transglutaminase from *Streptomyces moberaensis* is most often used in the food industry. Proteins in meat and fish, milk proteins in yoghurt or cheese, but also gluten proteins in bakery products are cross-linked by transglutaminase. In the case of meat products, transglutaminase can be used to produce restructured meat from meat scraps. It can also be used as a substitute for cutter adjuvants in the production of scalded sausage and to achieve faster cut strength in raw sausage. Another application in the meat sector is the cross-linking of sodium caseinate gels as a fat substitute in cooked and raw sausages. In fish products, transglutaminase is used for the production of surimi. Surimi is a fish gel made from fish meat scraps, water and salt, and transglutaminase. Transglutaminase is also used in the production of fish pies, and this reduces water loss during thawing of frozen fish by injecting or placing transglutaminase in a drum prior to freezing ("tumbling"). In the case of dairy products, transglutaminase is used to control the texture of creams made from skimmed milk powder and to make low-fat desserts. In the case of low-fat desserts, ice cream or yogurt, this creates a mouthfeel of full-fat products with a creamy consistency. This is created by cross-linking casein, which has

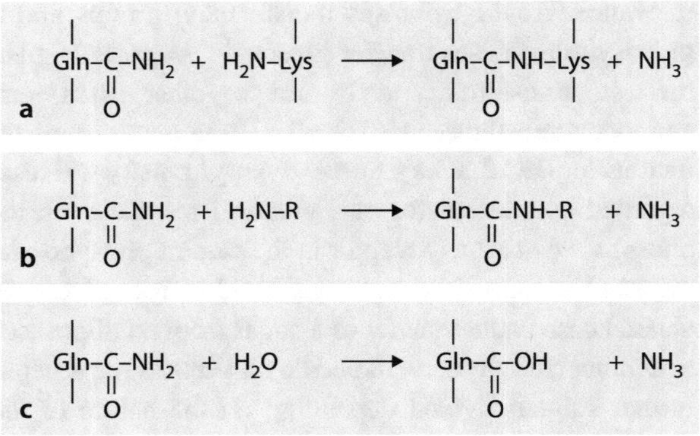

Fig. 16.2 Schematic representation of reactions catalyzed by transglutaminase. (**a**) Cross-linking; (**b**) acyl transfer reaction; (**c**) deamidation reaction (modified after De Jong and Koppelman 2002)

$$
\text{a} \qquad \text{Gln–C–NH}_2 + \text{H}_2\text{N–Lys} \longrightarrow \text{Gln–C–NH–Lys} + \text{NH}_3
$$

$$
\text{b} \qquad \text{Gln–C–NH}_2 + \text{H}_2\text{N–R} \longrightarrow \text{Gln–C–NH–R} + \text{NH}_3
$$

$$
\text{c} \qquad \text{Gln–C–NH}_2 + \text{H}_2\text{O} \longrightarrow \text{Gln–C–OH} + \text{NH}_3
$$

oil droplet-like structures, and thus serves as a fat substitute. In plant protein-based products, transglutaminase is used to reduce the allergenic potential of wheat flours as it reacts excellently with the β-gliadin fractions of wheat protein due to the high glutamine content.

In principle, oligosaccharides can be produced with two subclasses of enzymes. These are glycosyl transferases (EC 2.4.x.x) but also glycosyl hydrolases (glycosidases, EC 3.2.x. x; see Sect. 16.4.3) in special circumstances (see below). In the case of glycosyltransferases, a distinction is made between the Leloir and the non-Leloir glycosyltransferases. In cells, the synthesis of oligosaccharides occurs via the Leloir pathway. This means that a sugar-nucleotide complex, a so-called activated sugar, acts as a sugar donor and the latter is transferred to another sugar, the acceptor molecule. A new oligosaccharide and a free nucleotide (often UDP) are formed. The latter is regenerated again in the cell and will react again with a sugar to form a new sugar-nucleotide complex. This activated sugar is then available for further reactions with Leloir transferases again. The activated sugar is generated due to the hydrolysis of UTP (formed from ATP), which provides the thermodynamic energy needed to synthesize the glycosidic bond via a coupled reaction. Leloir transferases thus require activated sugars (sugar-UDP). They are selective for the sugar molecule. Moreover, these enzymes are quite unstable and poorly available. These characteristics make them ineffective for industrial applications. Starter cultures possessing high Leloir transferase activity are an alternative as whole cells for oligosac-charide synthesis during food fermentation.

However, in many plant and microbial cells are non-Leloir transferases that do not require activated sugars but use the special high-energy glycosidic bond of the disaccharide sucrose between its D-glucose and D-Fructose unit. The bond energy between both monosaccharides is similar to the one of the phosphate in UTP or ATP, respectively. Unfortunately, the transfer is limited to glucose or fructose units when using this compound as substrate. However, since sucrose is an inexpensive starting substrate, non-Leloir transferases are of high interest for biotechnological applications (Seibel et al. 2006). An example of oligosaccharides that could be produced with non-Leloir transferases are fructosides. Fructosides are produced by fructosyltransferases (EC 2.4.1.-) and are low-calorie and anticariogenic sweeteners for foods and cosmetics. They have a higher sweetening power than sucrose and are dietary, i.e., their absorption in the intestine is much slower compared to glucose, which allows them to be metabolized by intestinal bacteria and not by humans.

Questions
8. How can the texture of food enzymatically be generated or increased, respectively?
9. What are non-Leloir transferases and why is sucrose a unique starting compound for enzymatic polyglucans or polyfructan synthesis?

16.4 Hydrolases

16.4.1 Importance to the Food Industry

Hydrolases are enzymes that reversibly catalyze the hydrolysis (cleavage) of various types of bonds with the participation of water (Fig. 16.3). The reaction equilibrium is far on the right site at high free water activity values (aW values). However, under special circumstances, the hydrolases can be used to catalyze condensation or transfer reactions as well.

Within the EC classification they belong to the main class EC 3 and within this class they are further subdivided depending on which type of bond is cleaved. Of particular relevance for use in the food sector are hydrolases, which catalytically cleave ester bonds (lipases/esterases; EC 3.1.x.x), glycosidic bonds (glycosidases EC 3.2.x.x), and peptide bonds (peptidases/proteases EC 3.4.x.x). Hydrolases are coenzyme-independent, which facilitates their use, but care must be taken to ensure that the "correct" enzymes (preparations) are used. Depending on the food/food matrix, specific requirements are placed on the enzymes. This includes, for example, the pH and temperature range in which they are active/stable, the presence or absence of ions that are necessary for their activity or have an inhibitory effect and obviously their substrate-, regio-, and stereo-selectivities. Furthermore, other components may be present in the food that inhibit the activity of certain enzymes (e.g., trypsin inhibitors in soy). Hydrolases bearing the same EC number do catalyze the same reaction, but depending on the origin of the enzyme, one enzyme may have a narrower substrate spectrum and catalyze fewer reactions due to less substrate acceptance than another hydrolases having a broader substrate spectrum.

$$A\text{--}B + H_2O \rightleftharpoons A\text{--}H + B\text{--}OH$$

Fig. 16.3 Schematic representation of the catalytic reaction of hydrolases

16.4.2 Lipases/Esterases: Applications in the Food Industry

The difference between lipases and esterases is their activity properties at water/oil interfaces (Fojan et al. 2000). Lipases are depending on such an interface because they have a hydrophobic lid sealing the active center in water. In presence of a hydrophobic phase, the lid uncloses and the substrate can enter the active center. Esterases do not need such an interface as they do not have a sealing lid. Lipids play an important role in the taste and texture of foods. By means of transesterification (exchange of fatty acids; Fig. 16.4), the structure of fats (triglycerides) can be influenced accordingly. An example of this is designer margarine. Through the transesterification of triglycerides, on the one hand, the position distribution and on the other hand the fatty acid composition can be changed. If the modification of the fatty acid composition in the food is the goal, free fatty acids or triglycerides of other dietary fats are also used. Through this modification, the nutritional and technological properties of the fats can be specifically modified.

Another example of the use of lipases in food is the production of cheese such as Parmesan, Cheddar or Gouda. The use of enzymes saves 50% of the time needed for ripening. The reason behind this is as follows: Lipases hydrolyze the triglycerides (lipolysis) and thereby specifically release fatty acids. These serve as starting materials for the formation of aroma substances such as aldehydes or ketones. The fatty acids are metabolized accordingly by the starter cultures.

Lipases, together with other enzymes such as esterases, proteases and peptidases, can also be used to produce an *enzyme modified cheese* concentrate (EMC). For this purpose, fresh cheese or cheese curd is mixed with a saline solution and homogenized. The enzymes are then added, whereby the concentration and type of enzymes determine the subsequent aroma of the EMC. After a defined incubation (4–5 days, 30–40 °C), EMC can be used as a paste or, after drying, as granules. Due to the 15–30 times higher flavor intensity of EMC compared to naturally matured cheese, only 0.1–2% EMC is added to finished products such as sliced cheese or frozen pizzas. EMC is used especially in frozen meals and microwaveable products (convenience products), as it is sensory superior to conventional cheese in these cases.

As a further field of application, lipases can be used to produce butter flavors. For this purpose, partial hydrolysis of milk fat takes place. For example, butterfat is melted and emulsified together with an aqueous phase (e.g., phosphate buffer) with the aid of lecithin. This emulsion is partially hydrolyzed with lipases at 40 °C and the reaction is then stopped at a desired time by thermal treatment. The profile can be controlled by the reaction conditions chosen, the enzyme concentration and in particular by the lipase used

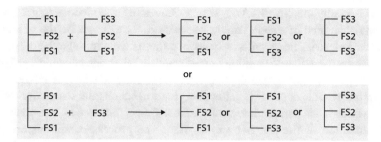

Fig. 16.4 Principle possibilities of two transesterification reactions of a triglyceride with a triglyceride or a fatty acid (stoichiometry not considered. Enzyme: 1,3-position-specific lipase; mod. according to Uhlig 1991)

(selectivity). Such butter flavor emulsions are used in the production of chocolate, baked goods, and convenience products, among others.

The production of imitated sheep and goat cheese products is also based on the partial hydrolysis of milk fat with lipases. Specific lipases are added to the cow's milk for this purpose, and the resulting release of special fatty acids produces a flavor similar to that of sheep's or goat's milk. This lipase-modified milk can then be used to produce cheeses.

16.4.3 Glycosidases: Applications in the Food Industry

The bond type, which is cleaved by glycosidases, is chemically an ether bond and thus has the structure R-O-R'. The residue R is the so-called glycone of the substrate molecule and determines the acceptance in the active site. The aglycone residue (R'), on the other hand, is not of particular importance for most glycosidases, although certain steric effects or charges may play a role in substrate recognition in the active site sometimes. Aglycone residues include sugars, lipids, or peptides. Most significant examples of glycosidases in the food sector are amylases, β-glucanases, xylanases, and β-galactosidases.

The polysaccharide starch is enzymatically degraded by amylases, whereby it can only be hydrolyzed in the gelatinized (swollen) state of starch. Starch consists of amylose (α-1,4-glycosidically linked D-glucose molecules) and amylopectin (α-1,4-glycosidically linked D-glucose and, after 15–30 glucose molecules, an α-1,6-glycosidically linked side chain). The ratio of both macromolecules to each other is plant species-specific and determines the properties of the particular starch. The starch-cleaving glycosidases are divided into endo- and exoamylases. The endoamylases include α-amylase (EC 3.2.1.1). It cleaves internal α-1,4-glycosidic bonds of amylose, but not terminal or α-1,6-glycosidic bonds. Exoamylases include β-amylase (EC 3.2.1.2), which cleaves one maltose molecule (disaccharide) at a time from the non-reducing chain end. The name β-amylase is misleading because this amylase is also selective for α-1,4-bonds, but the product is β-maltose. For historical reasons, the amylase was therefore named β-amylase. Another exoamylase is γ-amylase (EC 3.2.1.3), which cleaves off one glucose molecule from the non-reducing chain end of starch. Amylases are used, for example, in the so-called starch process. This industrial huge process describes the enzymatic degradation of starch to glucose, whereby more than 20 million tons of starch are hydrolyzed annually. The starting raw material is corn, from which the starch is extracted. The starch is mixed with water under the influence of heat (105 °C, jet-cooker) to form starch milk with a dry substance of 40% (w/v), which is thus microbiologically stable. This is followed by enzymatic liquefaction or dextrination of the starch by α-amylase. The process conditions are optimally adjusted to the enzyme (e.g., 95 °C, 2 h). Liquefaction is followed by saccharification. This is carried out enzymatically by γ-amylase. Saccharification can take several days (e.g., 60 °C, 24–72 h). The glucose is then refined and can be further processed, e.g., into glucose/fructose syrup (Sect. 16.6.2).

Endo-β-1,2(4)-glucanases (EC 3.2.1.6), for example, cleave β-1,3- and β-1,4-bonds of glucose in β-glucans and β-1,4-bonds in cellulose. All substances that consist of two or more glucose units and are connected via a β-glycosidic bond are referred to as β-glucans. They find applications, for example, in breweries to facilitate filtration and release of unmalted barley. The reason for this is that the cell walls of the barley endosperm cell of the malt consist of about 75% β-glucans. In Germany, the addition is not permitted due to the German purity law. β-glucanases are also used in wine clarification and filtration.

Xylanases, or more precisely endo-β-1,4-xylanases (EC 3.2.1.8), catalyze the cleavage of β-1,4-glycosidically linked xyloses into xylans (components of plant cell walls). Xylanases find application in the bakery industry for the hydrolysis of xylans, as these can bind up to ten times their own weight in water and thus interfere with the formation of the gluten network. The hydrolysis of xylans improves dough properties such as extensibility and stability.

The food industry is very interested in bringing more milk and whey products onto the market that are *lactose-free* and thus tolerated by lactose-intolerant people. For lactose hydrolysis, β-galactosidases are suitable (in industry often called "lactases"; EC 3.2.1.23). The substrate for β-galactosidases is lactose (4-O-(β-D-galactopyranosyl)-D-glucopyranose), which is present in cow's milk at approximately 4.5–4.8% (w/w). By the use of β-galactosidases, lactose is cleaved into its components glucose and galactose (Fig. 16.5a). The β-galactosidase preparations must be free of peptidases; otherwise, a bitter off-taste of the milk may result due to proteolysis. There are two principal manufacturing processes for producing lactose-free milk. One is to add the β-galactosidases to the milk before heat treatment, and the other is to add them to the already heat-treated milk. There are advantages as well as disadvantages to both methods. When added before heat treatment, high amounts of β-galactosidase must be added to the milk so that the lactose is hydrolyzed in as short a time as possible. However, since the heat treatment inactivates the enzyme, it does not have to be declared on the product package. In the other case, the β-galactosidase is added sterile to the milk after the heat treatment. Accordingly, lactose hydrolysis does not take place during the milk processing in the dairy, but during storage and transport of the milk. Thus, a longer time is provided for the lactose

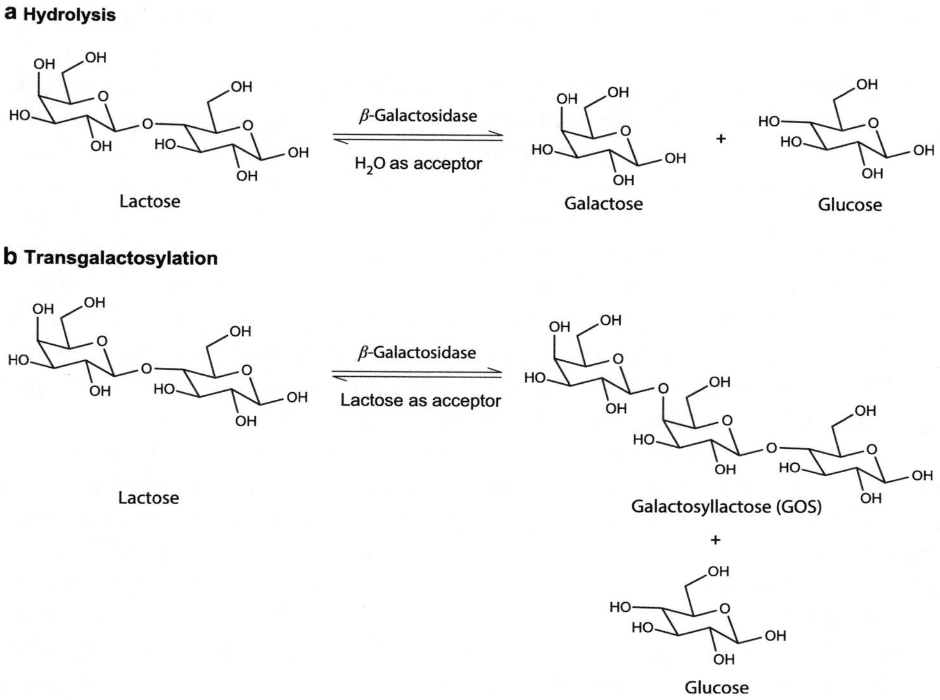

a Hydrolysis

Lactose β-Galactosidase / H$_2$O as acceptor   Galactose + Glucose

b Transgalactosylation

Lactose β-Galactosidase / Lactose as acceptor   Galactosyllactose (GOS) + Glucose

Fig. 16.5 Schematic representation of lactose hydrolysis (**a**) and transgalactosylation (**b**) with a β-galactosidase

hydrolysis, which reduces the amount of β-galactosidase required and thus the cost. However, it should be noted that the β-galactosidase must be added to the milk under sterile conditions, as there is no further thermal treatment of the milk. The β-galactosidase addition must also be declared on the packaging, as the enzymes are still actively present in the lactose-free milk.

In addition to cleavage, glycosidases can also be used for glycoside synthesis (formation of oligosaccharides). In general, oligosaccharides are defined as glycosides containing 2–10 covalently bound monomeric sugar units. One type of oligosaccharides are the so-called galactooligosaccharides (GOS). They can be formed from lactose with the help of β-galactosidases in a transglycosylation reaction. Transglycosylation occurs when another OH-group saccharide (from glucose, galactose, lactose, or GOS molecule) acts as a galactose acceptor (Fig. 16.5b). The kinetically controlled transgalactosylation is thus in competition with thermodynamically controlled lactose hydrolysis (Chaves e Souza et al. 2022). Depending on the reaction conditions (lactose concentration, water activity, temperature, pH), different GOS(-mixtures) can thus be generated. Of particular importance is that GOS are classified as prebiotic by EFSA and they must be present in infant food.

16.4.4 Proteases/Peptidases: Applications in the Food Industry

The enzyme name proteases or peptidases, respectively, are used synonymously and catalyze the cleavage of peptide bonds (amide bonds). They can be categorized based on their catalytic mechanism (serine, cysteine/thiol, carboxy and metallo peptidases), among others. Furthermore, they are distinguished based on the localization of the peptide bond in the protein/peptide they cleave (endopeptidases and exopeptidases). The peptide bond selectivity of peptidases can be described using the nomenclature of Schechter and Berger (1967), which considers the active site during cleavage. Specific regions in the active site (S for *sub-sites*) interact only with certain amino acid patterns of the peptide (P for *position*). The active site regions are named from the cleavage site towards the *N-terminus* ($S1, S2, ..., Sn$) and the *C-terminus* ($S1', S2', ..., Sn'$). The interacting regions of the peptide are correspondingly designated $P1, P2, ...Pn$ and $P1', P2', ...Pn'$, respectively.

Peptidases are used in a variety of ways in the production and processing of foods. For example, they can be used for the production of flavor-active hydrolysates or for the production of hypoallergenic baby food. In order for the hydrolysates to meet the necessary requirements, a high degree of hydrolysis must be achieved, i.e., as many peptide bonds as possible of the starting protein (e.g., casein, wheat gluten) should be cleaved. Due to the amino acid pattern selectivity of the individual peptidases, such a total hydrolysate cannot be achieved by using a single peptidase only. A combination of different peptidases is thus essential to achieve a high degree of hydrolysis. In Fig. 16.6, such a hydrolysis is exemplified for a theoretical protein. In the case where only one endopeptidase is used (1), the protein is hydrolyzed into longer peptides, but no amino acids are released. If only an exopeptidase is used (2), individual amino acids may be released, but hydrolysis stops if amino acids are present that do not fit into the substrate spectrum of the exopeptidase used. However, if endo- and exopeptidases are used together (3), the protein can be hydrolyzed almost completely. By possibly using another exopeptidase with ideally a complementary selectivity, the whole protein can theoretically be cleaved into its amino acids. De facto a mixture of different endo- and exopeptidases is necessary for such a hydrolysis result.

Probably the best-known application of a peptidase in food processing is in cheese production. Rennet from the calf's stomach was and is used for this purpose. It contains mainly the responsible enzyme chymosin (EC 3.4.23.4). Thickening of the milk occurs by a selective cleavage of the κ-casein at the Phe_{105}/Met_{106} position done by chymosin. This cleavage causes destabilization and thus coagulation of the casein micelles by removal of the hydrophilic glycopeptide (caseinomacropeptide). In addition to native rennet from the calf stomach, nowadays, chymosin is increasingly used, which was produced recombinantly. The yeast *Kluyveromyces lactis* and the mould *Aspergillus niger* are used as recombinant production strains. The advantage of recombinant chymosin production is that the preparation contains only chymosin and not pepsin as in Kälber-Lab (rennet), which can lead to the release of unwanted bitter peptides.

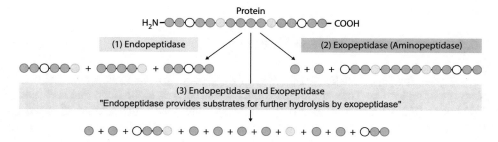

Fig. 16.6 Exemplary hydrolysis of a protein by an endopeptidase (1), an exopeptidase (2) and a combination of both (3). Preferred cleavage sites are indicated by green spheres (endopeptidase) and blue spheres (exopeptidase). Amino acids, which are represented by white spheres, cannot be cleaved if they are located at the P1 position (Stressler et al. 2015)

As already mentioned in Sect. 16.4.2, peptidases are also used for the production of EMC. Here, too, a combination of endo- and exopeptidases is required to achieve the highest possible degree of hydrolysis.

Peptidases are also used in the baked goods industry. Gluten is a collective term for a mixture of proteins found in the seeds of some types of cereals. It is the most important functional component in wheat flour. Any influence on the gluten network also means a strong influence on the dough and ultimately on the finished bread. Peptidases are used, among other things, to improve the machinability of doughs and to reduce mixing time. Furthermore, peptidases can be used to positively influence the color, taste, water absorption, dough extensibility, and much more.

Peptidases are also used in the beverage industry. For example, peptidases can be used to stabilize beer. *Beer haze* is usually due to cross-linking of high molecular weight proteins in the malt. Peptidases such as papain (EC 3.4.22.2) can be used to prevent this. Peptidases can also be used to produce gluten-free beers that can be consumed by people suffering from celiac disease. One enzyme that is used here is prolyl endopeptidase (EC 3.4.21.26) from *A. niger.*

Peptidases are also used in meat processing. For example, the peptidases papain (EC 3.4.22.2), bromelain (EC 3.4.22.32), and ficain (EC 3.4.22.3) are used for the softening of meat.

The peptidase thermolysin (EC 3.4.24.27) from *Bacillus thermoproteolyticus* can be used for the synthesis of the sweetener aspartame. Aspartame is a dipeptide consisting of L-aspartic acid and the methyl ester of L-phenylalanine. Due to the regio- and stereoselectivity of the enzyme, the condensation reaction to aspartame can be carried out without the need for complex protecting group chemistry required in chemical synthesis.

Question
  10. Why are hydrolases the most often used exogenous enzymes in food manufacturing?

16.5   Lyases

16.5.1 Importance to the Food Industry

Lyases form the fourth main enzyme class (EC 4.x.x.x) and catalyze a *non-hydrolytic* cleavage of molecules. They are further subdivided according to the bond that is cleaved. C-C bonds (EC 4.1.x.x), C-O bonds (EC 4.2.x.x), C-N bonds (EC 4.3.x.x), C-S bonds (EC 4.4.x.x), C-halogen bonds (EC 4.5.x.x), and P-O bonds (EC 4.6.x.x) are cleaved. In their reaction, they usually leave a double bond in the molecule. For the food industry mainly lyases of class EC 4.2.2.x are used, because they are catalytically active on polysaccharides (pectins). Pectins are plant polysaccharides consisting mainly of α-1,4-glycosidically linked D-galacturonic acid molecules. Within this class (EC 4.2.2.x), pectate lyase (EC 4.2.2.2), pectate disaccharide lyase (EC 4.2.2.9), and pectin lyase (EC 4.2.2.10) are primarily relevant. Pectate lyases are endoenzymes and catalyze a β-eliminating cleavage. Their preferred substrates are pectic acid and low-esterified pectins. Ca^{2+} ions are mandatory for their catalytic activity. Pectate disaccharide lyases, also called exo-polygalacturonic lyases, cleave digalacturonic acid from the reducing end of the substrate. Trigalacturonic acid is accepted as the smallest substrate. Again, Ca^{2+} ions have an activating effect on the enzyme, and the pH optimum is in the alkaline range (pH 8.0–9.5). Pectin lyases are also endoenzymes and preferentially cleave highly esterified pectins. They generate esterified unsaturated oligogalacturonic acids. They are the only enzymes that can directly cleave pectins.

16.5.2 Areas of Application in the Food Industry

Pectin lyases find their application in liquid products of fruits and vegetables. Therefore, they are among the most used technical enzyme preparations in food technology/processing. Pectin enzymes are also used in fruit juice clarification and mashing.

16.6   Isomerases

16.6.1 Importance to the Food Industry

Isomerases are listed in the fifth main enzyme class (EC 5.x.x.x). Isomerases are enzymes that catalyze an intramolecular isomerization reaction, i.e., the intramolecular conversion of a compound into an isomeric compound. Isomers are molecules that have the same molecular formula but a different structural formula. The most prominent isomerase in the food industry is the so-called xylose isomerase (EC 5.3.1.5) which natively converts xylose (aldose) into xylulose (ketose) and vice versa. In the food industry, this enzyme is used for the reversible conversion of glucose into fructose, which are also well-accepted substrates for it. That is why the enzyme is also called "glucose isomerase". Worthy of mention, although still more in the field of research, is an epimerase that is catalytically active on carbohydrates (EC 5.1.3.x). This is a so-called cellobiose-2-epimerase (EC 5.1.3.11). Also of interest is an intramolecular transferase (EC 5.4.x.x) called isomaltulose synthase (EC 5.4.99.11).

16.6.2 Applications in the Food Industry

An important intermediate product in the food industry is the so-called *high fructose corn syrup,* HFCS (Parker et al. 2010). Its properties in food include sweetening capacity, hygroscopicity, texture formation, browning, flavor formation and enhancement, crystallization inhibition with low viscosity, and freezing point depression. It is used, for example, in soft drinks, beverage powders, bakery products, snacks, confectionery, dairy products, ice creams, alcoholic beverages such as liqueurs and pet food additives. In the production of HFCS, the enzyme xylose isomerase (also called glucose isomerase) plays a decisive role. Xylose isomerase (EC 5.3.1.5) was discovered by Japanese scientists in the 1950s. Its natural substrate is D-xylose (hence xylose isomerase), which is converted into its keto form (D-xylulose). However, in vitro experiments revealed that the enzyme also accepts D-glucose as a substrate in addition to D-xylose. The conversion product of glucose isomerization is D-fructose. Accordingly, xylose isomerase is able to catalyze an intramolecular rearrangement of a pyranose into a furanose and vice versa. At process equilibrium (dependent on temperature and sugar concentrations), 84% D-xylose and 16% D-xylulose or 49% D-glucose and 51% D-fructose are present. The K_M value for glucose is about 100 mM, which is quite high. The Michaelis constant K_M describes the affinity of the enzyme to its substrate. It corresponds to the substrate concentration at which half of the maximum reaction rate is reached (see Chap. 4). This means that if the initial concentration of glucose is high enough, the yield of fructose is acceptable or, the longer the process continues, the more glucose is converted. Since isomerase belongs to the metalloenzyme group, metal ions have an effect on the activity of the enzyme. Table 16.1 shows the metal dependence of xylose isomerase from *Bacillus coagulans.*

Syrups from enzymatic starch hydrolysis (Sect. 16.4.3) with glucose contents of about 94–98% of total solids are used for the isomerization process. The reason for glucose isomerization is the nearly triple higher sweetness of fructose and lower glycemic index when compared with glucose. The xylose isomerase is used in its immobilized form and the resulting polymer particles are packed in huge columns for continuous HFSC production (24 h, 7 days). The final product of this enzymatic process contains about 42% fructose at equilibrium at about 55 °C (HFCS-42). At a temperature of 90 °C approx. 55.6%, fructose would be obtained. However, this temperature is industrially not relevant due to insufficient temperature stability of the isomerase. Higher fructose contents (55–90%) are obtained by enrichment using chromatographic means (HFCS-55; HFCS-90). The chromatographic step allows glucose to be partly separated, recycled in the enzymatic process and

Table 16.1 Metal dependence of xylose isomerase from *Bacillus coagulans*. The relative enzyme activities (%) in relation to the respective substrate are shown

Metal salt (10 mM)	D-Xylose	D-Glucose
Without	4	0
MnCl$_2$	100	16
CoCl$_2$	27	100
MgCl$_2$	40	15

isomerized again to fructose, resulting in a significant increase in fructose yield of the syrup.

The food-relevant disaccharide palatinose or isomaltulose is formed by intramolecular rearrangement from sucrose (α-1,2 linkage to α-1,6 linkage between glucose and fructose). This isomerization is catalyzed by the enzyme isomaltulose synthase (EC 5.4.99.11), also called sucrose-α-glucosylmutase, optimally at 40 °C and pH 5.8. Cells of *Protaminobacter rubrum* immobilized in Ca^{2+} alginate are used for this purpose and not the free enzyme. Alternatively, *Leuconostoc mesenteroides* or *Serratia polymuthica* can be used as whole cell catalyst. The yield of palatinose is 85% with a selectivity of also 85%. Trehalulose is formed as a by-product, i.e., the reaction has to be stopped when the corresponding conversion occurs. The substrate conversion rate is >99.5%. Palatinose can be used as a sweetener and has a similar taste to sucrose, but only about 40% sweetness at half the caloric value. It is a substitute for sucrose as only weak insulin stimulation is produced, making it suitable for diabetics. Palatinose is broken down more slowly than sucrose. This leads to a slower or lower increase in blood glucose and thus a lower insulin requirement. Palatinose (also its hydrogenated derivatives) has a low potential for bacterial plaque formation on teeth (caries prophylaxis). It is produced by Südzucker AG (Germany) and Mitsui Seito Co. Ltd (Japan). Palatinose has been approved as a novel food since 2005 and is produced on a scale of >100,000 t per year (since 1985). It is used in sweets, icing, puddings, chewing gum, sports drinks, toothpaste, chocolate products, and many more.

Cellulose 2-epimerases (EC 5.1.3.11) are not yet used in industry, but they show great potential in research. They catalyze the intramolecular epimerization of lactose (galactose-β-1,4-glucose) to lactulose (galactose-β-1,4-fructose) and epilactose (galactose-β-1,4-mannose). Both lactose isomers show a prebiotic effect in animal feeding experiments, but they are not yet approved as prebiotics by EFSA. The advantage of cellobiose-2-epimerases for lactulose/epilactose production is that, in contrast to chemical lactulose/epilactose production, the enzyme could be directly used in foods and needs lactose as the only substrate.

Question
11. What kind of reactions do isomerases catalyze and what is the most prominent isomerase in the food industry?

16.7 Ligases

16.7.1 Importance to the Food Industry

Ligases are the sixth main enzyme class (EC 6.x.x.x). They catalyze the covalent linkage of two molecules by the formation of C-O (EC 6.1.x.x), C-S (EC 6.2.x.x), C-N (EC 6.3.x.x), C-C (EC 6.4.x.x), C-phosphoric acid ester bonds (EC 6.5.x.x), mostly with simultaneous consumption of energy-rich compounds such as ATP.

16.7.2 Application in the Food Industry

While ligases find direct application in the synthesis of substances in other industrial sectors (chemical industry, pharmaceutical industry), they play a subordinate role in the food industry. They are used in the analysis of foodstuffs in the context of, for example, the animal species determination of sausage and meat products by PCR analysis (Rabag Farag et al. 2015). First, the DNA is isolated and then amplified by PCR. If no species-specific primers are used, but so-called universal primers, the PCR product must then be characterized in more detail by means of various procedures, such as RFLP (restriction fragment length polymorphism analysis), SSCP (single strand conformation polymorphism analysis), Southern blotting (hybridization), or sequencing. With the exception of sequencing, the latter analytical methods are used in combination with simple electrophoretic visualization of the resulting DNA fragments for qualitative PCR. The detection limit of animal species is <0.1% depending on the type of PCR method.

16.8 Conclusion

Enzymes or enzyme preparations, respectively, have become indispensable in modern food technology. In addition to the endogenously occurring enzymes in the raw materials of the food, primarily exogenously, i.e., separately, added enzymes play a decisive role in food processing. In this way, the diverse food products from the supermarket can be produced cost-effectively, as they are automated produced, and with high sensory and nutritional quality. The exogenous enzymes are usually not pure, but represent enzyme mixtures. This is due to the fact that the enzyme preparations are predominantly produced by bioreactor cultivation of food-grade microorganisms and the enzyme extracts obtained are not or only partially purified due to the costs incurred for purification. As a result, the enzyme preparations are advertised as having a certain major activity, but additional enzyme activities (side activities) are present. Users are often unaware of these side activities, but they can be important for product quality and process reproducibility. Therefore, industrial enzyme preparations should be examined by the user himself for side activities and characterized biochemically.

The economic importance of enzyme preparations is also evident from the sales figures: global sales in 2021 were valued at US\$ 11.47 billion and are projected to expand at a compound annual growth rate of 6.5% from 2022 to 2030. This increase is mainly attributed to the growing use of enzyme preparations in the food and beverage industry (Grand View Research 2022).

Answers
1. The endogenous enzymes belong to the cells of the plant or animal raw foods that is the starting material in the food industry. In addition, the enzymes from microorganisms are considered as endogenous enzymes if foods are fermented during food processing or are naturally contaminated before further manufacturing. Exogenous enzymes are separately produced by biotechnological means and added during food processing.
2. The names of exogenous food enzymes are created imaginatively by the companies under marketing considerations. It is hardly possible to know which scientific enzyme sub-subclass is behinds the name. The food enzyme preparations do have more than one enzymatic activity in nearly all cases but only the main activity is quantified and labeled for the customer.
3. The reason for use of exogenous enzymes is their significant contribution to the quality of foods regarding, e.g., taste, preservation, shelf life, appearance, digestibility, nutritional and technological properties. The latter comprised, e.g., texture, foaming, emulsion, water holding capacity, and viscosity.
4. The use of exogenous enzymes in food production is strictly regulated by laws and permissions of the European Parliament and the Council in the EU. In the moment, the national regulations of the individual countries are still in force. However, the enzyme has to be on a positive list in the near future.
5. Food enzymes have to be cheap and available in huge quantities. Thus, a microbial overproduction is most suitable. Wild-type strains are evolutionary adapted to optimum regulation of their enzymes and thus, do avoid any overproduction by effective repression and regulation on gene level. Thus, known molecular regulation systems in suitable host strains are used to achieve the overproduction of an enzyme economically by genetic engineering means in short time.
6. The food-grade concept aims a safe and environmental friendly production of food enzymes. The microbial production strains have to be uncritical to human health and, only species from risk class 1 are approved. The final genetic modifications in the production strain have to be free of antibiotic resistance and reduced to a minimum.
7. The catalyzed reaction by glucose oxidase converts selectively glucose and oxygen to produce gluconolactone and H_2O_2. Thus, several positive changes happen in the food. Firstly, Maillard reactions generating taste and color changes are prevented due to missing reduced ends from glucoses. Secondly, the gluconolactone is chemically hydrolyzed to gluconic acid and this lowers the pH making the food more stable against spoilage as the latter is also supported by generation of H_2O_2. Catalase is often co-added with glucose oxidase if H_2O_2 is not desired or interfering in the food.

8. The transglutaminase does form covalent bonds between different kinds of proteins reacting on glutamine and lysine residues of the respective proteins. In this way, a covalent and temperature stable protein network is built up in foods depending on the amount of accessible glutamine and lysine residues of the particular protein composition.

9. Non-Leloir transferases are enzymes which do not need activated substrates, such as UDP-glucose, to condensate monosaccharides and built up polyglucans or polyfructans. The thermodynamic energy for the condensation reaction is coming from cleavage of the high energy bond between glucose and fructose in the disaccharide sucrose which is equal to the energy bond of phosphate in ATP or UTP, respectively. Sucrose is quite unique in this sense.

10. Hydrolases are coenzyme independent and catalyze the hydrolytic cleavage of covalent bonds in poly-/oligosaccharides, proteins/peptides, and fats. These compounds are the most important ingredients in foods and their hydrolytic modifications do alter significantly the attributes of foods, e.g., taste, texture, flavor, technological functions, nutritional value etc. Hydrolases are widespread in nature and commonly have a broad substrate spectrum. Nearly for any food processing condition (pH, temperature, ionic strength etc.) a suitable hydrolase is available.

11. Isomerases catalyze intramolecular rearrangements of functional groups in a substrate. The most prominent isomerase in food is the so-called xylose isomerase. However, the substrate in food industry is not xylose but glucose that is intramolecularly rearranged to the much sweeter monosaccharide fructose. The xylose isomerase is used in its immobilized form in a continuous production process. The resulting syrup is a mixture of glucose/fructose whereas the fructose content varies between 42 and 90% depending on the subsequent downstream processing by column chromatography.

Further Reading

Kunz B. Lebensmittelbiotechnologie. 2nd ed. Hamburg: Behr's Verlag; 2016.
Lösche K, editor. Enzymes in food technology. 1st ed. Hamburg: Behr's Verlag; 2000.
Whitaker JR, Voragen AGJ, Wong DWS. Handbook of food enzymology. New York: Marcel Dekker; 2003.
Whitehurst RJ, van Oort M. Enzymes in food technology. Chichester: Wiley-Blackwell; 2010.

References

Bauer JA, Zámocká M, Majtán J, Bauerová-Hlinková V. Glucose oxidase, an enzyme "Ferrari": its structure, function, production and properties in the light of various industrial and biotechnological applications. Biomolecules. 2022;12(3):472–97.
Chaves e Souza AF, Gabardo S, Silva Coelho RDJ. Galactooligosaccharides: Physiological benefits, production strategies, and industrial application. J Biotechnol. 2022;359:116–29.

De Jong GAH, Koppelman SJ. Transglutaminase catalyzed reactions: Impact on food applications. J Food Sci. 2002;67(8):2798–806.

Fojan P, Jonson PH, Petersen MTN, Petersen SP. What distinguishes an esterase from a lipase: A novel structural approach. Biochimie. 2000;82(11):1033–41.

Grand View Research. Enzymes market size, share & trends analysis report by product (lipases, polymerases & nucleases, carbohydrase), by type (industrial, specialty), by source (plants, animals), by region, and segment forecast, 2022-2030. Report ID: 978-1-68038-022-4, 140 pages; 2022.

Kaushal J, Mehandia S, Singh G, Raina A, Kumar AS. Catalase Enzyme: Application in bioremediation and food industry. Biocatal Agric Biotechnol. 2018;16:109–99.

LFBG. Food and feed code; 2005. http://www.gesetze-im-internet.de/lfgb/

Mayolo-Deloisa K, Gonzáles-Gonzáles M, Rito-Palomares M. Laccases in food industry: bioprocessing, potential industrial and biotechnological applications. Bioeng Biotechnol. 2020;24:Article 222. (8 pages)

Merz M, Eisele T, Berends P, Appel D, Rabe S, Blank I, Stressler T, Fischer L. Flavourzyme, an enzyme preparation with industrial relevance: Automated nine-step purification and partial characterization of eight enzymes. J Agric Food Chem. 2015;63(23):5682–93.

Parker K, Salas M, Nwosu VC. High fructose corn syrup: production, uses and public health concerns. Biotechnol Mol Biol Rev. 2010;5(5):71–8.

Rabag Farag M, Alagawany M, Ezzat Abd El-Hack M, Tiwari R, Dhama K. Determination of different animal species in meat and meat products: trends and advances. Adv Anim Vet Sci. 2015;3(6):334–46.

Schechter I, Berger A. On the size of the active site in proteases. I. Papain. Biochem Biophys Res Commun. 1967;27(2):157–62.

Seibel J, Beine R, Moraru R, Behringer C, Buchholz K. A new pathway for the synthesis of oligosaccharides by the use of non-Leloir glycosyltransferases. Biocatal Biotransform. 2006;24 (1/2):157–65.

Stressler T, Ewert J, Merz M, Glück C, Fischer L. Functional protein hydrolysates—potentials of peptidases for protein modification in food. FOOD-LAB. 2015;3(15):20–4.

Uhlig H. Enzymes work for us. Munich: Hanser Fachbuchverlag; 1991.

Enzymes in Detergents and Cleaning Agents

17

Karl-Heinz Maurer

What Do You Learn in This Chapter?

Enzymes in laundry detergents and household cleaners make a significant contribution to enhancing the performance and improving the environmental properties of laundry detergents and household cleaners. They are also essential for differentiating premium brands in relation to standard and "value-for-money" products. Proteases, amylases, cellulases, lipases, mannanases, and pectate lyases are used in detergents. In automatic dishwashing detergents, it is proteases and amylases. In this chapter, performance characteristics of the enzymes and the typical requirements of the different enzyme classes for these products are discussed. The benefits of enzyme technology to the sustainability of washing are assessed, as are aspects of health, safety, and environmental relevance. The economic importance is touched upon and an outlook on the expected future is attempted.

17.1 History of Detergent Enzymes

As early as 1913, Otto Roehm had applied for a patent in which he claimed the use of proteolytic enzymes in laundry detergents in analogy to the use of proteases in the leather production. The only proteases known at that time came from the pancreas of slaughtered animals and, as we know today, are not ideally suited for detergents. Detergents at that time

K.-H. Maurer (✉)

CLIB - Cluster Industrial Biotechnology, Düsseldorf, Germany

© The Author(s), under exclusive license to Springer Nature Switzerland AG 2024
K.-E. Jaeger et al. (eds.), *Introduction to Enzyme Technology*, Learning Materials in Biosciences,
https://doi.org/10.1007/978-3-031-42999-6_17

consisted of perborate and silicate and were consequently too alkaline for the proteases to perform well. Accordingly, the only option was to use enzymes as a soaking or prewashing agent. It was not until proteases from bacterial sources became available in the 1960s that the performance of these bacterial proteases became usable in detergents, which by then had advanced significantly. It was not until this time that visible enzyme action on proteinaceous soils in the wash cycle was possible. These enzymes are secreted as extracellular enzymes by bacteria and fungi. They are naturally more stable than intracellular enzymes or the digestive enzymes of animals. Since then, the alkaline extracellular proteases from *Bacillus*-species, all of which belong to the subtilisin superfamily, have become the standard in detergents. Although 60 years have passed since then, during which the formulations of textile detergents have changed considerably, it is still the subtilisins that define the type of detergent enzyme.

Using the example of subtilisin proteases, it also became clear in the early days that enzymes as exogenous proteins in the human body can lead to allergies by inhalation. The phenomenon was first observed in employees of a detergent production plant. The first enzymes had been used as spray-dried powder preparations, which resulted in the workers being exposed to relatively large amounts of enzyme dust, the pulmonary effects of which were exacerbated by surfactants and bleach. Attributing allergy symptoms to enzymes was a real medical challenge due to lack of experience. The enzyme products, which initially consisted of spray-dried concentrate, were then confectionated into dust-free coated granules which, when properly handled and controlled, do not result in the release of enzyme dust into the air we breathe.

In recent decades, other classes of enzymes have become available for use in laundering: Amylases, lipases, cellulases, mannanases, and pectate lyases (Olsen and Falholt 1998). The possibility of fermentative production by means of recombinant microorganisms played a decisive role here. Without this option, technical use would be neither technically feasible nor economically viable. Since 1990, further fields of application were added in the form of machine and manual dishwashing detergents (Herbots et al. 2012). This resulted in the market for enzymes in this market segment representing approximately a quarter of the global market of enzymes, with an estimated value of 950 million € (2015). The market for detergents and cleaning agents is currently served essentially by two companies: Novozymes as the clear market leader and IFF.

17.2 Production of the Enzymes

All enzymes used in detergents and cleaning agents are produced by submerged large scale fermentation of bacteria and fungi. Depending on the production organism, the typical fermenter size is 30–120 m^3. Almost without exception, recombinant microorganisms are used here: for bacteria *Bacillus species*, for fungi *Aspergillus oryzae* and *Trichoderma reesei* (Aehle 2007). The production organisms are all categorized in risk group 1, which means that they pose no risk to humans, animals, or the environment. Nevertheless, the producing companies attach great importance to the fact that the production organisms are only kept in a closed system (containment) and no release into the environment takes place. This is done not only in order to comply with legal requirements, but also to prevent the production organisms, in the construction of which decades of research have been invested, from being made accessible to the competition. The fermentation media essentially contain complex (protein) or (ammonium) salt nitrogen sources, sugars, and trace elements. Fermentation can be carried out in batch or fed-batch and takes between 2 days (for bacteria such as *Bacillus*) and up to 14 days (for fungi such as *Aspergillus* or *Trichoderma*) depending on whether batch or fed-batch fermentation is used.

After termination of the fermentation, the first step of the processing is the quantitative separation of the biomass. This can be done by rotary drum filters, chamber filter presses, and separators/microfiltration. In the course of further processing, concentration and the separation of undesired accompanying substances, which can affect, for example, the color and odor of the enzyme product, takes place: Especially for liquid products, it is essential that the color of the liquid product desired by the detergent manufacturer is not altered or affected by brown or yellow color of the enzyme preparation. Similarly, the perfuming of the detergent should not be altered by odorous components of the enzyme product. One possibility for concentration and purification is the crystallization of the enzyme, which makes highly concentrated and very pure concentrates possible.

For liquid products, such concentrates have to be stabilized so that they survive transport and storage until the production of the liquid detergent or cleaning agent without damage. In the simplest case, this stabilization can consist of the addition of up to 50% propylene glycol (1,2-propanediol) or glycerol. The resulting reduction in water activity in the enzyme product may be sufficient to achieve the desired stabilization. In addition, liquid products for detergents and cleaning agents are preserved against microbial contamination. Parabens and sorbitol are used as preservatives, while benzoic acid or sodium benzoate are not permitted.

For powder products and tabs, the enzymes must be converted into dust-free granules. For this purpose, several technologies for granulation have been developed over time. In all these cases it is advantageous to start from a liquid concentrate containing at least 10% active enzyme. For the production of prills, which today have only historical significance, the liquid concentrate even had to be spray-dried.

The various forms of granulates that have been incorporated into powder products over the last 60 years consist of *high shear mixer granulates*, prills, extrusion granulates, and *fluidized bed* granulates (Fig. 17.1). Mixer granulates are formed by shaping enzyme, carrier substances, or additives such as salts or starch and polymers into homogeneous granulates in mixers at high shear force, analogous to the production of crumbles for a crumb cake.

Prills are produced from a mixture of dried enzyme and molten polyethylene glycol dropped over a rotating disc, the droplets solidifying by cooling during free fall in a cooling tower.

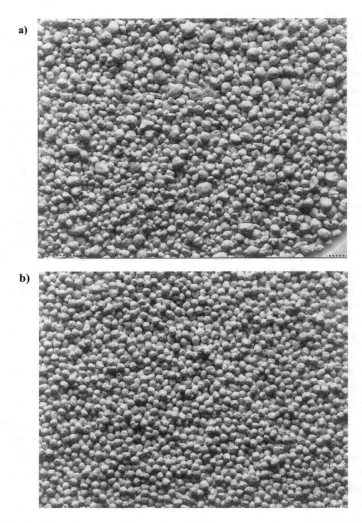

Fig. 17.1 Enzyme granulates for laundry and dish wash detergents. The average particle size is 0.5–0.6 mm. (**a**) Mixer granules; (**b**) fluidized bed granules

Extrusion granules are produced by extruding a mixture of concentrated enzyme and additives such as starch, waxes or salts. This process can be compared to the production of extremely short spaghetti. The extrusion granules are rounded during cooling, as the edges of these granules otherwise provide a starting point for mechanical damage.

Fluidized bed granules are produced by spraying the enzyme solution onto predetermined core particles that are kept floating by an airstream in a fluidized bed coater. The core particles can consist of (coated) sugar or salt. The advantage here is that very defined particle sizes or distributions are possible depending on the core particle. Also, by the amount of enzyme concentrate sprayed on, the activity of the granules can be adjusted very well and, if required, also very high. In addition, these particles are usually very round.

All these granules are coated with a coating layer of waxes such as polyethylene glycol (PEG) or polyvinyl alcohol (PVA) in the final stage of their manufacture to protect them against mechanical abrasion. All granules for detergents must maintain a diameter of 0.3 mm to a maximum of 0.8 mm to prevent segregation in the detergent during transport and storage. In the case of compacted detergent products with a higher average particle size (megaperls®), larger granules may be useful and necessary. In principle, there are high requirements for stability in case of impact on hard surfaces *(impact)*, stability against shearing forces *(shearing)*, and abrasion *(attrition)*. The evaluation of the corresponding physical stability has long been a major problem between enzyme manufacturers and enzyme users. Within the framework of a consortium (Enzyme Dust Consortium), work has been carried out for more than 15 years on the development of analytical methods for these properties.

Questions
1. How are enzymes for detergents manufactured?
2. What are the requirements for enzymes to be used in detergents and what measures taken to make enzymes fit for use in detergents?

17.3 Production of Detergents and Cleaning Agents Containing Enzymes

Detergents for cleaning textiles are offered worldwide in many forms: from bars of soap to multi-chamber sachets. In essence, however, the forms of supply can still be divided into powder and liquid products. Geographically, there have been clear preferences in the past, which are currently weakening at different rates: Asia, the Middle East, and Africa with a preference for powders, in North America preference for liquid products and Europe moving away from powders to a mix of both. The advantage of powders is the ability to use bleach in the form of, for example, percarbonate and bleach activators such as TAED or NOBS, allowing bleaching by peracetic acid even at temperatures such as 40 °C. The formulation of bleach-active components in liquid formulations proved to be extremely difficult despite a variety of efforts using chemical, biochemical and packaging technology approaches.

For powder products as well as for liquid products, the incorporation of enzymes requires considerable effort in order to meet the requirements for worker health and safety and for stability in the formulation. The effort required for enzyme granulates relates essentially to the method of dosing, which is usually carried out from big bags via belt weighers, where the granulate is dosed immediately before packaging in order to minimize mechanical stress on the granulates. For liquid products, the safety aspect is easier to ensure by closed pipelines and by following the process instructions. Here it is usually necessary

Table. 17.1 Examples for laundry detergent formulations (% w/w)

Function/substance	Liquid detergent	Concentrated liquid detergent	Powder detergent Low temperature / No bleach USA/JP	Powder detergent with bleach Europe	Compacted detergent Europe	Tablets
Surfactants (anionic surfactants, soaps, nonionic surfactants)	10–30	10–40	7–22	10–20	10–20	15–25
Complexing agents (citric acid, phosphonates)	0–5	5–10	10–50	10–20	20–40	15–30
Bleach (Percarbonate)	0	0	0–5	11–27	13–28	10–25
Enzyme[a]						
Protease	0.01–0.06	0–0.09	0–0.04	0.01–0.06	0.02–0.06	0.02–0.07
Amylase	0.002–0.02	0.002–0.025	0–0.013	0.003–0.04	0.002–0.04	0.02–0.04
Cellulases	0–0.003	0–0.003	0–0.004	0–0.007	0–0.007	0–0.005
Lipase	0–0.006	0–0.01	0–0.002	0–0.002	0–0.002	0–0.002
Mannanase	0–0.005	0.005	0–0.003	0–0.004	0–0.004	0–0.005
Enzyme-stabilizing agents (Boric acid/boronic acid)	0.2–1/4–6	0.2–1/4–6				
Propylene glycol						
Water	ad 100%	ad 100%				
pH (1% solution)	7.5–10	7.5–9	9.5–11	9.5–11	9.5–11	9.5–11

[a]Enzyme is given as active substance

to adapt the formulation of the detergent to the needs of the enzyme. Unless multi-chamber sachets are used to ensure that aggressive chemicals do not directly affect the enzymes, anionic surfactants, complexing agents, alkalinity and water activity, for example, must be adapted to the stability of the enzymes. Otherwise, in the homogeneous aqueous solution, the enzymes may be denatured by these substances during storage, and thus the consumer cannot be guaranteed stable performance of a product. When formulating proteases, additional care must be taken to prevent this class of enzymes from proteolytically degrading themselves or other enzymes by suitable stabilization.

Examples of powder and liquid formulations of detergents are given in Table 17.1.

In the case of automatic dishwashing detergents, there are uniformly more powder and above all tabs users, but liquid products are gaining in importance. The formulation of automatic dishwashing detergents differs significantly from that of laundry detergents. A key feature of machine dishwashing detergents is the lower dosage of surfactants. Bleach is of high importance in solid machine dishwashing products. Again, formulation is a challenge despite dual or multiple chamber processes. Tabletting poses additional challenges for the granulation of enzymes, as extremely high enzyme loading of the granules and their mechanical robustness are important here.

Examples of powder and liquid formulations of automatic dishwashing detergents are given in Table 17.2.

17.3.1 Manual Dishwashing Detergents

For a few years now, proteases and amylases have also been used in manual dishwashing products. The advantage here depends on the type of manual dishwashing. Very different procedures exist worldwide for manual dishwashing. Naturally, the use of enzymes is particularly advantageous when longer dwell times are used, such as those that occur when the dishes are soaked. In this case, too, the recipes must be adapted to the enzymes' requirements for stable storage behavior.

17.4   Proteases

17.4.1 Molecules

Subtilisin-type proteases still characterize the detergent enzymes segment today (Maurer 2015). Originally, they were found only in *Bacillus species*, but have since been identified in many other organisms. They are characterized by a pH profile that has a maximum in the alkaline or highly alkaline range, which coincides with the isoelectric point of the respective subfamily. Accordingly, subtilisins are grouped into two subfamilies (IS-1: pH 8—alkaline and IS-2: pH 10–11—highly alkaline). They are quite stable at temperatures up to 55 °C, against complexing agents and surfactants and sufficiently stable against oxygen

Table 17.2 Ingredients of automatic dish washing detergents

Function/substance	Classical automatic dish detergent		Compacted automatic dish detergent	Phosphate-containing	Phosphate-free
Alkali	Metasilicate/ disilicates	30–70	Soda	0–40	0–40
			Sodium bicarbonate		0–40
	Sodium carbonate	0–10	Disilicates	0–40	0–40
Complexing agents and disperging substances	Phosphates	15–40	Phosphate	>30	
			Citrate		>30
	Polymers	0–10	Phosphonate	0–2	0–2
			Polycarboxylates	0–5	0–15
Bleaching system	Chlorine bleach	0–2	Oxygen bleach	3–20	3–20
			TAED	0–6	0–6
			Mangan catalyst		>1
Wetting agent / surfactant	Surfactant	0–2	Surfactant	0–4	0–4
Enzymes		0	Enzyme[a] Amylase, protease	<6	<6
	Paraffin oil	1	Perfume	<0.5	<0.5
			Paraffin oil	<1	<1
			Silver protection	<1	<1
pH (1% solution)		12–13		<11	11–12

[a]Enzyme is given an enzyme granulate

bleach. Only against chlorine bleach they are highly sensitive, in the presence of hypochlorite they are denatured immediately.

Subtilisins like all detergent enzymes are secreted as extracellular enzymes from the bacterial cell. They are initially formed as an inactive pre-pro enzyme. This has a length of approximately 380 amino acids, whereby the pre-pro sequence with a length of 105–111 amino acids serves as an intramolecular chaperone for correct folding. The prosequence is autocatalytically cleaved after secretion. It can serve to restore the correct 3D structure after denaturation. The active subtilisin has a molecular weight of approximately 27 kDa. The spatial structure of subtilisins is one of the densest enzyme structures overall. The structure is stabilized by up to three calcium binding sites of different binding strength.

The active site of subtilisins is composed of the amino acids aspartic acid, histidine, and serine. After formation of the enzyme-substrate complex, the reaction mechanism of the serine proteases proceeds via a tetrahedral covalent transition state at the serine of the active site, after which the new amino end of the hydrolyzed protein exits as a leaving group and the new carboxyl end remains bound to serine as an acyl enzyme before this is finally also released via a second, negatively charged tetrahedral transition state in interaction with histidine and water as reaction partners.

As extracellular enzymes, subtilisins are not only relatively stable but also relatively non-specific in their choice of substrates. They act as endoproteases on a broad number of peptide bonds.

According to the European chemicals legislation REACH, subtilisins have been notified (https://echa.europa.eu/de/registration-dossier/-/registered-dossier/14104/1) as one of the enzymes produced in a quantity of more than 1000 tons per year in the European Union. This tonnage does not necessarily describe the amount of active enzyme, but the dry matter of the highest concentrated intermediates. Like all enzymes, they have been notified and evaluated according to the enzyme nomenclature under their EC number EC 3.4.21.62.

Virtually all subtilisins relevant for washing and cleaning agents can be assigned to one of the following subfamilies: Subtilisin Carlsberg (Ex: Alkalase®) alkaline IS-1, Subtilisin lentus (Savinase™, Purafect™, BLAP) high alkaline IS-2 and Subtilisin BPN′ (Purafect Prime™) alkaline IS-1.

The various subfamilies can also be differentiated on the basis of their antigenic properties. Subtilisins crystallize easily above a concentration of 10% active enzyme. For this reason, crystal structures were also known at an early stage and were used for the first time in 1984 for targeted mutagenesis in order to generate protease variants that could not be oxidized by hydrogen peroxide in borate buffer by replacing methionine in position 222. The corresponding products came onto the market in 1993. However, they only became successful there when they were able to prove their potential in automatic dishwashing detergents. In textile detergents the bleaching stability of the proteases proved to be irrelevant even in universal detergents containing bleach. This was all the more surprising as these products at that time contained perborate as bleach and TAED as bleach activator.

Since then, thousands of genetically engineered variants of the various subtilisins have been generated, investigated, and patented in the research departments. The procedures of *site-directed mutagenesis* were soon supplemented by *random mutagenesis* and *directed evolution*. However, the extensive patent landscape for each amino acid position in subtilisins made it very difficult to market the molecules generated in this way on a widespread basis.

17.4.2 Enzyme Action

As non-specific endoproteases, the subtilisins prefer denatured proteins as substrates for hydrolysis. This ideally generates water-soluble peptides. However, washing is not a defined enzyme-catalyzed reaction as in the test tube. The substrate in washing is uncharacterized proteins that are present as a mixture with sugars, natural polymers and fats/lipids on a textile carrier. These mixtures come from our food (egg, milk), our body (blood), and the environment (grass, pigments). They are used not only to test for enzyme-relevant soiling, but also to test for surfactant action or bleaching. In order to test and

Fig. 17.2 Manufacturing of natural soils for testing the performance of laundry detergents. Various soils are applied to the textile (cotton or cotton-polyester blend) by machine and afterwards air-dried at room temperature (by courtesy of Henkel AG & Co KGaA)

evaluate the effect of proteases, soils corresponding to actual life are therefore prepared by machine or manually (Fig. 17.2). For reasons of durability, these soils are sometimes also heat-treated, i.e., denatured. As in real life, such heat treatment can lead to chemical reactions that further modify a protein. An example of this is the formation of reaction products that correspond to the Amadori reaction in the Maillard reaction between proteins and sugars. This usually makes it more difficult for proteases to attack. In addition, the immobilization of protein-containing dirt on the textile support makes the hydrolysis of the proteins even more difficult.

In order to determine the effect and performance of proteases, natural (native) and denatured soils are consequently used. Such soils are produced under standard conditions and can also be obtained in part from specialized companies. In contrast to industrial applications of enzymes, the detergent manufacturer does not know which substrate the enzyme is to act on or which soils are to be removed, so a large number of different soils are tested. The effect is determined by determining the enzyme performance as a function of the enzyme dosage under standard conditions and, for example, with the defined concentration of a specific detergent. Since the detergent as such has a soil-removing effect, this effect must be subtracted by appropriate controls. The degree of whiteness is determined optically, for example, either by using the color of the soil or by adding a pigment (carbon black, India ink) as an indicator to an insufficiently colored soil (Fig. 17.3). This then results in the combinations known in the industry, e.g., blood-milk-ink. As a further variant, the textile carriers are varied (cotton, cotton-polyester blends and pure synthetic fibers such as polyester).

Simplified, it can be said that washing tests are carried out analogously to the investigation of an enzyme-substrate curve, except that a variety of substrates are used in washing by

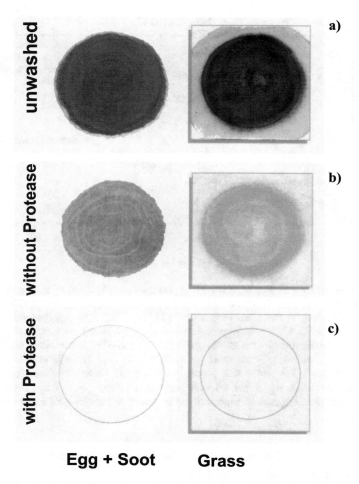

Egg + Soot   Grass

Fig. 17.3 Two typical stains for testing proteases (**a**) before washing; (**b**) after washing without protease, and (**c**) with protease. The washing process was carried out with a bleach containing heavy duty detergent at 40 °C

means of which the enzyme action is measured, that the washing or dishwashing machine serves as the reaction vessel and the greying (remission) is measured as a sum parameter. As a rule, the control value of the enzyme-free detergent is subtracted and the sum of the differences of all tested soils is plotted (Fig. 17.4).

The determination of enzyme performance can begin on the scale of microtitre plates and then proceeds to the scale of 50 ml in so-called Launder-O-Meters, in which 6–8 different preparations are run in parallel in metal beakers in a water bath, up to series of 6–10 washing machines of the same type, in which the fuzzy logic is usually inactivated in order to create completely identical conditions (Fig. 17.5) In order to evaluate different

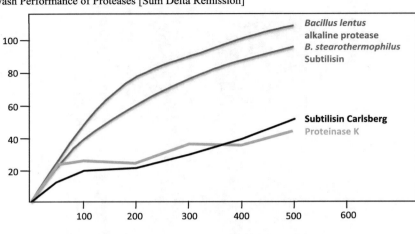

Wash Performance of Proteases [Sum Delta Remission]

Fig. 17.4 Wash performance of different serine proteases as a function of enzymatic activity. The enzyme specific washing performance was tested by using a large number of different protease-specific stains in washing machines or launder-O-meters using a bleach containing, heavy duty powder detergent at 40 °C. The difference in whiteness (remission) with and without protease is plotted as a function of protease activity. Subtilisin 309 (Savinase®) is a member of the *Bacillus lentus* alkaline proteases. Subtilisin Carlsberg is produced by *Bacillus licheniformis* and commercially available as Alkalase®. Proteinase K is presently not used in detergents. The specific activity of different subtilisins determined at pH 8.5 with casein as a substrate can easily differ by a factor of 2. Thus, the figure does not represent the wash performance as a function of the amount of protease protein

geographical washing habits and to secure the results, washing machines of completely different geographical origin are also used.

The water used for washing must be of defined and consistent water hardness and composition. As a rule, synthetic tap water of medium hardness is used (15° dH).

17.4.3 Stability, Toxicity, and Environmental Aspects of Proteases

The stability requirements affect the enzyme product as well as the detergent produced with it. On its way to production, the enzyme product is sometimes transported over long distances and stored for long periods under challenging conditions (temperature, humidity). Nevertheless, it is expected that the enzyme activity will not suffer as a result.

Once the enzyme product has been incorporated into the detergent or cleaning agent, the expectation of both the manufacturer and the consumer is that the performance of the product and thus also of the enzymes will remain constant over a long period of time without any problems during storage, both at the retailer and at the consumer. Proteases

Fig. 17.5 Experimental set-up of washing machines for testing different detergent formulations (varying ingredients including different enzymes or different enzyme concentrations) (by courtesy of AB Enzymes GmbH)

present a particular problem to the industry in this regard because of their effect on proteins in aqueous solution. This challenge is solved by the reduction of the water activity, and in addition by reversible inhibition of the proteases. This inhibition has to be reversed by the dilution factor when rinsing in the washing machine. Suitable reversible inhibitors are compounds analogous to the tetrahedral transition state of the substrate during catalysis. Diols and boric acid can form corresponding complexes. In practice, 1,2-propanediol (propylene glycol) and boric acid or boronic acids such as 4-formylphenylboronic acid (FPA), but also peptide-based reversible inhibitors are used for this purpose.

Using the example of subtilisins, a wide range of investigations were also carried out into the sustainability aspects of detergent enzymes. By investigating the washing performance of a formulation with different protease concentrations, detergent dosages, and washing temperatures, it was possible to show that the proteases make a decisive contribution to reducing detergent quantities and washing temperatures. Several life cycle assessments have also shown that the production of detergent proteases by recombinant microorganisms is far superior to the production by classical production strains.

Due to their effect on proteins, proteases like the subtilisins have an additional toxicological effect which consists in the fact that they may have an irritating effect on mucosa and skin. Concentrated enzyme solutions or enzyme products must therefore be declared accordingly as irritating to the skin and eye, in contrast to consumer products. The same properties also lead to the fact that in the assessment of environmental toxicological properties, subtilisin proteases have been determined to be toxic to certain algae, water fleas (*Daphnia*) and fish fry. For this reason, concentrated subtilisin products must be declared as environmentally toxic. However, the concentration of proteases in consumer products is too low to require specific labelling.

Questions
3. What is the lead molecule for proteases used in detergents?
4. What is the standard type of production organism for these proteases and what is the reason for that decision?

17.5   Amylases

17.5.1 Molecules

Just as practically all detergent proteases relevant to date belong to the subtilisin family and are produced with *Bacillus species*, the main α-amylases used in detergents and cleaning agents to the basic molecule of α-amylase from *Bacillus lichenifomis*. Characteristics of this molecule are its pH profile with a maximum in the neutral pH range (6–7) and its high thermostability (up to 90 °C depending on calcium concentration). Since 2003, other *Bacillus amylases* have been added, characterized by a pH optimum in the alkaline (pH 8–10) range and somewhat lower thermostability (up to 60 °C).

The molecular weights of the *Bacillus amylases* are significantly higher than those of the proteases, at 50 kDa. All *Bacillus*-α-amylases are characterized by three domains and belong to family 13 of the glycosyl hydrolases. The A domain is the catalytic domain, characterized by a barrel structure formed by eight β-sheets, which in turn are surrounded by eight α-helices. The B domain consists of a β-sheet containing a calcium ion binding site, and the C domain also consists of a β-sheet. All members of family 13 possess aspartate residues as catalytic bases or nucleophiles and glutamate residues as catalytic acids or proton donors.

In the experiments on *protein engineering* and the directed evolution of amylases for detergents and cleaning agents, the initial focus was on stability against oxygen bleaching. In automatic dishwashing detergents, the bleach content is very high and both the storage and process stability of classical amylases are not excellent here. By replacing oxidation-sensitive methionine residues with non-oxidizable amino acid residues, it was possible to achieve a significant increase in stability, although this came at the price of reduced performance.

17.5.2 Enzyme Action

The α-amylases had been known and available for a long time, but their breakthrough came with the development of liquid laundry detergents in the 1980s and enzyme-containing automatic dishwashing detergents in the 1990s. In combination with the proteases, they show real synergy in both detergents and cleaners, i.e., a stronger joint effect than would be possible with each enzyme class in isolation. The fact that the amylases harmonize excellently with subtilisins is not surprising, since both are co-expressed by the same or related *Bacillus species*.

The action of amylases is based on the removal of starch, which in textile laundry comes either from finishes or from food residues such as porridge or pudding. The effect can also be differentiated according to the origin of the starches (potato, rice, corn, wheat, or tapioca). Amylases act as endoamylases and produce soluble dextrins of 2–6 glycosyl residues as well as limit dextrins at the branching sites of amylopectin. The use of pullulanases, i.e., amylases that hydrolyze such α-1,6-linkages, has been the subject of intensive patent work, but there are no relevant products on the market.

The importance of starch as a typical soil in washing has increased with its increased use in processed foods. Although starch as such is colorless and consequently actually uncritical, it has a supporting effect on textiles when dirt is re-deposited on the textile (redeposition) also from the wash liquor.

When washing dishes by machine, the repeated application of starch can lead to dulling of the decoration and lower gloss. The amylases maintain or restore the original gloss. This effect is particularly appreciated in commercial kitchens.

A range of artificially soiled test fabrics is also available for testing amylase performance. These consist, for example, of cocoa/sugar/potato starch, starch/charcoal on cotton or cotton/synthetic blend fabrics. Pure starch-loaded fabrics loaded with different natural starches are also in use. In these cases, the textile must be dyed on starch after washing to make the effect optically measurable.

For the determination of the amylase effect in automatic dishwashing detergents, different soiling and test methods have been developed with the support of industry associations. In the case of automatic dishwashing, the composition of the ballast soil plays a major role for both proteases and amylases.

Questions
 5. What is the lead molecule for amylase used in detergents and what is the manufacturing organism?

17.6   Cellulases

17.6.1 Molecules

While proteases and amylases are almost exclusively derived from Bacilli and can be clearly assigned to uniform superfamilies (subtilisins, resp. family 13 glycohydrolases), the variance of molecules of cellulases for detergents is high. All cellulases used in detergents are β-1,4-endo glycohydrolases or endoglucanases, but they originate from different donor organisms and are also assigned to different glycohydrolase families.

The first cellulases used in detergents by Kao in Japan in 1986 were obtained from an alkaliphilic *Bacillus* strain. This molecule is characterized by a high molecular weight. The second cellulase offered for detergents consisted of a complete fungal cellulase mixture of *Thermomyces lanuginosus* (at that time *Humicola insolens*). This was a mixture of β-1,4-endoglucanases, cellobiohydrolases, and cellobiases. A β-1,4-endoglucanase from this mixture was cloned in *Aspergillus oryzae* and marketed as a monocomponent enzyme in 1993. Later, other molecules from alkalophilic Bacilli and from fungi *(Trichoderma reesei, Melanocarpus albomyces)* were marketed by DuPont/IFF and AB Enzymes as monocomponent products.

17.6.2 Enzyme Action

Cellulases act exclusively on cellulose-based textiles such as cotton or Tencel. Their effect can either be based on the detachment or prevention of the re-deposition of particulate dirt (dust, tennis ash, soot) (e.g., *Bacillus* cellulases) or on a fiber effect. Why cellulases help to remove particulate dirt from cellulosic fibers but not from polyester fibers has not yet been conclusively clarified. The effect on the fiber may be the smoothing of cotton fibers by breaking down protruding microfibrils and/or the prevention or removal of pilling. This keeps the color fresh in colored textiles or prevents pilling in knitted textiles. The textiles thus remain attractive in their color and the appearance of their surface for longer (Fig. 17.6).

However, the effect on the fiber is bought with the risk of fiber damage. This is particularly critical when using natural and complete cellulase systems from fungi. After all, these systems always aim at the complete hydrolytic degradation of cellulose to glucose, which is definitely not the goal of the washing process. For this reason, it is only with the cloned and genetically engineered monocomponent cellulases that enzymes have become available that allow safe and controlled enzyme action. Moreover, these enzymes are dosed very carefully. The optimum fiber effect of cellulases is usually set to be achieved over several washes or wash-carry cycles. Proof of the effect is then obtained after 5, 10, or 20 washes.

All cellulases that achieve fiber effects require binding to the fiber, which is achieved via cellulose binding domains (CBDs). A cellulase can also have several binding domains. The linkers connecting the CBD to the catalytic domain are usually highly flexible and can serve as targets for proteases. Through such proteolytic degradation, a cellulase can completely lose its effect, although the catalytic domain shows measurable activity.

Incidentally, similar cellulases with a fiber effect are used in textile technology. Here, cotton textiles are treated once with cellulases to prevent pilling (bio-polishing®).

The cellulases used to prevent redeposition or detachment of pigment soil generally have no effect on the fiber. The effect can be measured either directly by removing pigment dirt (dust, soot, tennis ash) or by the lack of transfer of pigment dirt from soiled textiles to white textiles by determining the degree of whiteness. The effect they achieve, also known

Fig. 17.6 Fiber effect of cellulases for color refreshment and fiber smoothing/antipilling. Cotton textile after ten wash cycles using a liquid detergent at 40 °C without (top) and with (bottom) celullase (by courtesy of AB Enzymes GmbH)

as inhibition of greying, does not require binding to the cotton fiber. The exact mechanism of the graying-inhibiting or antiredeposition effect has not been published. It is interesting to note that the effect is not inhibited but, on the contrary, enhanced by addition of carboxymethylcellulose, which has the same effect and is a substrate of these cellulases.

Questions
 6. What are the effects of cellulases used in laundry detergents?

17.7   Mannanases

In the last 20 years, the food industry has increasingly used natural polymers to create a certain texture, viscosity, or mouthfeel in processed foods. These foods can be, for example, ice cream, sauces or desserts. The polymers used are substances such as mannan (guar gum, locust bean gum), carrageenan, pectin, or xylan. In addition to the desired effects in the food, these substances also have the effect of ensuring good binding to textiles. Enzymes that are able to hydrolytically degrade these substances therefore significantly support the detergent in removing the corresponding soiling. This is especially true if the food is colored by chocolate, fruit juice, or by brown reaction products of the Maillard reaction, which are typical for frying or roasting processes, e.g., in meat sauces.

These polymers can also bind dirt particles and thus contribute to the greying of textiles. The use of mannanases and carrageenases was already patented in the 1990s.

17.7.1 Molecules

Novozymes entered the market in 2000 with a mannanase (Mannaway™). The product is based on a mannanase gene from a *Bacillus alkalophilus* strain expressed in *Bacillus licheniformis*. The product was initially only used for the premium brand of a detergent manufacturer, but since 2006 mannanases have been used on a broad front.

17.7.2 Enzyme Action

Mannanases are able to hydrolyze the β-1,4-mannose bond in galactomannan. Due to the *endo effect* on the polymer, its adhesion to the textile fiber is significantly reduced. As a result, the bond to other particulate substances such as cocoa is also lost. The use of mannanases can be clearly demonstrated by the effect on specific soils, such as chocolate ice cream, which was produced using guar gum.

17.8 Lipases

Lipases hydrolyze fat and oils (triglycerides) to diglycerides, monoglycerides, and glycerin. By this, they may support the emulsion and detachment of fatty soils by surfactants, especially because mono- and diglycerides themselves may act as emulsifiers. However, the detachment of the fatty soil also depends on the washing temperature which is necessary for the melting process. Triglycerides of long-chained, saturated fatty acids are difficult to remove at temperatures below their melting point even with lipases.

Lipases are only used in detergents and not in automatic dishwashing detergents, as the formation of lime soaps must be expected in this application.

Similar to proteases, lipases show a strong interaction with the surfactants used, both in terms of performance and stability. Both have to be optimized by a clever ratio of different anionic and non-ionic surfactants, the so-called surfactant block.

17.8.1 Molecules

The first lipase for detergents was launched in 1986 by Novo Nordisk under the name Lipolase®. It is the lipase from *Thermomyces lanuginosus* (then *Humicola insolens*) cloned and expressed in *Aspergillus oryzae*. The molecule has since been optimized several times by protein engineering (Lipex®, Lipoclean®) and is still the guiding principle for all detergent lipases. The aim of further development was to achieve a stronger effect in the first wash cycle.

In 1995, a detergent lipase derived from *Pseudomonas pseudoalkaligenes* and produced in a *Pseudomonas strain* was launched by Gist Brocades. It was later described to be expressible in *Bacillus amyloliquefaciens*. This molecule is not currently on the market.

Lipases are among the technical enzymes that would not be available without recombinant production strains.

17.8.2 Enzyme Action

Ever since lipases have been on the market, they have had their fixed niche in the enzyme portfolio of detergent manufacturers. Like cellulases and mannanases, they are used to a large extent to differentiate premium products. Due to their effect in removing greasy soils, they replace some of the non-ionic surfactants. Typical soils for the detection and testing of lipase activity are triglyceride based lipstick soils and sebum (e.g., shirt collars). As already mentioned, lipases usually require several washings until the optimal result is achieved. Higher lipase concentrations are usually not a good approach as high dosages of lipases may cause undesirable side effects: Lipases accumulate at the interface between fat and water. If, in the case of heavily soiled textiles, the fat cannot be completely hydrolyzed, emulsified, and removed, the lipase may remain active on the textile after washing and continue to be hydrolytically active on the traces of fat after drying in the cupboard with the air humidity present. The result in this case is an unpleasant odor of the washed laundry.

However, if the lipase dosage is correct and well adjusted, this case can be excluded.

17.9   Pectate Lyases

17.9.1 Molecules

Pectin is a plant polymer used to thicken and solidify jams and preserves. Pectins are essentially composed of linear α-1,4-linked D-galacturonic acids interrupted at regular intervals by α-l-rhamnose. Additional oligomeric side chains of arabinose, galactose, or xylose may be coupled to the rhamnose. The acid function of galacturonic acid may be esterified to varying degrees with methanol. Pectin is water soluble depending on temperature, depending on its origin and composition. At lower washing temperatures and in the case of intensively colored jams, the removal of soiling based on berry jams is quite difficult without enzymes. For this reason, different pectinases were tested early on for their removal. Pectinases are produced by microorganisms which use them to break down the pectin content of plant cell walls. Pectinases include pectin esterases, endo- and exopolygalacturonases as well as pectin and pectate lyases. So far, only pectate lyases have been found to have an effect in detergents.

Although most known pectinase products are produced with fungi such as *Aspergillus species*, they are not ideally suited for detergents due to their pH optimum in the weak acidic range. The pectate lyase genes used in detergents are of bacterial origin and are derived from *Bacillus species*. They are produced with the aid of *B. licheniformis* or *B. subtilis*.

17.9.2 Enzyme Action

Pectate lyases do not act via a hydrolytic reaction mechanism, so they are not hydrolases. The reaction mechanism of the lyases (β-elimination) begins with the removal of a proton by arginine and leads in a three-step process to the departure of the non-reducing end in the form of a 4-deoxy-galacturonyl residue unsaturated between C4 and C5.

Pectate lyases only act in alkaline pH, which facilitates their use in detergents. By cleaving the polymer chain of pectin, which ensures that soiling is otherwise difficult to remove in color-strong jams such as wild berries, they can lead to significant effects depending on the type of pectin.

17.10 Sustainability Aspects

The quantity in which detergents and cleaning agents are used in households and in commercial use means that these products have an environmental relevance. For this reason, life cycle assessments, also known as life cycle analyses (LCA) have been drawn up for the production and use of these products.

Enzymes in detergents and cleaning agents have played a major role in significantly reducing the amount of detergent or cleaning agent needed for a standard wash or rinse cycle in recent decades (Table 17.3; Dreja et al. 2014).

A decisive advantage here is that only a small quantity of enzymes is required as catalysts. At the same time, it was possible to significantly improve the performance on standard soils, so that more output was possible with reduced chemical use. At the same time, it was possible to reduce the average temperature in both washing and machine rinsing. However, dosing, like the choice of temperature, is in the hands of the consumer. For this reason, dosing instructions and dosing cups were changed for washing. In the case of automatic dishwashing detergents, tabs have been launched on the market which prescribe an optimum dosage. To reduce temperatures, initiatives have been in place for some time to make consumers aware of the possibility of reduced temperatures (*"I prefer 30"*). This gives the consumer the opportunity to save private money through reduced energy consumption. However, at the same time, for hygiene reasons, it is recommended to run each washing machine once a month at high temperature (90 °C) with a heavy duty detergent containing bleach, in order to control micro-organisms, e.g., in the sump of the washing machine.

Another aspect of sustainability is the fate of the components in the environment. As a rule, in Central Europe it is assumed that the wastewater is processed via a multi-stage sewage treatment plant. In any case, biodegradability is an absolutely crucial criterion. As proteins, enzymes are rapidly and completely biodegradable. For this reason, they are also excellent ingredients in this respect. The biodegradability of detergent enzymes optimized by protein engineering was investigated particularly intensively, since enzymes with improved stability had to be tested for their biodegradability. However, these variants

Table 17.3 Comparison of the amounts of detergent and automatic dish detergent needed for a 5 kg wash (moderately stained and at moderate water hardness) resp. for a dishwasher over a period of 30 years

	1986	2016
Laundry detergent for 5 kg laundry	200 g	69 g
Add (powder/tabs)	30 g[a]	18 g

[a]Consisting primarily of chlorine bleach and metasilicate

were also just as rapidly and completely biodegradable as all other enzymes investigated to date.

Another aspect is the effect on aquatic organisms. Except for proteases, enzymes show no effects here either. However, proteases in concentrated and active form may harm water fleas and fish fry due to their intrinsic effect on proteins. For this reason, enzyme products must be labelled accordingly. After their use in washing or cleaning, however, these enzymes are only present in small quantities or activity in the wastewater of a dishwasher or washing machine due to autoproteolysis and are no longer present at all in a sewage treatment plant.

Long before the new European chemicals legislation, the results of studies on the health and environmental relevance of detergent ingredients such as enzymes were published in the so-called HERA project of the AISE (International Association for Soaps, Detergents and Maintenance Products) (www.hera-project.com).

Questions
  7. What are sustainability aspects of enzymes used in detergents?

17.11 Conclusion and Outlook

In the past, a lot of research has also been done on enzymes that can have a direct or indirect bleaching effect on bleachable soils. Hydrogen peroxide, for example, which has been used in detergents for some time in the form of the addition compound percarbonate, has a direct bleaching effect. However, hydrogen peroxide from precursors like percarbonate or perborate can only be used in powder detergents. Oxidases such as glucose oxidase would in principle be able to produce hydrogen peroxide in liquid products as well. Oxidases require a substrate to generate hydrogen peroxide. However, this would require the enzyme and its substrate (in the case of glucose oxidase, glucose) to be prepared separately. This, of course, is accompanied by the question of how much substrate is needed to produce the required concentration of hydrogen peroxide. The great advantage of hydrolytic enzymes, that they only require water for their reaction on the substrate, does not apply to oxidases.

In addition to the generation of hydrogen peroxide, other enzymes are also capable of generating medium- to long-chain peracids. The reaction is based on the perhydrolytic conversion of esters with hydrogen peroxide instead of water to peracids. This side reaction has previously been demonstrated in lipases, certain subtilisin proteases, esterases, and acyltransferases. The advantage of this enzyme reaction is seen primarily in liquid products, although a two- or multi-chamber system is required. Although there are a number of patents on this, no commercial use has yet been made of it.

> **Take Home Message**
> - Enzymes used in detergents are proteases, amylases, cellulase, lipases, and pectate lyases in laundry and proteases and amylases in dishwashing.
> - Detergent enzymes are produced by large scale fermentations (30–120 m^3) and formulated specifically for the application.
> - The production organisms in general are genetically modified to allow economically feasible cost of goods.
> - The use of enzymes in detergents allows for lower washing/cleaning temperatures, reduction of the chemicals consumed, and better performance of the detergent—thus, they represent a major contribution to the sustainability of detergents.

Answers
1. Enzymes for detergents are produced by genetically modified microorganisms in large scale submersed fermentations. The enzymes are secreted by the production strains and are concentrated and purified from the supernatant of the fermented broth.
2. Detergent enzymes are required to be performant in the process, and stable in the formulation during storage and during application. In addition they are required not to change the color or the odor of the formulation. In powder or tablet formulations non-dustiness is required not to release enzyme dust in manufacturing or application. Performance is based upon the enzyme molecule, the stability can be based upon intrinsic stability of the molecule and stabilizers used in the formula (low water content/activity, and the use of reversible inhibitors, The non-dusting properties for powders are ensured by granulation and coating of enzyme particles.
3. Proteases of the Subtilisin family are the lead molecule for detergent proteases. In general they originate from *Bacilli*, where they are secreted.
4. For manufacturing of the proteases, *Bacillus species* are used due to the ability to secrete the enzyme with high efficiency.
5. The lead molecule of detergent amylase is the *Bacillus licheniformis* alpha-amylase. It does especially fit as the molecule originates from *Bacillus* and is highly compatible with the Subtilisin proteases.
6. Cellulases in laundry detergent are used for two reasons on cotton/cellulosic textiles: (1) Some molecules demonstrate an anti-greying/antiredeposition effect. (2) Some molecules show an antipilling and color refreshing effect.

7. The use of enzymes in detergents represents a major contribution to sustainability, as it allows for lower wash temperatures, less consumption of chemicals, and improved washing/cleaning performance including longer lifetime of cotton textiles (cellulases).

References

Aehle W, editor. Enzymes in industry. 3rd ed. Weinheim: Wiley-VCH; 2007. p. 154–192.

Dreja M, Vockenroth I, Plath N, Schneider C, Martinez M. Formulation, performance and sustainability aspects of liquid laundry detergents. Tenside Surfactants Detergents. 2014;51: 108–112.

Herbots I, Kottwitz B, Reilly PJ, Antrim RL, Burrows H, Lenting HBM, Viikari L, Suurnäkki A, Niku-Paavola M-L, Pere J, Buchert J. Enzymes—non-food application. In: Ullmann's encyclopedia of industrial chemistry. Weinheim: Wiley-VCH; 2012.

Maurer K-H. Detergent proteases. In: Grunwald P, editor. Industrial biocatalysis. Singapore: Pan Stanford; 2015.

Olsen HS, Falholt P. The role of enzymes in modern detergency. J Surfactants Detergents. 1998;1: 555–567.

Enzymes and Biosensor Technology

18

Michael J. Schöning and Arshak Poghossian

What You Will Learn in This Chapter

This chapter describes the use of enzymes in biosensors. Enzyme-based biosensors have enjoyed a prosperous growth market for more than five decades and are increasingly being used in biotechnological processes. Typically, they comprise a biological recognition element together with a physicochemical transducer. The main advantages of biosensors are their easy-to-use operation, cost-effective manufacturing, and their ability for sensitive and highly accurate detection of specific analytes. A brief overview on "biosensor technology" is given, including relevant sensor parameters, followed by major developments in electrochemical enzyme biosensors with a more detailed focus on possible applications in the field of biotechnology. Looking beyond the "end of one's nose" will offer insight into alternative transducer principles and biomolecules.

According to the definition, the term **sensor** (from Latin *sentire*, to feel or sense) describes a detector or measured variable transducer that conveys knowledge about a measured variable (e.g., the temperature in a bioreactor) and converts this information into a suitable measurement signal (e.g., a voltage that can be measured with a voltmeter). As a technical component, the sensor thus represents the primary element of a measuring device that can (quantitatively) detect certain physical or chemical/biological properties and then convert these into a suitable measurement signal by means of a so-called transducer.

M. J. Schöning (✉) · A. Poghossian
Institute for Nano- and Biotechnologies (INB), FH Aachen – Jülich, Jülich, Germany
e-mail: schoening@fh-aachen.de

© The Author(s), under exclusive license to Springer Nature Switzerland AG 2024
K.-E. Jaeger et al. (eds.), *Introduction to Enzyme Technology*, Learning Materials in Biosciences,
https://doi.org/10.1007/978-3-031-42999-6_18

Sensors are generally divided into different types. Depending on the measurand to be determined, these can be mechanical sensors, thermal sensors, acoustic sensors, sensors for optical signals and radiation, sensors for electrical and magnetic radiation, and sensors for chemical and biological measurands. The latter group—the **biosensors**—will be dealt with in more detail in this chapter.

18.1 Chemo- and Biosensors

18.1.1 Definition: Chemo/Biosensor

According to IUPAC (International Union of Pure and Applied Chemistry), a chemosensor is a miniaturized device that selectively and reversibly detects chemical compounds or ions and provides an analytically useful signal. Biosensors are a subgroup of chemosensors, i.e., they are also miniaturized sensors, but they use biological recognition mechanisms for analyte recognition (Thevenot et al. 1999).

Miniaturized in this context means a closed, ideally integrated transducer in which the chemical or biological recognition element—the receptor—is in direct spatial contact with the transducer (Fig. 18.1).

A biosensor—or more precisely, a biosensor chip—consists of three essential core components: the receptor layer, the transducer, and the signal processing.

The receptor layer, the actual sensor-active layer, is in direct contact with the sample to be examined, the analyte. The analyte can be gaseous or liquid. The interaction between receptor layer and the analyte molecule is selective, i.e., comparable to the simplified "lock and key principle" in enzyme-substrate reactions; at the same time, this mechanism must be reversible.

In the next step, the interaction between receptor and analyte molecule is transformed into a physical output signal (a measurand). The transducer is responsible for this. The transducer converts the chemical or biological information at the input of the chemo/biosensor into a metrologically accessible signal.

The third part of such a chemo/biosensor represents the final signal processing. This is directly linked to the corresponding display unit. Here, the electrical output signals are adapted by means of signal amplification, impedance matching, compensation of disturbance variables, or characteristic curve correction, for example. This part also represents the interface to the environment and the user, e.g., for input/output media, open-loop or closed-loop control systems or the actuators (control elements such as pumps, valves, nozzles, etc.).

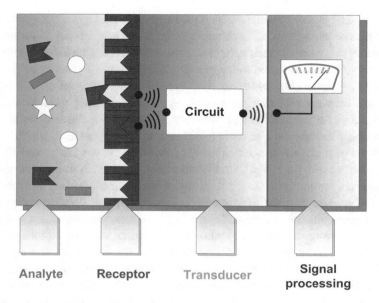

Fig. 18.1 Schematic diagram of a chemo/biosensor

18.1.2 Classification of Biosensors

Biosensors are often classified depending on either their receptor layer or the transducer used. Receptor layers can be present at different levels of integration: In the simplest case, they can be dielectric or semiconducting layers that provide binding sites at the interface to the analyte with which the molecules to be detected interact. Synthetically produced organic molecules with a defined, three-dimensional cage structure—so-called ionophores—allow the detection of monovalent (e.g., Na^+) or polyvalent (e.g., Ca^{2+}) ions. The ions can diffuse specifically into the cavities of the ionophores and change the physical properties there.

For the detection of more complex molecules, such as glucose or urea, enzymes are often used; the products resulting from the biocatalytic reaction with the substrate to be detected can then be measured. In addition to enzymes, antibody-antigen reactions, for example, can also function as a receptor principle, or even parts, i.e., fragments, of a DNA molecule can be used to "fish" the complementary, matching DNA counterpart out of the analyte solution.

In addition to ionophores for the precise detection of ions, synthetic receptors are increasingly being used for larger biomolecules, including viruses or even complete cells; these are known as MIPs *(molecularly imprinted polymers)*. They are constructed in such a way that they have replicated (at least) parts of the biomolecule to be detected in their structure as a negative imprint—similar to a fingerprint (Schirhagl 2014). The biomolecule can diffuse into this three-dimensionally designed receptor structure and interact there. Cell-based biosensors exhibit the highest complexity and integration density: Here, complete microorganisms, bacteria, cells, or tissue sections act as receptor layers. Such receptors are present in their natural environment and thus have their highest activity and sensitivity. However, uncontrolled growth ("proliferation") often takes place on the surface of the biosensor chip, which can have a strong negative impact on its functionality.

In general, there are two principal procedures for immobilizing biomolecules (such as enzymes) on the sensor surface as a receptor layer: physical and chemical immobilization. When immobilizing, care must be taken to ensure that the enzyme molecules can be stably coupled to the chip surface on the one hand, so that they are not easily washed off again directly on contact with the analyte. On the other hand, the immobilization must not be too stringent, so that, for example, the functionality of the biomolecules is negatively affected. Figure 18.2 shows a schematic overview of the common immobilization methods.

Physical immobilization essentially involves the three processes: adsorption, membrane or gel entrapment. Porous or hydrophobic surfaces are often sufficient to couple enzymes adsorptively, e.g., via van der Waals forces. On the one hand, these are weak bonds, but on the other hand, they are a very moderate immobilization method, in which the functionality of the receptor molecules remains almost undisturbed compared to their unbound state in solution. Comparable is the case for inclusion in a dialysis membrane. If the enzymes do not adhere sufficiently to the chip surface, they can still be fixed to the chip surface by a membrane. However, in this case it must be ensured that the pore size of the membrane allows the unhindered penetration of the substrate to be detected and, conversely, that the enzyme cannot diffuse out. Confinement in a gel (e.g., agar-agar or polyacrylamide) has comparable properties to membrane confinement, but the functionality of the enzyme may suffer. Common to all physical immobilization methods is that a high loading density of receptor molecules per area can be achieved. This can ultimately have an advantageous effect on the detection sensitivity of the biosensor.

In contrast, **chemical immobilization** focuses more on the directed and locally addressable immobilization of the enzymes on the surface of the sensor chip. Here, a distinction is made between three common methods: covalent binding, crosslinking or crosslinking together with inert proteins. In the case of covalent binding between enzyme and sensor chip, the chip surface is modified by certain chemical formulations to such an extent that direct chemical binding (e.g., via amino groups) with functional groups of the enzyme can take place. In the case of crosslinking, an additional crosslinking of the enzyme molecules with each other (e.g., by glutardialdehyde) and with the chip surface can optionally be carried out. In this case, however, there is a risk that the enzyme molecules are crosslinked too densely or that those sites of the enzyme molecules are crosslinked that are actually supposed to react with the substrate to be detected. To circumvent this, another possibility is to introduce additional inert proteins (e.g., *bovine serum albumin*, BSA) as "spacers" between the receptor molecules. Another approach to covalent binding is avidin/biotin linkage with one of the strongest known bonds between reaction partners. In this approach, avidin is first covalently attached to the sensor surface and then, in the next step, the biotin-functionalized receptor molecule is coupled to the avidin via the biotin bridge.

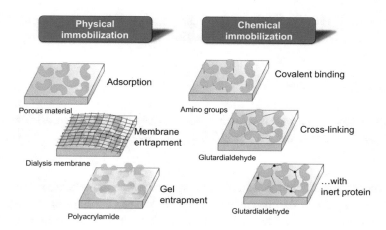

Fig. 18.2 Immobilization strategies for coupling enzymes on sensor chips (schematic)

Which immobilization procedure can ultimately be used for which enzyme must be optimized on the basis of the experimental setup (Mulchandani and Rogers 1998; Wollenberger et al. 2003). This is also the commonly chosen approach for the development and commercialization of biosensors on the part of industry.

As already listed above, the signal converter (transducer) is responsible for the physical conversion of the (bio-)chemical information at the input of the biosensor into a metrologically accessible signal. In general, there are six different classes of transducers for chemo/biosensors:

- **Electrochemical transducers:** systems in which electrons, ions, and phase boundaries between electron and ion conductors play a decisive role. Electrochemical transducers can be divided into potentiometric (Sect. 18.2.1), amperometric (Sect. 18.2.2), coulometric, and conductometric transducers.
- **Electrical transducers:** systems in which electrical (non-ionic) conductivity plays a role.
- **Optical transducers:** these systems use the change in electromagnetic radiation due to interaction with one or more substances of the analyte (Sect. 18.3.1); quantities may include absorption (attenuation), luminescence or fluorescence (light emission), reflection, refraction, or diffraction.
- **Thermal transducers:** systems based on a change in heat due to a chemical or bio/chemical conversion (calorimetry) of the analyte molecule to be detected.
- **Mass-sensitive transducers:** systems which use the change in mass loading on their surface by interaction with the analyte (Sect. 18.3.2); a distinction is made between piezoelectric transducers—often called quartz crystal microbalances—and SAW *(surface acoustic wave)* sensors.

- **Magnetic transducers:** systems that use the (super-)paramagnetic properties of certain analyte molecules (e.g., O_2) for their detection.

18.1.3 Sensor Parameters

The requirements placed on biosensors depend largely on their intended use. The most important criteria are sensitivity, selectivity, stability, response time, application, and design.

- **Sensitivity:** By definition, sensitivity describes the slope of the biosensor's calibration curve, i.e., the ratio of the change in the output signal (e.g., voltage or current) when the substrate concentration is changed as the input signal. This is also accompanied by the subsequently usable (dynamic) measuring range, which is limited by the lower and upper detection limits. The slope of the calibration curve also makes a statement about the resolution (measurement accuracy), i.e., which smallest change in the analyte concentration can still be detected. It is often argued (as with the lower detection limit) that the change in the measurement signal must be three times greater than the natural "*noise level*" of the sensor signal.
- **Selectivity** provides an indication of how well discrimination can be made between the substrate molecule to be detected and other (interfering) components in the analyte; low cross-sensitivity is generally desired.
- The **stability** defines the time period during which the biosensor provides a constant output signal under constant conditions. Stability—ideally a long lifetime—can be adversely affected by the drift and hysteresis of the sensor signal. Drift means the change of the sensor signal at constant conditions over time; hysteresis describes the undesired difference of the sensor signal at (repeatedly) the same analyte concentration, depending on whether measurements are made in the direction from high to low concentration or *vice versa*. Biosensors must provide stable measurement signals under harsh conditions in bioprocesses (shear forces, sterilization, media composition), for example.
- The **response time** is the time required for the sensor signal to reach a stable final value. Often one refers to the $t_{90\%}$—or $t_{95\%}$—final value of the sensor signal after the analyte concentration has been varied. Real-time analyses, i.e., short response times, are preferred in practical use.
- **Application** and **design** are two aspects that should not be underestimated, especially against the background of commercial biosensors. The end user often wants the sensor to be miniaturizable, compact in design and robust; it should be user-friendly and able to automatically compensate for disturbance variables. In addition, acquisition and operating costs, self-calibration, product safety, and user acceptance play a decisive role. Depending on whether the biosensor is to be used in a biotechnological process (e.g., for in-line analysis in a bioreactor) or even in the human body for in-vivo

measurement, additional factors such as product safety, biocompatibility (i.e., compatibility of the sensor in the body), or sterilizability must be included in the sensor development.

Questions

1. What is the definition of a biosensor and what are the main components of a chemical sensor / biosensor?
2. Comment on typical immobilization strategies to couple enzymes onto sensor surfaces.
3. Which transducers are commonly utilized in enzyme biosensors?
4. Name the most important parameters that indicate the behavior of biosensors.

18.2 Electrochemical Enzyme Biosensors

Electrochemical enzyme biosensors often use either a voltage change (potentiometry) or a current change (amperometry) as the output signal (sensor signal), which depends on the substrate concentration to be detected, or more precisely on the product concentration resulting from the biocatalytic conversion. In the following, the two most commonly used transducer principles in practice are introduced and application examples are presented.

18.2.1 Potentiometry

18.2.1.1 Definitions

Potentiometry determines potential differences that occur at phase boundaries between electron conductors (electrodes) and ion conductors (electrolyte solutions) as a function of the activity of a specific ion (ion-selective potentiometry) in a solution. Electrodes are, for example, pH glass electrodes, electrolyte solution in this context means the analyte solution. Whenever an electrically conductive, solid material (e.g., a metal electrode) is brought into contact with a conductive liquid (electrolyte), an electrochemical double layer is formed at the phase boundary, which has electrical charge-separating properties. Various models exist for describing these electrochemical processes at the electrode/electrolyte phase boundary, such as the originally developed Helmholtz model and its refinement on the basis of the Gouy-Chapman theory, as well as its further development based on the Stern and Grahame model (Bard and Faulkner 2001).

In practice, the potentiometric measuring chain shown in Fig. 18.3 is used as the measuring setup for **ion-selective potentiometry**. This consists of two electrochemical half-cells: an ion-selective electrode (ISE) or measuring electrode and a potential-constant reference electrode (reference electrode). Both electrodes are in contact with each other via the analyte and are read out electrically by a high-impedance voltage meter. At its lower end, the ISE has a so-called ion-selective membrane which can interact with the ion to be

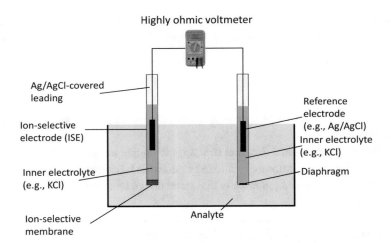

Fig. 18.3 Schematic diagram of a potentiometric measuring chain consisting of measuring electrode (ISE), electrolyte (analyte), reference electrode and voltmeter

detected in the analyte. The reference electrode is often an Ag/AgCl reference electrode with an internal electrolyte; it is in contact with the analyte via a diaphragm (e.g., a porous frit).

The measuring principle of the ISE is based on the potential change at its phase boundary to the analyte as a result of a change in the activity of the free, non-bound ions of a certain type in the analyte. This sounds very abstract, so let give you an example: A pH or potassium electrode can detect the pH value or the potassium ion concentration accordingly by means of its H^+ - or K^+ -selective membrane. This change in potential can then be detected without current using a high-impedance voltage meter with respect to a potential-constant reference electrode. In terms of formula, this change in potential can be described by the **Nernst equation** (Eq. 18.1):

$$U = U_0 \pm 2.3 \frac{R \times T}{z_i \times F} \log a_i \tag{18.1}$$

with slope S:

$$S = \frac{2.3 \times R \times T}{z_i \times F} \tag{18.2}$$

where U: electrode potential in V; U_0: standard electrode potential at $a_i = 1$ in V; R: universal gas constant, $R = 8.31447 \text{ J mol}^{-1} \text{ K}^{-1}$; T: absolute temperature in Kelvin; z_i and a_i: valence and activity of the potential-determining ion, respectively; F: Faraday constant, $F = 96{,}485.34 \text{ C mol}^{-1}$.

The Nernst equation can therefore be used to determine the activity of an ion species in the analyte as a measurand. This is initially referred to as the activity and not the concentration, since the activity takes into account the mutual influence of the ions in electrolyte solutions as a result of the Coulomb interaction at high electrolyte concentrations ($>10^{-2}$ M). This is the theory, but in practice one often works with ion concentrations rather than activities. The relationship between activity a_i and ion concentration c_i can be established via the activity coefficient f_A :

$$a_i = f_A \times c_i \quad \text{with} f_A \leq 1 \tag{18.3}$$

The activity coefficient is not only determined by the ion concentration to be measured, but also by the total ion concentration of the solution (i.e., its ionic strength). In diluted analyte solutions (ion concentration $\leq 10^{-3}$ M), f_A is so close to 1 that the ion activity can be directly replaced by the "ion concentration of interest."

Thus, for the Nernst equation (Eq. 18.1) at room temperature (298.15 K), a theoretical slope for monovalent ions (e.g., H^+ ions) of 59.16 mV/decade and for divalent ions (e.g., Ca^{2+} ions) a slope of 29.58 mV/decade follows, provided that the corresponding ISE is used.

Ion-selective electrodes are distinguished on the basis of the nature and function of their ion-selective membrane into: glass membrane electrodes (e.g., pH glass electrode), crystalline solid-state membrane electrodes (e.g., fluoride ISE), and liquid membrane or PVC-based electrodes with ionophores (e.g., for K^+-, Li^+-, Ca^{2+}-, Mg^{2+}-detection etc.). There are also gas-selective electrodes (e.g., for CO_2 or NH_3 determination) and enzyme electrodes (Sect. 18.2.1.2).

In contrast to the ISE, the **reference electrode** is an electrode with a constant equilibrium potential and is used as a reference point for the measurement. The Ag/AgCl electrode shown schematically in Fig. 18.3 is the most commonly used reference electrode nowadays due to its ease of use, high reproducibility as well as its wide usable temperature range (up to 150 °C). With this type of electrode, an Ag wire coated with AgCl is usually immersed in a solution containing Cl^- ions (internal electrolyte, e.g., 3.5 M KCl). The electrically conductive connection of the reference electrode to the measuring solution is formed via a ceramic diaphragm.

In real measurements, in addition to the primary ion to be detected (e.g., Na^+), interfering ions (e.g., K^+) are also present in the analyte solution, which can have a negative effect on the measurand, i.e., the electrode potential. This relationship is taken into account via the modified Nernst equation, the so-called Nikolsky-Eisenmann equation.

18.2.1.2 Potentiometric Enzyme Electrodes

The potentiometric enzyme electrode is—like the gas-selective electrode—a secondary ISE. This uses an ISE, e.g., a pH glass electrode, as the basic sensor: the enzyme layer is immobilized on the pH-sensitive glass membrane (e.g., by means of a dialysis membrane; Sect. 18.1.2, Immobilization procedure). This then allows detection of one or more reaction

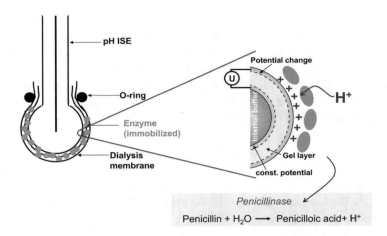

Fig. 18.4 Enzyme electrode for the detection of penicillin and reaction principle (schematic)

products formed in the analyte solution as a result of the catalytic reaction between enzyme and substrate (to be detected). Figure 18.4 schematically shows such an enzyme electrode for the detection of penicillin.

The enzyme used here is penicillinase (β-lactamase), which hydrolytically cleaves the β-lactam ring of penicillin. The enzymatic reaction produces H^+ ions, which leads to an acidification at the interface between the enzyme layer and the surface of the pH glass electrode. The acidification—i.e., one of the products—is detected by the pH glass electrode in a concentration-dependent manner. The more penicillin there is in the analyte solution, the greater the potential difference compared to the initial state (without penicillin).

In addition to penicillin, a large number of other analytes can be detected with potentiometric enzyme electrodes (Table 18.1). For this purpose, the appropriate enzymes must be immobilized on the pH glass electrode. For example, in the case of urea determination, the enzyme urease is used. The biocatalytic reaction results in a pH shift towards alkaline pH values when the urea concentration in the test sample increases. At the same time, however, the concentration of ammonium ions (NH_4^+) also increases, so that this increase can also be measured, for example, with a gas-selective ammonia electrode.

The response times of potentiometric enzyme sensors can vary (greatly) depending on the enzyme activity and the immobilization method: they are usually in the range of 1–5 min. The detection range is usually between 0.1 and 10 mM. The lifetime also depends on the enzyme activity, the type of immobilization as well as leaching effects.

Modern chip technologies, such as silicon planar technology, enable the integration of several potentiometric enzyme electrodes on a single sensor chip. At the same time, due to the smaller area of the sensor membranes (in the range of square micrometers), less enzyme quantity is required compared to the above-mentioned ISE, which is accompanied by a cost saving.

Table 18.1 Practically usable ISE and EnFET with immobilized enzymes and substrates to be detected (selection); a change in pH value is usually detected as the sensor signal

Analyte	Enzyme/enzyme system
Glucose	Glucose oxidase
	Glucose dehydrogenase
	Glucose oxidase/MnO_2 nanoparticles
Penicillin	Penicillinase
	Penicillin-G-acylase
Urea	Urease
Glutamine	Glutaminase
Creatinine	Creatinine deiminase
Glutamate	Glutamate oxidase
Formaldehyde	Ethanol oxidase
Acetylcholine	Acetylcholine esterase
Maltose	Maltase/glucose dehydrogenase
Sucrose	Invertase/mutarotase/glucose oxidase
	Invertase/glucose dehydrogenase
Ascorbic acid	Peroxidase
Ethanol	Ethanol oxidase
Phosphorous pesticides	Acetylcholine esterase
	Organophosphate hydrolase
Cyanide	Peroxidase
Lactate	Lactate oxidase
Glycerol	Glycerol oxidase
Urea	Urease
Inosine	Xanthine oxidase
ATP	H^+-adenosine triphosphatase
Cephalosporin C	Cephalosporinase

Figure 18.5b shows schematically the structure of such an **enzyme FET** (field-effect transistor) and a photo of a fabricated sensor chip (Fig. 18.5a), which is inserted into a catheter and encapsulated.

In an enzyme FET (or EnFET), the space charge region in the semiconductor chip between source and drain—the channel in Fig. 18.5b—is modulated by the voltage U_G applied to the gate insulator SiO_2 as well as the drain-source voltage U_{DS}. This behavior corresponds to that of a conventional field-effect transistor in microelectronics, whereby such a typical FET always consists of a p- or n-doped semiconductor, into which two oppositely conducting regions, source and drain, are diffused accordingly (n- and p-conducting, respectively). If there is now an additional pH-sensitive layer on the gate insulator (e.g., made of Ta_2O_5) and an enzyme layer on top of it, the pH value change resulting from the enzymatic reaction can be recorded accordingly. In this case, an additional surface potential is created, which then additionally modulates the space charge region, depending on the substrate concentration. In other words, one ultimately has a

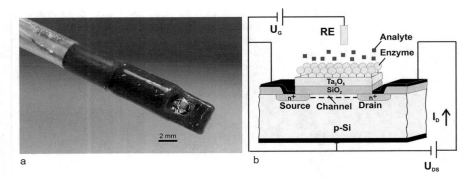

Fig. 18.5 Photo (**a**) and schematic (**b**) of an EnFET (enzyme field-effect transistor); RE, reference electrode; I_D, drain current; U_G, gate voltage; U_{DS}, drain-source voltage

comparable behavior to that of the pH glass electrode. Examples of this (immobilized enzymes, substrates to be detected) can be found in Table 18.1 (see also Mulchandani and Rogers 1998; Poghossian and Schöning 2021).

18.2.2 Amperometry

18.2.2.1 Definitions
In amperometry (as a subgroup of voltammetry), a voltage is applied to an electrochemical cell to cause a Faraday reaction (a substance turnover) and the resulting current is measured. This increases with increasing analyte concentration; however, the basic requirement for this is that the substance to be detected must be electroactive.

Figure 18.6 schematically shows the setup of an amperometric electrode required for the measurement. In the simplest case, this consists of two electrodes—a working electrode (WE) and a counter electrode (CE), which are in contact with each other via the analyte solution—as well as a controllable voltage source (the potentiostat) and a current measuring device (amperemeter). Commonly used materials for the working electrode are Pt, Au, Ag or graphite (carbon); the counter electrode is, for example, an Ag/AgCl reference electrode (Sect. 18.2.1.1).

The **electrode reaction** (oxidation or reduction) takes place with the electroactive substance to be detected in the analyte. In order for such a reaction to start, an energy input in the form of a voltage must take place from "outside"; a defined potential is applied between WE and CE by means of the potentiostat. Ultimately, this results in a charge transfer process at the interface between the working electrode and the analyte. The potential that must be applied externally to the measuring cell ultimately depends on the electroactive component to be detected; the two most important electroactive molecules in amperometry are oxygen (O_2) and hydrogen peroxide (H_2O_2). The relationship between measured current and the concentration of the electroactive substance to be detected at a constant (polarization) voltage is described by Eq. (18.4):

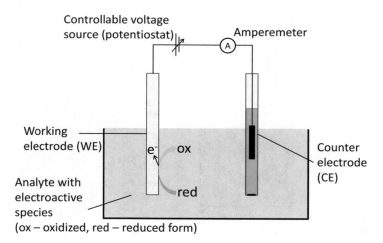

Fig. 18.6 Schematic diagram of an amperometric measuring chain consisting of working electrode (WE), analyte solution, counter electrode (CE), potentiostat, and current meter (ox, red: oxidized or reduced form of the electroactive substance in the analyte)

$$I_{\mathrm{Diff}} = z \times F \times D \times A \times \frac{c}{d} \qquad (18.4)$$

where I_{Diff}: diffusion-limited current in A; D: diffusion coefficient $\mathrm{cm}^2\,\mathrm{s}^{-1}$; A: area of the working electrode in cm^2; d: thickness of the diffusion layer in cm; z: number of electrons involved in the reaction; F: Faraday constant; c: concentration of the analyte in $\mathrm{mol}\,\mathrm{L}^{-1}$.

In the case of **O$_2$ detection** using a **Clark electrode** (Fig. 18.7), the oxygen diffuses from the analyte through the membrane (e.g., Teflon) into the measuring chamber according to its partial pressure. The voltage applied between the WE and CE across the potentiostat is approximately $-600\,\mathrm{mV}$; at this potential, the redox process takes place in a diffusion-controlled manner, i.e., depending on the O$_2$ concentration: thus, the WE becomes the cathode and the CE becomes the anode. Equations (18.5) and (18.6) show the electrochemical reactions that occur:

$$O_2 + 4e^- + 2H_2O \rightarrow 4OH^- \qquad (18.5)$$

$$4Ag + 4Cl^- \rightarrow 4AgCl + 4e^- \qquad (18.6)$$

The oxygen molecules are reduced to hydroxide ions (OH$^-$) at the cathode and at the same time the oxidation of silver to silver chloride takes place at the anode.

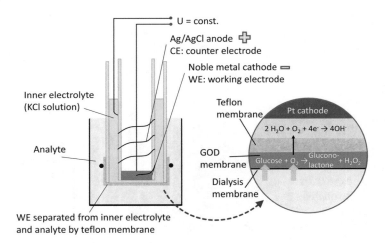

U = const.

Ag/AgCl anode ⊕
CE: counter electrode

Noble metal cathode ⊖
WE: working electrode

Inner electrolyte
(KCl solution)

Analyte

WE separated from inner electrolyte
and analyte by teflon membrane

Teflon
membrane   Pt cathode

$2 H_2O + O_2 + 4e^- \rightarrow 4OH^-$

GOD
membrane $Glucose + O_2 \rightarrow$ Gluconolactone $+ H_2O_2$

Dialysis
membrane

Fig. 18.7 Design of an oxygen electrode (Clark electrode) with Teflon membrane and extension as an amperometric enzyme electrode (magnification right, schematic)

The fact that the liquid is retained at the membrane makes it possible to measure the oxygen partial pressure in the test sample: only the gas molecules can diffuse to the electrode (cathode)—four electrons are involved in the reaction per O_2 molecule. With increasing O_2 concentration, I_{Diff} increases accordingly.

Due to its design as a liquid electrolyte electrode (inside) separated from the environment by a gas-permeable membrane, the Clark electrode can also be used for the detection of O_2 in the gas phase. The advantages of the gas-permeable membrane are the reduction of contamination inside the electrode, the low influence of the flow rate of the measuring solution, the internal solution is optimally matched to the WE and the CE and calibration in air is possible. The distance between the membrane and the WE represents a diffusion barrier and should therefore be kept as small as possible (approx. $5-10$ μm).

The advantage of amperometry over potentiometry is the linear dependence of the diffusion-limited current on the concentration (in contrast to the logarithmic dependence), i.e., smaller changes in concentration can be recorded more accurately. A disadvantage is the dependence on the area of the working electrode: small (micro-) working electrodes supply smaller currents, which in turn must then be amplified via electronic circuits (with as little noise as possible). Furthermore, the temperature dependence of the diffusion coefficient and the oxygen solubility must be taken into account in the measurement (e.g., by additional temperature compensation).

To increase the measuring accuracy, a three-electrode measuring arrangement is often used in practice instead of the two-electrode measuring arrangement shown in Fig. 18.6: in addition to the WE and Ag/AgCl reference electrode (RE), an additional counter electrode made of, e.g., Pt or Au is then used (which then ensures that the current flows via the CE, the RE thus always remains currentless and supplies a stable reference potential).

18.2.2.2 Amperometric Enzyme Electrodes

Amperometric enzyme electrodes are usually based on the principle of the Clark electrode presented in Fig. 18.7, exemplified for glucose determination in the right part of the figure. Analogous to the potentiometric enzyme electrode in Sect. 18.2.1.2 (in which the enzyme layer was immobilized on, for example, a pH-sensitive glass electrode), here the enzyme layer (e.g., GOD: glucose oxidase) is coupled to an O_2 electrode. One possible approach is to fix the enzyme GOD onto the gas-permeable Teflon membrane using a dialysis membrane (physical immobilization). As an oxidoreductase, GOD can catalyze redox reactions, which is exactly the basic requirement for the amperometric transducer.

If the electrode is in contact with the test sample, glucose and O_2 from the analyte first diffuse into the dialysis membrane and are catalytically converted there according to the enzyme-substrate reaction; the GOD binds glucose (in the form of β-D-glucose) and O_2 and converts them into gluconolactone and H_2O_2. This means that the more glucose present in the analyte, the more gluconolactone and H_2O_2 are formed, but also the more O_2 is consumed in this reaction. This O_2 consumption, or more precisely, the remaining amount of O_2 molecules not needed for the catalytic reaction, can then diffuse through the Teflon membrane to the Pt cathode in the next step. There they are then reduced at -600 mV (Eq. 18.5). A high glucose concentration in the analyte thus results in less O_2 arriving at the Clark electrode and thus the I_{Diff} decreases with increasing glucose concentration.

The disadvantage of this measuring principle is that the sensor signal depends on the dissolved O_2 concentration in the measuring solution and the sensor reacts relatively sluggishly due to the Teflon membrane, the response time is in the range of minutes.

An alternative approach is to record not the O_2 consumption (decrease) but the H_2O_2 production (increase) during this reaction (Fig. 18.7). For this purpose, the corresponding redox potential for H_2O_2 has to be set, which is $+600$ mV compared to the Ag/AgCl electrode. In this configuration, "only" the sign of the potential then changes; this is referred to as the Pt anode and Ag/AgCl cathode. The associated (oxidation) reaction of H_2O_2 at the Pt anode is shown in Eq. (18.7) and the reaction at the cathode is shown in Eq. (18.8):

$$H_2O_2 \rightarrow O_2 + 2H^+ + 2e^- \tag{18.7}$$

$$4H^+ + 4e^- + O_2 \rightarrow 2H_2O \tag{18.8}$$

The majority of amperometric enzyme electrodes with oxidoreductases used today make use of this measuring principle of H_2O_2 detection: The sensor signal starts with a "low" base current (when there is no glucose in the measuring solution and thus no H_2O_2 is formed) and increases steadily with increasing glucose concentration in the measuring solution; there is no limitation as with the detection of O_2 consumption (where there is practically no O_2 left at high glucose concentrations).

Figure 18.8a shows an exploded view of an example of the construction of such an amperometric enzyme sensor based on H_2O_2 determination, as used in practice. The enzyme layer (GOD) is immobilized by means of a crosslinker (glutardialdehyde together with BSA) on a carrier film (e.g., made of polyurethane) between two membrane layers. The polycarbonate layer limits substrate diffusion to the enzyme layer (in order to be able to detect even high substrate concentrations in the bioreactor, for example) and at the same time blocks high-molecular substances that could negatively influence enzyme activity. The substrate to be detected (here: glucose) diffuses to the enzyme and is converted there (reaction 1), producing H_2O_2. The H_2O_2 diffuses through the cellulose acetate membrane to the Pt electrode where it is oxidized according to Eq. (18.7) (reaction 2), the resulting current is proportional to the glucose concentration. The cellulose acetate membrane is permeable only to small molecules, such as H_2O_2, thus eliminating other electroactive substances (anti-interference) that could interfere with the measurement. Such a setup is typically used for **electrochemical analyzers** (Sect. 18.2.3), where—depending on the stability of the enzyme—up to about 1000 measurements are possible before the enzyme membrane needs to be replaced. The response time of the sensors is <1 min.

Another way of recording the substrate or product concentration during the enzymatic reaction is to use (redox) **mediators** as auxiliary molecules. Mediators can assume two different states, they can be present in oxidized and reduced form, e.g., the redox pairs ferrocene/ferricinium and ferricyanide/ferrocyanide, each representing the oxidized/reduced form.

The mediator can transport the electrons resulting from the catalytic reaction from the active site of the enzyme to the electrode surface. It thus acts as an electron acceptor instead of oxygen and passes from the oxidized to the reduced state. After reaching the electrode surface, it then releases the electrons again and thus returns to the oxidized initial state. The game can then start all over again with the "mediator shuttle." One advantage here is that the enzymatic reaction is now no longer dependent on the oxygen concentration.

In contrast to the electrochemical analyzers, **disposable enzyme biosensors,** such as those used as test strips (Fig. 18.8b) for glucose detection in diabetics, are usually manufactured in thick-film technology using *screen printing*. WE, RE, and CE are deposited as an approx. 10−40 μm thick paste through a stencil on the carrier material (e.g., plastic) and then cured; the GOD is then immobilized and the surface is protected by, e.g., another plastic cover. This leaves only the test field open for the analyte solution. In the meantime, such disposable sensors can be miniaturized to such an extent that sample volumes of a few 100 nL (!) are sufficient, and the response time is sometimes only in the range of a few seconds. Thick-film technology enables the sensors to be produced in large quantities, with high reproducibility and at very low cost, even if the low manufacturing prices ultimately do not reach the consumer.

Analogous to the potentiometric ones, modern chip technologies are also used for amperometric enzyme electrodes, e.g., to integrate several electrodes in the form of **sensor arrays** on a single sensor chip. This enables the simultaneous detection of several analytes and/or coupled enzyme reactions (Bäcker et al. 2013).

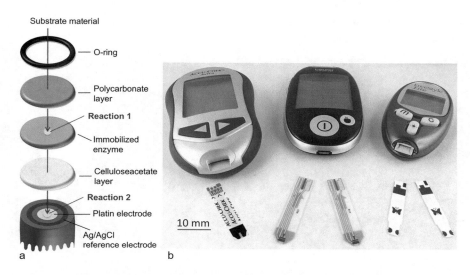

Substrate material

O-ring

Polycarbonate layer

Reaction 1

Immobilized enzyme

Celluloseacetate layer

Reaction 2

Platin electrode

10 mm

Ag/AgCl reference electrode

a

b

Fig. 18.8 Exemplary structure of an enzyme electrode based on H_2O_2 detection (**a**) and various glucose measuring devices with disposable biosensors, produced using thick-film technology (**b**)

In such microelectrodes, at least one dimension is always so small that its properties (such as mass transport) become a function of its size; in practice, the critical dimension range is between about 0.1 and 50 μm. The advantage of microelectrodes is the hemispherical diffusion profile (in contrast to macroscopic electrodes, which essentially allow only planar diffusion of the analyte molecules), i.e., considerably more electroactive molecules reach the electrode surface per unit of time and area and can be converted there. This results in a higher current density (i.e., current per unit area) and thus a higher detection sensitivity for low concentrations. Lower detection limits down to the picomolar range are discussed in the literature. The disadvantage associated with such microelectrodes, that the I_{Diff} in Eq. (18.4) is area-dependent, is often circumvented by forming a large number of electrodes (up to several thousand) into arrays (so-called **microelectrode arrays**, MEAs): The individual currents are then added up so that the total current finally recorded (in the microampere range) is again of the same order of magnitude as for the original macroelectrodes (Wang 2006).

Amperometric enzyme sensors are currently the largest and most important group among biosensors. Common examples of amperometric enzyme electrodes and immobilized enzymes or substrates and products to be detected can be found in Table 18.2 and in Mulchandani and Rogers (1998) and Gründler (2004).

18.2.3 Application Examples

Electrochemical enzyme sensors are used in a variety of applications in clinical and sports medicine diagnostics (e.g., blood glucose measurement for diabetics or lactate

Table 18.2 Examples of amperometric enzyme sensors, and the immobilized enzymes and products to be detected (selection not complete)

Analyte	Enzyme/enzyme system	Product
Glucose	Glucose oxidase	O_2, H_2O_2
Lactate	Lactate monooxygenase	H_2O_2
Cholesterol	Cholesterol oxidase	H_2O_2/ ferrocene
Polyphenol	Polyphenol oxidase	o-chinone
Pesticides	Acetylcholine esterase	H_2O_2
Glycerol	Glycerol oxidase	H_2O_2
Oxalate	Oxalate oxidase	O_2, H_2O_2
Ethanol	Ethanol oxidase	H_2O_2
Glutamate	Glutamate oxidase	O_2, H_2O_2
Lactose	Galactose oxidase/glucose oxidase	O_2, H_2O_2
Sucrose	Glucose oxidase/mutarotase	O_2, H_2O_2
Galactose	Galactose oxidase	O_2, H_2O_2
Glutamine	Glutaminase/glutamate oxidase	O_2, H_2O_2
Choline	Choline oxidase	O_2, H_2O_2
L-amino acid	L-amino acid oxidase	H_2O_2
Catecholamine	Catechol oxidase	H_2O_2

measurement to monitor the fitness of athletes), food technology, the pharmaceutical industry and environmental monitoring, but also increasingly in biotechnology.

In the field of medicine in clinical laboratories, metabolic products such as blood glucose, cholesterol, lactate or urea can be detected by means of commercial biosensors integrated in analyzers. The detection of glucose was one of the first applications of biosensors, and thus the first commercial glucose analyzer from Yellow Spring Instruments used Clark's patent and triggered a "boom": companies such as Fuji Electric, ZWG, Eppendorf immediately followed suit. In the meantime, glucose measurement is taken into account by all well-known manufacturers of blood gas analyzers such as Siemens, Roche, Radiometer, Instrumentation Laboratory, EKF, Abbott, and Nova Biomedical.

The second (even larger) group of enzymatic glucose biosensors are **disposables** based on **test strips**. These are suitable as self-tests for diabetics for self-monitoring and are used

Table 18.3 Overview of current commercial analyzers for biomedical applications as well as portable devices for PoC applications (exemplary selection)

Company	Model	Analyte	Reference
Yellow Springs Instrument	Multiparameter Bioanalytical System YSI 7100 MBS	Glucose, glutamate, glutamine, lactate	www.ysi.com
Analox Instruments	Fast Multi-Assay Analyzer GL6	Ethanol, glucose, glycerol, lactate, methanol, sucrose, lactose	www.analox.com
TRACE Analytics	MultiTRACE	Glucose, lactate, methanol, ethanol	www.trace.de
Abbott Point-of-Care Inc.	i-STAT	Glucose, urea, lactate, creatinine	www.abbottpointofcare.com
Alere GmbH	epoc® Blood Analysis System	Glucose, creatinine, lactate	www.alere.com

for *point-of-care* (PoC) analysis in hospitals directly on the ward, at the bedside or in doctors' practices. The pioneer here was the company Medisense (1986) with the ExaTech "Glucose Pen." In the meantime, the market in this sector is literally "rolling over," with new companies advertising daily with even better, easier to use, faster, and more accurate glucose meters; the market leaders here are currently the companies Roche and Abbott. Glucose monitoring still accounts for more than 80% of the global biosensor market.

Modern PoC analyzers, which often use chip-based sensor arrays (for single use), now allow the simultaneous determination of several analytes (Sect. 18.4.1). Table 18.3 lists two examples from Abbott and Alere.

For continuous process control in **biotechnology** (e.g., for the control of fermentation processes), analytical systems using enzyme biosensors are offered for the determination of, e.g., glucose, urea, lactate, sucrose, ethanol, methanol, creatinine, glutamate, cholesterol, and glycerol. Table 18.3 shows a selection of commercial multiparameter systems for biotechnological applications (companies Yellow Spring Instruments, Analox Instruments, TRACE Analytics).

In contrast to medical technology, biosensor technology is currently less widespread in the field of biotechnology. This can be explained primarily by the fact that the sample matrices in medicine (blood, serum, saliva, urine) are more clearly defined than in biotechnology: here, a large number of (sometimes viscous) samples with a wide variety of compositions and interfering substances can be found. This still poses a challenge to biosensor development. At the same time, in the case of electrochemical off-line or on-line analyzers, sterile sampling must be ensured so that the bioprocess itself is not contaminated; in addition, matrix effects must not influence the biosensor signal. In the case of in-line measurement, even the complete biosensor chip must be completely sterilizable.

The limited shelf life of enzyme biosensors and the resulting often insufficient stability of the enzymes is still one of the challenges in the commercialization of such sensors. Against this background, the disproportionate use of single-use biosensors or the replacement of the biosensors after a certain number of measurements performed can also be explained.

Questions
5. Compare the theoretical slope from Nernst equation in potentiometry for monovalent and divalent ions.
6. Discuss the principle of an enzyme-modified field-effect transistor (EnFET).
7. Which quantities determine the diffusion-limited current in amperometry?
8. Discuss the difference between an amperometric and a potentiometric measurement chain.

18.3 Looking Beyond the "End of One's Nose": Alternative Transductor Principles and Biomolecules

Optical and mass-sensitive transducer principles represent a further possibility for "translating" the (bio)chemical input signal into a physically accessible output variable. Even if their widespread use—especially in the field of enzyme-based sensor technology—cannot keep pace with the electrochemical transducer principles presented in Sect. 18.2, common sensor types are nevertheless presented below.

18.3.1 Optical Biosensors

Optical chemo/biosensors use electromagnetic radiation in the infrared (0.5 mm−760 nm), visible (750−400 nm), or ultraviolet (400−100 nm) wavelength range. The optical properties of the radiation are changed by interaction with the analyte and detected by a detector. Optical quantities that can be used for measurement are, for example, absorption, emission, fluorescence, reflection, refraction, diffraction, or polarization. Optical biosensors with enzymes are often based on absorption or fluorescence measurements.

Absorption is the attenuation of monochromatic light when it passes through a (homogeneous) analyte solution of defined layer thickness in the visible or UV (ultraviolet) range. The absorption behavior can change when the analyte molecule to be detected reacts with the receptor layer. The relationship between light absorption and concentration of the substance to be detected is described by the **Lambert-Beer law** in Eq. (18.9).

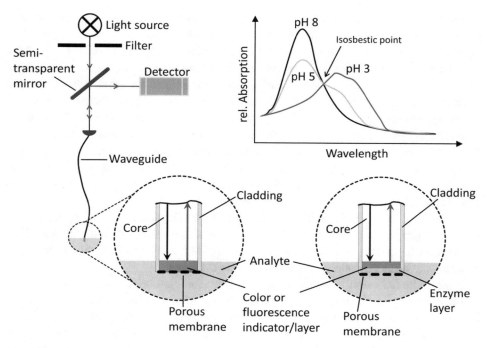

Fig. 18.9 Optical chemo/biosensor based on glass fibers with receptor layer (absorption, fluorescence indicator) and, as an embodiment, with additionally immobilized enzyme membrane; the graph shows the pH value change in the absorption spectrum as an example

$$\text{Abs} = \log \frac{I_0}{I_{in}} = \varepsilon \times c \times d \qquad (18.9)$$

where Abs: absorption, I_0: intensity of transmitted (or reflected) light in W m^{-2}; I_{in}: intensity of incident (irradiated) light in W m^{-2}; ε: decadic molar extinction coefficient in L mol^{-1} cm^{-1}; d: thickness of the irradiated layer in cm; c: concentration of the absorbing substance in mol L^{-1}.

In contrast, **emission** is the emission of light after energy absorption due to chemical (chemiluminescence) or biochemical (bioluminescence) reactions resulting from the interaction between receptor and analyte molecule; spontaneous emission is called **fluorescence**. Materials in which fluorescence occurs are fluorophores, and the emitted light is always lower in energy than the previously absorbed light.

The "simplest" optical sensor that uses the physical effect of absorption is the pH indicator test strip: a paper or plastic strip coated at the end with a combination of different pH color indicators. These change their color behavior (absorption properties) depending on the pH of the test solution in which they are immersed. A qualitative comparison is made between the coloring of the pH indicator test strip and an enclosed color scale; typical color indicators are litmus, bromothymol blue, or phenolphthalein.

The technically refined embodiment for the exact (quantitative) recording of the change in the absorption properties is carried out with an optrode (Fig. 18.9).

The sensor consists of an optical fiber (light guide) at the end of which (end face) a receptor layer is immobilized (e.g., in a porous matrix) and is in contact with the test sample via a membrane. The optical fiber consists of a core (which transports the light) and a cladding; the light is coupled into the optical fiber cable by means of, for example, a laser via an appropriate optical system (filter, aperture, semi-transparent mirror). The coupled light is transported along the glass fiber to the receptor layer, where the interaction between the analyte and, for example, the pH indicator layer (color indicator) takes place—a change in absorption occurs. The back-transported (reflected) light reaches the detector (e.g., an absorption spectrometer, a so-called photometer) via the semi-transparent mirror, which records the intensity as a function of the wavelength.

The absorption spectrum in Fig. 18.9 shows how the relative absorption (plotted against the wavelength) changes depending on the pH value of the analyte solution. The shift of the maxima on the y-axis can, for example, be calibrated as a sensor signal; in principle, evaluation at a single, defined wavelength is also sufficient.

While color indicators are used for pH determination, fluorophores are used as fluorescence indicators for O_2 determination (as the second important parameter). Here, the fact that O_2 can quench the fluorescence of such fluorophores is exploited; this is referred to as "quenching"—the more oxygen present in the analyte, the more the fluorescence signal decreases.

If an enzyme membrane is immobilized in the immediate vicinity of the color or fluorescence indicator layer (Fig. 18.9, bottom right), an **optical enzyme biosensor (enzyme optrode)** can be constructed in this way. Analogous to the electrochemical enzyme electrodes (potentiometric, amperometric) introduced in Sects. 18.2.1 and 18.2.2, the products generated and/or starting substances consumed as a result of the biocatalytic reaction are also measured here: the pH change resulting in the case of hydrolases is detected by means of a color indicator, the oxygen consumption in the case of oxidoreductases is detected by means of the fluorescence indicator. Typical substrates such as penicillin, glucose, or lactate can therefore be detected both electrochemically and optically (Tables 18.1 and 18.2).

The main advantage of such optical biosensors is that long distances (several hundred meters) can be bridged potential-free (e.g., in potentially explosive atmospheres) using fiber optic cables. Furthermore, no reference electrode is required (the absorption spectra in Fig. 18.9 all pass through the so-called isosbestic point as a reference value). However, the disadvantage is that the turbidity of the analyte solution can falsify the measurement signal and the dynamic measurement range for color and fluorescence indicators is (significantly) smaller than for electrochemical sensors.

In addition to optical enzyme biosensors, completely different optical detection principles are becoming increasingly important in the field of biosensors today: **DNA chips**—often also referred to as gene probes, microarrays, or biochips—enable the detection of up to 500,000 genetic markers on a sensor chip with the size of a fingernail. Using largely automated processes, a specific (short) section of the complementary single-stranded DNA to be detected is immobilized on the sensor surface (capture DNA); the target DNA present in the analyte solution, e.g., fluorescently labelled (also present as a single strand), can then be read optically if hybridization is successful. Typical applications are the detection of genetic diseases, paternity tests, forensics, food monitoring, and SNP *(single nucleotide polymorphism)* analyses.

As a second example, so-called **SPR biosensors** *(surface plasmon resonance)* enable highly accurate investigations of the kinetics of binding processes between affinity partners (e.g., antigen-antibody, DNA-DNA, protein-protein); the detection sensitivity is in the range of about 1 pg mm^{-2}. Such sensitivities are necessary, for example, in the detection of infectious diseases (e.g., anthrax, smallpox, plague, Ebola, Marburg viruses), in the monitoring of pesticides in agriculture and in the detection of toxins or pathogens in homogenized foodstuffs.

18.3.2 Mass-Sensitive Biosensors

A mass-sensitive biosensor detects changes in the "mass loading" of its surface. A distinction is made between two main designs, the *quartz crystal microbalance* (QCM) and the *surface acoustic wave* (SAW) sensor.

Quartz crystal microbalances are often called "molecular balances." They exploit the piezoelectric properties of quartz crystals, because their resonance frequency is strongly dependent on whether molecules are adsorbed (or not) on their surface. When the molecules adsorb on the surface of such a vibrating quartz crystal, its resonance frequency changes according to the following modified form of the **Sauerbrey equation** (Eq. 18.10). The oscillation is damped by the adsorption, and this damping increases with increasing analyte concentration. If the measurements are carried out in liquids, the viscosity and density of the medium must also be taken into account.

$$\Delta f = - \left[\sqrt{\frac{\eta_{Fl} \times \rho_{Fl}}{4\pi f_o}} + 2 \times f_o^2 \times \frac{1}{\sqrt{\mu_Q} \times \sqrt{\rho_Q}} \times \frac{\Delta m}{A} \right] \quad (18.10)$$

with Δf: change of resonance frequency in Hz, f_o: output (resp. resonance) frequency in MHz; Δm: mass change in g; A: vibrating area of quartz in cm^2; μ_Q: shear modulus of quartz, $\mu_Q = 2.947 \times 10^{11}$ g cm^{-1} s^{-2}; ρ_Q: density of quartz, $\rho_Q = 2.648$ g cm^{-3}; η_{Fl}: viscosity of liquid in N s m^{-2}; ρ_{Fl}: density of liquid in g cm^{-3}.

Figure 18.10 (top) schematically shows the structure of a QCM sensor and the **thickness shear oscillation** underlying the principle which undergoes a damping (decrease) during the interaction between analyte and receptor molecule: commercial QCM sensors have an Au electrode fabricated using thin-film technology on each side of the quartz wafer. The electrode in contact with the analyte is functionalized with a receptor layer, e.g., an

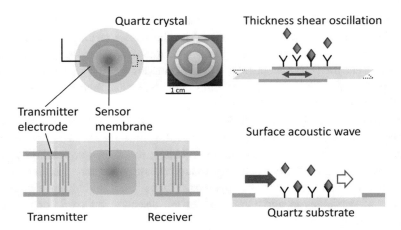

Fig. 18.10 Schematic diagram of a mass-sensitive biosensor as a quartz crystal microbalance (top) and SAW sensor (bottom)

antibody, which reacts with the corresponding antigen. It is therefore possible to measure the detection in real time without an additional marker. This is particularly suitable for immune reactions (e.g., the binding of biotin/streptavidin) or the detection of biomarkers (e.g., for the prostate-specific antigen). This method can also be used for highly sensitive detection of the adsorption of proteins, surfactants, polymers, cells, or bacteria on surfaces. Biolin Scientific, for example, offers such a system (Q-Sense) on the market, which can also be used to detect cells or proteins.

Typically, the detection limit of a QCM sensor (f_0 is, e.g., 10 MHz) is about 400 ng mm^{-2}; this corresponds (converted) to about a 0.02 monolayer copper of weight gain! To minimize interferences due to non-specific adsorption of molecules on the sensor surface, electronic fluctuations, and temperature dependence, differential measurement arrays are used in practice: a crystal pair consists of two oscillating crystals, one of which is designed with a receptor layer and one without.

As can be seen from Eq. (18.10), a higher resonance frequency implies a higher detection sensitivity. SAW sensors use the propagation of acoustic waves along the (planar) surface of a piezoelectric material. Surface acoustic waves are a special form of physical volume oscillations; due to the low restoring forces on a free surface, they remain bound there and can thus be guided along it (Fig. 18.10, bottom).

The chip contains two comb-shaped electrodes (e.g., made of Au) designed as interdigitated structures, which serve as transmitter and receiver. If radio waves are applied to the transmitter electrodes, they generate a synchronous, mechanical load in the crystal; this leads to the generation of an acoustic surface wave. This wave propagates along the surface of the piezoelectric crystal and is picked up by the receiver electrodes; the mechanical oscillations are converted into an electrical voltage. Proportionally to the mass loading, there is again a change in frequency or the change in the propagation speed of the surface acoustic wave. With this arrangement, higher resonance frequencies up to the GHz range are possible. As a result, a higher sensitivity can be achieved, which is in the femtogram per square millimeter range and corresponds approximately to the weight of a single *Escherichia coli* cell.

Questions
9. Which optical quantities can be used for the analyte detection with optical biosensors and which relationship exists between light absorption and concentration of the substance to be detected?
10. What is the working principle of a QCM biosensor (including Sauerbrey equation)?

18.4 Conclusion and Outlook

The use of enzymes in biosensor technology as receptor molecules now has a history of more than 50 years: the starting point was the idea of Leland C. Clark Jr. in 1962 at the Childrens' Hospital in Cincinnati (USA), who combined the oxygen electrode named after him with the enzyme GOD, thus heralding the birth of biosensor technology. His idea triggered a boom in recent decades, especially with regard to a large number of different enzyme-based glucose biosensors; global blood glucose meters market size is expected to grow from US\$ 15.8 billion in 2022 to US\$ 33 billion in 2030. On the other hand, the use of enzyme-based biosensors is also becoming increasingly important for monitoring, controlling and regulating biotechnological processes, e.g., during the fermentation of microorganisms or cells; typical parameters to be detected here are the concentrations of glucose, glutamate, lactate, ethanol, methanol, sucrose, or glutamine.

A current development trend focuses on actively integrating biosensors in particular into measurement and automation technology, which up to now has essentially been reserved for physically measured variables such as pressure or temperature (Sect. 18.4.1). In contrast, molecular gates with enzyme biosensors in the field of "biocomputing" are still largely at the research stage (Sect. 18.4.2), although in the medium term the relevance for applications will also gain in importance here.

18.4.1 Microsystems

Nowadays, the demand for simple and miniaturized sensor systems for chemical and biological quantities is growing: Ideally, a complete analytical measurement laboratory is realized in pocket size or even on a single chip. Such complete microsystems are often referred to as **lab-on-chip** system, MEMS *(micro electro mechanical system)* or μTAS *(micro total analysis system)*.

Figure 18.11 shows such a microsystem, which is made up of three essential blocks: the sensors, the actuators, and the electronics. The sensor technology usually consists of an

Fig. 18.11 Schematic structure
of a microsystem

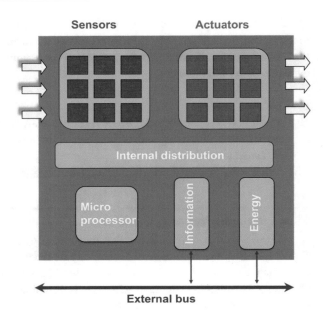

array of many individual sensors (e.g., also of different enzyme biosensors), with which the corresponding chemical/biological parameters can be recorded. In addition, the actuator system performs all the necessary mechanical and optical functions of the microsystem: for example, it ensures that the analyte solution is transported to the sensor system via micropumps, microswitches, or micropositioning valves. Actuators can also include other functionalities such as micromotors, membranes, filters, mirrors, or similar. The third block, the electronics, enables communication between sensors and actuators and performs typical electrical/electronic functions such as signal amplification, filtering, A/D (analog to digital) conversion, signal processing, and data output.

The best-known example of such a "lab-on-chip" analysis system is used in particular in the field of medical technology. It consists of a handheld device in which a cartridge about the size of a matchbox is inserted (Fig. 18.12), in which all the sensors and actuators are integrated. The measurement process is largely fully automated via a pressure mechanism that activates the microfluidics and initially pre-calibrates the sensor system via an additional calibration solution in the cartridge before the actual measurement process starts. Depending on the cartridge type and their combinations, this platform technology allows the detection of up to 25 parameters (e.g., pCO_2, pO_2, pH and various ions, glucose, urea, lactate, etc.). Especially the detection of e.g., glucose, urea and lactate is done with potentiometric and amperometric enzyme sensors.

Fig. 18.12 Sensor cartridge of
a "lab-on-chip" system,
including various amperometric
and potentiometric enzyme
sensors at the upper edge
(Abbott); size of cartridge:
4.5 cm × 2.8 cm

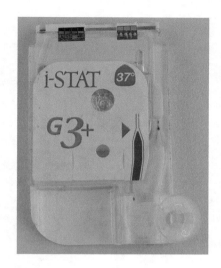

18.4.2 Molecular Gates with Enzyme Sensors

Biocomputing,, i.e., the representation and implementation of logical operations (Boolean logic)—as known from microelectronics—with the aid of biomolecules such as proteins, DNA, enzymes, antibodies, etc. opens up unimagined possibilities for the development of new technologies due to the high specificity and myriad possible combinations of the receptor molecules with each other (Katz 2012).

Recent examples are biomolecular keyboard locks, networks based on immunoreaction cascades, arithmetic operators, or so-called *logic gates* (adders, subtractors, AND, OR, XOR, XAND circuits, etc.). Against the background of a possible application, however, such biomolecular logic elements only make sense if they can be integrated accordingly at chip level: this is the only way to ensure the transfer of information to microelectronics and the corresponding data processing.

One research focus is currently molecular gates with enzyme sensors, which are combined, for example, with the electrochemical transducers from Sect. 18.2, e.g., with miniaturized field-effect sensors (Poghossian et al. 2015). As an example, Fig. 18.13 schematically explains such an AND gate: four enzymes (invertase, mutarotase, glucose oxidase, urease) are simultaneously immobilized on the sensor chip (left part of fig.). These are responsible for substrate recognition and logical further processing. If the substrate required for the respective enzyme is present, this corresponds to a logical "1," if not present, to a logical "0." The enzyme cascade on the sensor chip is triggered by the two input signals sucrose and O_2 (middle part of the figure). If both are present in the analyte— this corresponds to the combination (1,1)—then sucrose is catalyzed to α-D-glucose and fructose by invertase; in the next step, α-D-glucose is converted to β-D-glucose by mutarotase and then in the last step β-D-glucose is converted to gluconic acid by glucose oxidase (with O_2 consumption). The formation of gluconic acid ultimately induces a pH

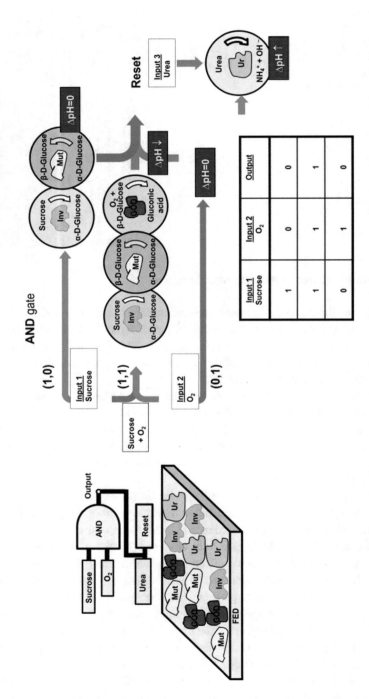

Fig. 18.13 Example of a molecular gate with enzyme biosensors. Mut, mutarotase; GOD, glucose oxidase; Inv, invertase; Ur, urease; FED, field-effect device

change (pH decreases) at the sensor surface, which can be detected. The reaction cascade is not started if only one of the two input variables is present: sucrose without O_2 (1.0) or O_2 without sucrose (0.1); in this case, the pH remains unchanged. The corresponding truth table can be found below in Fig. 18.13. To return the enzymatic gate to its initial state again (reset function), urea is added: the catalytic reaction with urease causes an increase in pH back to the initial value (right in fig.).

In this way, complex biochemical reaction processes can be converted into electrically usable output signals ("0" or "1") without having to follow all the individual processes in detail: only if all required input signals are present, a corresponding output signal results. Such a chip should have the ability to simultaneously and in "real time," ideally fully automatically, detect biochemically induced signal changes (analyte concentration) as input signal, analyze them and in the next step convert them into electronic output signals by each individual sensor unit performing the corresponding logical operation (OR, AND etc.). The logical operation performed can then be used to specifically address and stimulate the actuator system (e.g., the addition of a substance).

Possible areas of application are in the field of disease diagnostics (e.g., for the detection of biomarkers), which, according to a mathematical algorithm, form a resulting pattern from logical "0" and "1" that correlates with the disease pattern. Such a programmable "diagnostic chip," inserted in the human body, would react independently to metabolic changes with the sensor technology, diagnose disease symptoms and dose appropriate measures (e.g., a required medication) promptly on the basis of the system-integrated actuator technology. This could be done individually, in accordance with a tailor-made "personalized medicine." Conversely, the use of such an intelligent sensor chip in the field of biotechnology would also be conceivable, especially when it comes to being able to better assess and evaluate complex processes, such as in biogas processes; many biochemical reactions in living cells are catalyzed by the presence of the corresponding enzymes.

Question

11. Describe the theory of operation of an enzyme-based molecular logic gate.

Take-Home Message
- A chemical sensor is a miniaturized measurement device that selectively and reversibly detects chemical compounds or ions and provides an analytically useful signal. Biosensors are a subgroup of chemical sensors utilizing biological recognition mechanisms for analyte recognition. The chemo/biosensor consists of three main components: the receptor layer, the transducer and the signal processing unit.
- The receptor layer is coupled to the transducer surface by means of various immobilization strategies. Commonly applied immobilization methods for

(continued)

- enzymes include adsorption, membrane or gel entrapment, layer-by-layer technique, covalent binding, crosslinking, affinity coupling, or the use of nanoparticles as nanocarriers.
- Depending on their transducer principle, biosensors are often classified as electrochemical, electrical, optical, thermal, mass-sensitive, and magnetic biosensors.
- Ion-selective potentiometry belongs to the electrochemical transducer principles, enabling to determine the concentration (activity) of ions in an analyte solution. The relation between the ion concentration (activity) and the potential change as sensor output signal is described by the Nernst equation.
- Potentiometric enzyme electrodes means that the enzymatic reaction with its substrate delivers a product—often a pH change—which will be detected by the underlying ion-selective electrode (e.g., a pH glass electrode).
- The functioning principle of an EnFET is based on the detection of local pH changes near the gate surface, derived from the enzyme-substrate reaction. The amplitude of the output signal of the EnFET correlates with the substrate concentration in the solution.
- Amperometric biosensors are based on the measurement of the current between working and counter electrode, resulting from the oxidation or reduction of an electroactive substance to be detected in the analyte. Amperometric enzyme biosensors are currently the most widespread and successfully commercialized biosensor devices.
- Optical biosensors with enzymes are often based on analyte-dependent changes of absorption or fluorescence measurements.
- The QCM belongs to the class of mass-sensitive biosensors and exploits the piezoelectric properties of quartz crystals to detect changes in the mass loading of analytes on its surface.
- Interfacing molecular logic gates with electrochemical enzyme biosensors, in particular with field-effect devices, facilitates novel digital biosensors with a logic output signal in a YES/NO format, logically triggered actuators as well as closed-loop intelligent sense/act/treat biochips.

Answers

1. According to IUPAC, a chemical sensor is a miniaturized device that selectively and reversibly detects chemical compounds or ions and provides analytically useful signals. Biosensors are a subgroup of chemical sensors and use biological recognition mechanisms for analyte recognition. A chemical sensor / biosensor consists of three main components: the receptor layer, the transducer, and the signal processing unit.
2. Typically used enzyme immobilization methods include adsorption, membrane or gel entrapment, layer-by-layer technique, covalent binding, crosslinking, affinity coupling, and the use of nanoparticles as enzyme nanocarriers.

3. In general, there are six different classes of transducers for chemo/biosensors: electro-chemical, electrical, optical, thermal, mass-sensitive, and magnetic transducers.

4. The most important parameters of biosensors are sensitivity, selectivity, stability, limit of detection, response time, reproducibility, reliability, application, and design.

5. According to Nernst equation, at room temperature (298.15 K), the theoretical slope for monovalent (e.g., H^+) and divalent (e.g., Ca^{2+}) ions is calculated as 59.16 mV/decade and 29.58 mV/decade, respectively.

6. Most EnFETs are built-up of a pH-sensitive FET with immobilized enzyme onto its gate surface. The functioning principle of such EnFET relies on the detection of local pH changes near the gate surface due to the catalytic reaction between the enzyme and the substrate. This local pH change modulates the surface charge of the pH-sensitive layer, which in turn, will modify the space-charge distribution in the semiconductor. Consequently, the amplitude of the output signal of the EnFET correlates with the substrate concentration in the analyte.

7. The diffusion-limited current depends on the thickness of the diffusion layer, the analyte diffusion coefficient, the number of electrons involved in the electrochemical reaction, the surface area of the working electrode, the Faraday constant, and the concentration of the analyte.

8. In amperometry, the simplest setup consists of two electrodes, the working electrode and the counter electrode, which are in contact with each other via the analyte solution—as well as a controllable voltage source (the potentiostat) and a current measuring device (amperemeter). In contrast, the potentiometric measurement chain consists of an ion-selective electrode, which detects the ion of interest, and the potential-constant reference electrode. Both electrodes are in contact with each other via the analyte and are read out electrically by a high-impedance voltmeter.

9. Optical quantities that can be used for measurement are, for example, absorption, emission, fluorescence, reflection, refraction, diffraction, or polarization. Optical biosensors with enzymes are often based on analyte-dependent changes of absorption or fluorescence measurements. The relationship between light absorption and concentration of the substance to be detected is described by the Lambert-Beer law (see Eq. 18.9).

10. A quartz crystal microbalance (QCM) belongs to the class of mass-sensitive sensors. The QCM biosensor exploits the piezoelectric properties of quartz crystals to detect changes in the mass loading of analytes on its surface. The QCM measures a mass variation per unit area by measuring the change in frequency of this quartz crystal resonator. The Sauerbrey equation (Eq. 18.10) includes the change of resonance frequency, the output frequency, the mass change, the vibrating area of quartz, the shear modulus, and density of the quartz as well as the viscosity and density of the liquid (for aqueous samples).

11. Enzyme-based molecular logic gates (with field-effect devices) directly convert bio-chemical input signals (i.e., the analyte concentration) into processed electrical output signals by performing the corresponding logical operation (AND, OR, etc.). Such

systems allow the creation of novel digital biosensors with a logic output signal in a YES/NO format, logically triggered actuators as well as closed-loop intelligent sense/ act/treat biochips.

Acknowledgements The authors thank H. Iken for assistance in preparing the illustrations.

References

Bäcker M, Rakowski D, Poghossian A, Biselli M, Wagner P, Schöning MJ. Chip-based amperometric enzyme sensor system for monitoring of bioprocesses by flow-injection analysis. J Biotechnol. 2013;163:371–6.

Bard AJ, Faulkner RL. Electrochemical methods: fundamentals and applications. Weinheim: Wiley; 2001.

Gründler P. Chemical sensors: an introduction for scientists and engineers. Heidelberg: Springer; 2004.

Katz E. Biomolecular information processing: from logic systems to smart sensors and actuators. Weinheim: Wiley; 2012.

Mulchandani A, Rogers K, editors. Enzymes and microbial biosensors. New York: Humana; 1998.

Poghossian A, Schöning MJ. Recent progress in silicon-based biologically sensitive field-effect devices. Curr Opin Electrochem. 2021;29:100811.

Poghossian A, Katz E, Schöning MJ. Enzyme logic AND-Reset and OR-Reset gates based on a field-effect electronic transducer modified with multi-enzyme membrane. Chem Commun. 2015;51: 6564–7.

Schirhagl R. Bioapplications for molecularly imprinted polymers. Anal Chem. 2014;86:250–61.

Thevenot DR, Toth K, Durst RA, Wilson GS. Electrochemical biosensors: recommended definitions and classifications. Pure Appl Chem. 1999;71:2333–48.

Wang J. Analytical chemistry. 3rd ed. Weinheim: Wiley-VCH; 2006.

Wollenberger U, Renneberg R, Bier FF, Scheller FW. Analytical biochemistry. Weinheim: Wiley-VCH; 2003.

Therapeutic Enzymes

19

Christoph Syldatk

What You Will Learn in This Chapter

Enzymes can be applied therapeutically both externally, orally and intravenously, or even inhaled. In addition to the long-established application of animal, plant, and microbial enzymes for nutritional support, their mostly supportive therapeutic use in infectious and cancerous diseases, in connection with blood clotting and in various rare genetically caused metabolic diseases, is becoming increasingly important. In contrast to many drugs, enzymes generally show few or no side effects, but on the other hand, intravenous application places very high demands on their purity and quality in order to avoid allergenic reactions, for example. This chapter describes the use of enzymes as therapeutics for a wide variety of applications ranging from nutritional support to the treatment of cancer and of rare genetic diseases. An overview is given on production and use of various animal, plant, and microbial enzymes for nutritional support and of human enzymes and their therapeutic use in cancerous diseases and in various rare genetically caused metabolic diseases. In all cases, enzymes should be non-toxic and non-allergenic or themselves meet the requirements placed on food, if ingested orally by patients. Microbial enzymes should be derived from GRAS *(generally regarded as safe)* organisms. Classical

(continued)

C. Syldatk (✉)
Institute of Process Engineering in Life Sciences II - Electro Biotechnology, KIT - Karlsruhe Institute of Technology, Karlsruhe, Germany
e-mail: christoph.syldatk@kit.edu

© The Author(s), under exclusive license to Springer Nature Switzerland AG 2024   417
K.-E. Jaeger et al. (eds.), *Introduction to Enzyme Technology*, Learning Materials in Biosciences,
https://doi.org/10.1007/978-3-031-42999-6_19

production methods for enzymes of human origin for intravenous use start with their isolation from human urine or from blood plasma, modern methods use animal cell cultures for their production. Intravenous application of enzymes of human origin or recombinantly produced by animal cell culture requires the highest enzyme purity and quality in order to avoid allergenic reactions. PEGylation of therapeutic enzymes can be used to influence their solubility, immunogenicity, and pharmacokinetics, e.g., by reducing proteolysis and increasing stability in serum.

Introduction

While a number of animal, plant, and fungal enzymes have long been used orally for nutritional support (O'Connell 2006), the mostly supportive therapeutic use of enzymes in infectious and cancerous diseases, in connection with blood clotting and in various rare genetically caused metabolic diseases, is becoming increasingly important (Fig. 19.1). All applications take advantage of the high regio-, substrate-, and reaction-specificity of the enzymes used and the property that they are generally non-toxic and readily biodegradable. In contrast to many drugs, enzymes generally show few or no side effects when used therapeutically, although on the other hand, very high demands must be made on their purity and quality, especially in intravenous applications.

Enzymes can be used therapeutically both externally, orally and intravenously, or even inhaled. Table 19.1 provides an overview of the most important therapeutically used enzymes and their applications (Shanley and Walsh 2006).

Established external applications of enzymes for therapy are the fight against bacterial or fungal infections by the addition of lysozyme to eye drops for the treatment of conjunctivitis or of chitinases to creams and ointments for the external treatment of fungal diseases.

Examples of oral applications of enzymes are the support of digestive functions for an improved degradation of fats, proteins, and starch by ingestion of animal or vegetable enzymes such as pepsin dissolved in wine or bromelain from pineapple, the cleavage of lactose in dairy products in cases of lactose intolerance with β-galactosidase, or enzymatic sucrose cleavage, which is important in *congenital sucrase-isomaltase* deficiency (CSID).

The oral application of recombinant phenylalanine ammonium lyase (PAL) from yeast for phenylalanine degradation in the hereditary disease phenylketonuria (Vellard 2003).

Examples of intravenous applications include the use of urokinase in the treatment of heart attacks, pulmonary embolisms, and other thrombotic vascular occlusions, of thrombin as an important enzyme in blood coagulation and of asparaginase in the treatment of acute lymphoblastic leukemia (ALL).

In addition, since the mid-1980s, there has been an increasing number of other enzymes used in the treatment of rare diseases (Table 19.3). These are counted among the so-called *orphan drugs* and, by definition, concern diseases that affect fewer than 200,000 patients each. (Fig. 19.1 and Table 19.3).

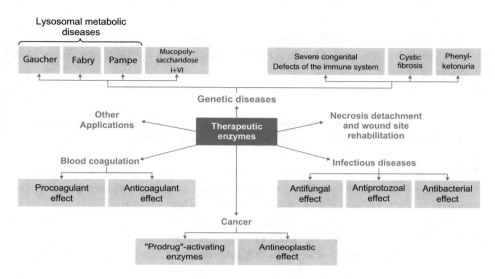

Fig. 19.1 Therapeutic use of enzymes in the treatment of diseases (after Vellard 2003)

Table 19.1 Overview of therapeutically used enzymes (after Shanley and Walsh 2006)

Enzyme	Therapeutic use
Tissue plasminogen activator (tissuePA)	Use as thrombolytic agent
Urokinase	Use as thrombolytic agent
(Activated) Protein C	Treatment of severe sepsis
DNase	Treatment of cystic fibrosis
Glucocerebrosidase	Treatment of Gaucher's disease
α-Galactosidase	Treatment of Fabry disease
Uric acid oxidase	Treatment of hyperuricemia
Asparaginase	Treatment of cancer, mainly childhood leukemia
Factor VIIa	Use in hemophilia
Factor IX (protease zymogen)	Use in hemophilia
α-Iduronidase	Treatment of mucopolysaccharidosis I (MPS I)
Hyaluronidase	Use in the treatment of heart attacks, in cancer therapy and in eye surgery
Superoxide dismutase	Degradation of oxygen radicals, use as an anti-inflammatory agent
Various proteases, including papain, collagenase and trypsin	Used for necrosis detachment and wound bed sanitation

For the treatment of cystic fibrosis, (CF) there are now sprayable and inhalable DNase-enzyme preparations that can be administered as sprays and inhaled, which are intended to support the breakdown of mucus in the lungs of affected patients (Vellard 2003).

Questions
1. What advantages does the therapeutic use of enzymes have over other drugs?
2. In what form can enzymes be applied therapeutically? Which are their main applications?
3. What should be considered when using enzymes therapeutically?

19.1   External Applications of Therapeutic Enzymes

The best known enzyme for external use in patients is lysozyme. This is used therapeutically in dissolved form in eye drops, as well as orally as a chewable or lozenge for the supportive treatment of bacterially caused purulent sore throats (angina).

Lysozyme, also called muramidase with 14.3 kDa and 129 amino acids, is a comparably small enzyme that attacks the bacterial cell wall and cleaves the β-1,4-glycosidic bonds between *N-acetylmuramic acid* and *N-acetylglucosamine residues* in the peptidoglycan backbone. cleaves.

Lysozymes already occur naturally as part of the innate immune system in animals and are also found in the tear fluid in humans. In addition to the so-called defensins, lysozyme protects the conjunctiva of the eye from bacteria, which is why deficiencies can lead to conjunctivitis.

Lysozymes are also found in plants, fungi, bacteria, and bacteriophages. Industrially, it is estimated that more than 100 t of lysozyme are extracted annually, mainly from the albumen of hen's eggs, and processed by freeze-drying. Alternatively, lysozyme can also be produced in larger quantities by fermentation with the bacterial strain *Streptomyces coelicolor* (Vellard 2003).

Lysozyme is also used in the food industry for preservation purposes. As a preservative, it is approved in the EU as a food additive under the number E 1105 for ripened cheese and outside Germany also for the preservation of beer. In cheese production, it prevents the formation of cracks in the cheese crumb caused by clostridia (so-called late bloating).

Also established for the external application in patients are creams and ointments containing chitinase-containing creams and ointments which can be used for the therapy of fungal diseases. The cell walls of many fungi causing skin diseases, so-called dermatophytes, consist of chitin, a polymer of *N-acetyl-1*,4-β-D-glucopyranosamine units. Chitinases (and also some of the lysozymes mentioned above) are capable of cleaving the β-1,4-glycosidic bonds they contain, and are widely distributed in nature. Mammals and plants also already naturally possess chitinases for pathogen defense, and in fish and insectivores, they also serve to digest insect cell walls. The recombinant production of chitinases has been established for a long time.

Furthermore, studies on the external use of enzyme cocktails for the detachment of necrotic skin after burns and for wound bed rehabilitation are also described in the literature (Shanley and Walsh 2006).

Table 19.2 Enzymes for nutritional support (after O'Connell 2006)

Enzyme	Substrate
Amylase	Carbohydrates, starch
Cellulase	Cellulose
Invertase	Sucrose
α-Galactosidase	α-Galactosides
Papain, pepsin, bromelain	Proteins
Superoxide dismutase	Superoxide
Lactase	Lactose
Pancreatin	Fats, carbohydrates and proteins

19.2 Oral Applications of Enzymes

Table 19.2 lists a number of enzymes used orally for nutritional support (O'Connell 2006).

Since these enzymes are ingested orally by patients either in liquid form (pepsin wine) or as tablets or capsules, they should themselves meet the requirements placed on food. A number of enzymes are therefore of animal origin such as the digestive enzyme pepsin from the gastric juice or pancreatin from the pancreas of cattle or pigs. Other enzymes come from plants such as the proteases papain from the milk juice of the melon tree (papaya) or bromelain from the flesh of the pineapple.

If appropriate microbial enzymes are used, they should be derived from GRAS *(generally regarded as safe)* organisms (Chap. 9). α-Galactosidases originate, for example, from the microorganisms *Lactobacillus acidophilus or Aspergillus oryzae* used in the food sector.

In all cases, it should be noted that the enzymes should be non-toxic and non-allergenic. After the production, the respective enzyme should furthermore be portioned in the necessary quantity in such a way that it can be easily absorbed by the patient himself. Enzymes which are not supposed to unfold their effect in the stomach but only in the intestine have to be stabilized or encapsulated accordingly.

Questions
  4. Give examples of external and oral uses of therapeutic enzymes and their manufacture for this purpose.

19.3 Intravenous Use of Therapeutic Enzymes

If enzymes are to be used intravenously, high demands must be made on their purity in order to avoid allergenic reactions. Conventional production methods therefore frequently use human urine or blood plasma as the starting point for isolating the enzymes in question, while modern methods use animal cell cultures for their production. The subsequent purification usually involves a large number of steps (Chap. 10). In all cases, virus inactivation must also be taken into account.

Since 2000, a derivatization with polyethylene glycols of different chain lengths has been very frequently applied to intravenously used therapeutic enzymes in order to minimize possible allergenic reactions in the patients and to positively influence the pharmacokinetics.

Currently, the most commonly used intravenously administered therapeutic enzymes are urokinase, thrombin and asparaginase.

19.3.1 Urokinase

Urokinase or urokinase-type plasminogen activator (abbreviated uPA) is an enzyme from the group of peptidases (also proteases) which can be used as an adjuvant in the treatment of heart attacks, pulmonary embolisms, and other thrombotic vascular occlusions as well as for the therapy and prophylaxis of thrombotic catheter occlusions.

Urokinase was first detected in human urine and described as a protease. Physiologically, this enzyme circulates in the bloodstream converting plasminogen to plasmin, a serine proteinase that dissolves various proteins in plasma and in particular fibrin clots, a process known as fibrinolysis.

For the preparation, urokinase is obtained from human urine by fractional precipitation with ammonium sulfate and further chromatographic steps such as gel filtration and affinity chromatography. The aim is to obtain an enzyme fraction with the prescribed molecular weight.

However, urokinase preparations from human urine could be contaminated by pathogenic viruses. Therefore, urokinase is heated to 50–70 °C for 8–12 h to inactivate viruses. To prevent denaturation, heat stabilizers such as glycocoll, lysine, arginine, sucrose, mannitol, sodium chloride, albumin, and gelatin are added. Alternative production methods now also allow recombinant production with cell cultures or with *Escherichia coli*.

19.3.2 Thrombin

Thrombin (Factor IIa) is the most important enzyme in blood clotting in vertebrates. Thrombin belongs to the group of serine proteases and cleaves fibrinogen to fibrin and the fibrinopeptides. In addition, it activates the blood clotting factors V, VIII, XIII during blood clotting. Together with thrombomodulin or protein C, it also plays an important role in inflammation and wound healing.

Thrombin is formed in the liver and is found continuously in the blood plasma as an inactive precursor or precursor with the designation "prothrombin." Genetic defects in the synthesis can lead to a deficiency here and to an increased risk of stroke.

Thrombin is produced from prothrombin-containing plasma fractions. Here, too, virus inactivation must take place which occurs with the addition of heat-stabilizing agents.

Heparin from pig intestine or bovine lung and hirudin from the leech are not enzymes but proteins used therapeutically as thrombin inhibitors which inhibit blood clotting. Both compounds are now produced recombinantly.

19.3.3 Asparaginase and Pegasparagase

Asparaginases (L-asparagine amidohydrolases) are enzymes that catalyze the hydrolysis of asparagine to aspartic acid, providing a means of degrading asparagine. Asparaginases are found in all living organisms; they are found in large amounts in the serum of guinea pigs and in the liver of several vertebrate species, as well as in fungi and several strains of bacteria. In humans, asparaginase is expressed in the brain, kidneys, testes, and intestines.

Therapeutically, asparaginase is used in its PEGylated form (Sect. 19.3.5) as so-called pegaspargases for the treatment of acute lymphoblastic leukemia (ALL) and subtypes of non-Hodgkin's lymphoma and is therefore classified as a cytostatic drug.

Healthy cells are capable of producing their own asparagine. However, leukemic cells (lymphoblasts) in acute lymphoblastic leukemia (ALL) can no longer do this and are dependent on the asparagine circulating in the blood. If asparaginase is added, it catalyzes the cleavage of asparagine to aspartic acid and ammonium, thus ensuring a low asparagine concentration in the blood. In this way, the leukemic cells no longer receive sufficient asparagine and can no longer divide.

Asparaginase is produced recombinantly with bacteria, primarily *Escherichia coli*. For therapeutic use, it is PEGylated, which has led to the name "pegaspargase." The aim here is to significantly increase the stability of the enzyme in the serum.

Questions
5. What is the importance of the enzymes urokinase, thrombin, and asparaginase for intravenous applications?

19.3.4 Therapeutic Enzymes for the Treatment of Rare Diseases

The possibility of recombinant production of peptides, proteins, and enzymes which are normally only available in very low concentrations and which would therefore be very costly to obtain from natural sources has led to the fact that, in addition to the above-mentioned therapeutic enzymes which are often still obtained conventionally, since the mid-1980s there have also been an increasing number of other enzymes which are used for the supportive treatment of rare diseases. Examples are "Gaucher's disease," "Fabry's disease," "Pompe's disease," or mucopolysaccharidosis MPS I and MPS IV. These are lysosomal metabolic diseases, some of which are genetically caused by a single point

Table 19.3 Overview of therapeutic enzymes for the treatment of rare diseases (after Veronese and Nero 2008)

Trade name	Generic name	Developed	Approved	Indication	Company
Adagen1	Pegademase	1984	1990	Enzyme replacement therapy in patients with severe immunodeficiency	Enzon Inc.
Ceredase1	Alglucerase	1985	1991	Enzyme replacement therapy in patients with Gauche type I disease	Genzyme Corporation
Pulmozyme1	Dornase A	1991	1993	Reduction of mucus viscosity and airway clearance in patients with cystic fibrosis	Genentech Inc.
Cerezyme1	Imiglucerase	1991	1994	Enzyme replacement therapy in patients with Gaucher disease type I–III	Genzyme Corporation
Oncaspar1	Pegaspargase	1989	1994	Treatment of acute lymphoblastic leukemia	Enzon Inc.
Sucraid	Sacrosidase	1993	1998	Treatment of congenital sucrase isomaltase deficiency (CSID)	Orphan Medical, Inc.
Elitek1	Rasburicase	2000	2002	Treatment of tumor- or chemotherapy-induced hyperuricemia	Sanofi-Synthelabo Research
Fabrazyme1	Agalsidase beta	1988	2003	Treatment of Fabry disease	Genzyme Corporation
Aldurazyme1	Laronidase	1997	2003	Treatment of mucopolysaccharidosis I	BioMarin Pharmaceutical Inc.
ReplagalTM	α-Galactosidase A	1998	2003	Long-term enzyme replacement therapy in the treatment of Fabry disease	Transkaryotic Therapies Inc.

mutation. The loss of a certain enzyme leads to disturbances in the fat or carbohydrate metabolism and consequently to the accumulation of toxic metabolites, which then cause severe nerve or muscle diseases, sometimes only after a long period of time. One possibility for supportive therapy of these diseases is the targeted administration of the missing enzymes. Examples of this are summarized in Table 19.3. The corresponding enzymes are produced recombinantly and are often also PEGylated before their therapeutic application. As these are rare diseases affecting fewer than 200,000 patients worldwide, these enzymes are classified as *orphan drugs*. group.

19.3.5 The PEGylation of Proteins

The term PEGylation generally describes a method first developed in the 1970s for the modification of biomolecules by covalent conjugation with polyethylene glycol (PEG) of different molecular weights. Polyethylene glycol is a non-toxic and non-immunogenic polymer. If it is used to derivatize biomolecules, it is possible not only to change the molar mass (e.g., to enable membrane retention of the coenzyme PEG-nictionamide adenine dinucleotide, PEG-NAD, after derivatization) but also to specifically influence important physical and chemical properties such as conformation, surface charge, electrostatic properties, and hydrophobicity. In the case of biopharmaceuticals, PEGylation is generally intended to influence their solubility, immunogenicity and pharmacokinetics, and pharmacokinetics, e.g., by reducing proteolysis and increasing stability in the serum. Since the 1990s, this technique has also been approved for the derivatization of therapeutic enzymes, which are then usually given the name "Peg-."

The first chemical step in the PEGylation of proteins is the derivatization of the free amino groups on the surface. A number of different derivatization reactions can be used for this purpose (Fig. 19.2). The first PEGylated therapeutic enzyme to be approved for the market was pegademase in the 1990s (Table 19.3). This is a bovine adenosine deaminasewhich removes toxic adenosine metabolites by deamination in the blood. Derivatization with 5 kDa polyethylene glycol reduced the immunogenicity and increased the half-life in plasma by a factor of 7 (Veronese and Nero 2008).

Pegaspargase (Table 19.3) was approved for therapy in 1994. Here, too, the half-life in plasma could be increased by a factor of 3 by PEGylation. The essential criterion for successful PEGylation should be that no loss of biological activity occurs (Veronese and Nero 2008).

Questions
  6. Give examples of therapeutic enzymes used to treat rare genetic diseases.
  7. What is the importance of PEGylation of therapeutic enzymes?

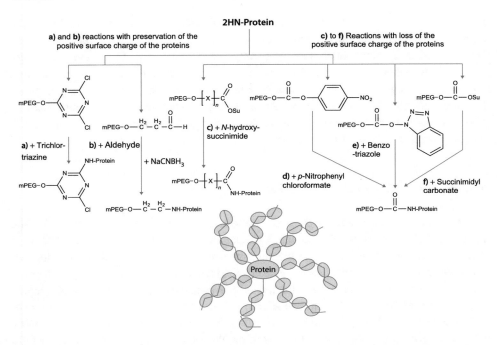

Fig. 19.2 PEGylation of enzymes with possible chemical reactions for prior derivatization of amino groups on the protein surface (**a–f**; after Veronese and Nero 2008). (**a, b**) Reactions with preservation of the positive surface charge of the proteins. (**c–f**) Reactions with loss of the positive surface charge of the proteins

19.4   Conclusion and Outlook

Even if the use of therapeutic enzymes is currently still limited to a manageable number of applications, this field of work in combination with the possibility of recombinant production of enzymes nevertheless opens up a wide range of possibilities for the future therapy of rare, genetically caused diseases. The possibility of using the high regio-, substrate-, and reaction-specificity of enzymes together with the property that, in contrast to many other drugs, they are non-toxic, biologically well degradable and usually show no or few side effects is very advantageous.

The possibility of optimizing the properties of enzymes and positively influencing their immunogenicity, stability, and pharmacokinetics through PEGylation has high potential. Therefore, further progress, important developments, and new products can be expected in this field of application in the future.

Of interest in the future may be, for example, therapeutic enzymes coupled with monoclonal antibodies, which could achieve a further increase in specificity.

As is the case with other biopharmaceuticals, the disadvantage is not only the high development costs but also the very long approval procedures, which are mainly caused by the necessary clinical tests.

Answers

1. Applications of therapeutic enzymes take advantage of their high regio-, substrate-, and reaction-specificity and the property that they are generally non-toxic and readily biodegradable. In contrast to many drugs, enzymes generally show few or no side effects when used therapeutically.
2. Enzymes can be applied therapeutically externally, orally and intravenously, or even inhaled. In addition to the long-established application of animal, plant, and microbial enzymes for nutritional support, their mostly supportive therapeutic use in infectious and cancerous diseases, in connection with blood clotting and in various rare genetically caused metabolic diseases, is becoming increasingly important.
3. Therapeutic enzymes should be non-toxic and non-allergenic or themselves meet the requirements placed on food, if ingested orally by patients. Microbial enzymes should be derived from GRAS *(generally regarded as safe)* organisms. Intravenous application of enzymes of human origin or recombinantly produced by animal cell culture requires the highest enzyme purity and quality in order to avoid allergenic reactions.
4. The best known enzyme for external use in patients is lysozyme. This is used therapeutically in dissolved form in eye drops, as well as orally as a chewable or lozenge for the supportive treatment of bacterially caused purulent sore throats (angina). Also established for the external application in patients are creams and ointments containing chitinase-containing creams and ointments which can be used for the therapy of fungal diseases. Examples for enzymes for oral use are the digestive enzyme pepsin from the gastric juice or pancreatin from the pancreas of cattle or pigs. Other enzymes come from plants such as the proteases papain from the milk juice of the melon tree (papaya) or bromelain from the flesh of the pineapple. Microbial enzymes used as amylases or α-galactosidases are derived from GRAS *(generally regarded as safe)* organisms as from the microorganisms *Lactobacillus acidophilus or Aspergillus oryzae* used in the food sector.
5. Urokinase or urokinase-type plasminogen activator (abbreviated uPA) is an enzyme from the group of peptidases (also proteases) which can be used as an adjuvant in the treatment of heart attacks, pulmonary embolisms, and other thrombotic vascular occlusions as well as for the therapy and prophylaxis of thrombotic catheter occlusions.

   Thrombin (Factor IIa) is the most important enzyme in blood clotting in vertebrates. Thrombin belongs to the group of serine proteases and cleaves fibrinogen to fibrin and the fibrinopeptides. In addition, it activates the blood clotting factors V, VIII, XIII during blood clotting.

   Asparaginases (L-asparagine amidohydrolases) are enzymes that catalyze the hydrolysis of asparagine to aspartic acid, providing a means of degrading asparagine. Therapeutically, asparaginase is used in its PEGylated form as so-called pegaspargases for the treatment of acute lymphoblastic leukemia (ALL) and subtypes of non-Hodgkin's lymphoma and is therefore classified as a cytostatic drug.

6. Examples for rare genetic diseases are "Gaucher's disease," "Fabry's disease," "Pompe's disease," or mucopolysaccharidosis MPS I and MPS IV. These are lysosomal metabolic diseases, some of which are genetically caused by a single point mutation. The loss of a certain enzyme leads to disturbances in the fat or carbohydrate metabolism and consequently to the accumulation of toxic metabolites, which then cause severe nerve or muscle diseases, sometimes only after a long period of time. One possibility for supportive therapy of these diseases is the targeted administration of the missing enzymes, which are produced in recombinant form.

7. The term PEGylation generally describes a method first developed in the 1970s for the modification of biomolecules by covalent conjugation with polyethylene glycol (PEG) of different molecular weights. In the case of biopharmaceuticals, PEGylation is generally intended to influence their solubility, immunogenicity and pharmacokinetics, and pharmacokinetics, e.g., by reducing proteolysis and increasing stability in the serum.

Take Home Message

Already for many years, enzymes are applied therapeutically both externally, orally and intravenously, or even inhaled. The application of animal, plant, and microbial enzymes has been long-established for nutritional support or to aid in the treatment of bacterial infectious diseases; their mostly supportive therapeutic use, in connection with blood clotting, in cancerous diseases and in various rare genetically caused metabolic diseases as "orphan drugs," has become important since the 1980s. As a great advantage compared to conventional drugs, enzymes generally show few or no side effects, but on the other hand especially intravenous application places very high demands on their purity and quality. For external and oral use, enzymes should be non-toxic and non-allergenic or themselves meet the requirements placed on food, if ingested orally by patients; microbial enzymes should be derived from GRAS *(generally regarded as safe)* organisms for this purpose. Intravenous application of enzymes of human origin or recombinantly produced by animal cell culture requires the highest enzyme purity and quality in order to avoid allergenic reactions. As a result, production and cleaning of this group of enzymes usually is very complex and expensive. While in the past therapeutic enzymes for intravenous application were often obtained from human blood plasma or human urine, they are now usually produced recombinantly using animal cell cultures. In all cases, virus inactivation is an important part of the cleaning processes. A PEGylation of intravenous applied therapeutic enzymes is a very common method to influence their solubility, immunogenicity, and pharmacokinetics, e.g., by reducing proteolysis and increasing stability in serum. Therapeutic enzymes modified in this way are given the suffix "Peg-" (e.g., "Pegasparagase").

References

O'Connell S. Additional therapeutic enzymes. In: McGrath BM, Walsh G, editors. Directory of therapeutic enzymes. Boca Raton: Taylor and Francis; 2006. p. 261–90.

Shanley N, Walsh G. Applied enzymology: an overview. In: McGrath BM, Walsh G, editors. Directory of therapeutic enzymes. Boca Raton: Taylor and Francis; 2006. p. 1–16.

Vellard M. The enzyme as drug: application of enzymes as pharmaceuticals. Curr Opin Biotechnol. 2003;14:1–7.

Veronese FM, Nero A. The impact of PEGylation on biological therapies. Biodrugs. 2008;22:315–29.

Enzymes in Molecular Biotechnology

20

Habibu Aliyu, Anke Neumann, and Katrin Ochsenreither

What You Will Learn in This Chapter

Molecular biotechnology uses a wide variety of enzymes. These can be grouped either to their reaction mechanism (enzyme class) or to the intended use in the manipulation of nucleic acids or to their composition of either only protein, a combination of protein and nucleic acid, or an enzymatic active nucleic acid. In this chapter they will be grouped according to their function into (1) enzymes for the *synthesis of polynucleotides,* like DNA polymerases, RNA polymerases, and reverse transcriptase. (2) *Enzymes cutting nucleic acids,* like restriction enzymes, DNases, and RNases, and the new type of CRISPR-CAS9. (3) Existing polynucleotides can be modified by the group of *nucleic acid-modifying enzymes* that cover, e.g., the ligases, methyltransferases, and phosphatases.

This chapter describes enzymes used in the field of molecular biotechnology, also called genetics. You will find an overview of the enzymes used to synthesize, hydrolyze, and manipulate nucleic acids. In addition to the reaction principles of single enzymes/enzyme classes, application examples in which combinations of

(continued)

H. Aliyu · A. Neumann
Institute of Process Engineering in Life Sciences II - Electro Biotechnology, Karlsruhe Institute of Technology-KIT, Karlsruhe, Germany
e-mail: habibu.aliyu@kit.edu; anke.neumann@kit.edu

K. Ochsenreither (✉)
Department of Biotechnological Conversion, Technikum Laubholz GmbH, Göppingen, Germany
e-mail: katrin.ochsenreither@technikumlaubholz.de

© The Author(s), under exclusive license to Springer Nature Switzerland AG 2024
K.-E. Jaeger et al. (eds.), *Introduction to Enzyme Technology*, Learning Materials in Biosciences,
https://doi.org/10.1007/978-3-031-42999-6_20

different enzymes are involved leading to valuable products will explain the power of these enzymes/demonstrate a typical workflow in molecular biology.

Historic Perspective

Current molecular biology advances and associated biotechnological applications date back to the mid-nineteenth century. It all began with the milestone discovery of nucleic acid in 1869 by Friedrich Miescher in pus cell extracts. Miescher named this material "Nuklein" because of its association with the cell nuclei. In 1909 and 1929, Phoebus Aaron Theodore Levene discovered ribose and deoxyribose sugars, respectively. Levene also coined the term nucleotide describing the unit of deoxyribonucleic acid (DNA). Using pneumococcal cell-derived deoxyribonucleases (DNases), Oswald Theodore Avery demonstrated in 1944 that DNA carries genetic information. By contrast, the ubiquitous nature of ribonucleic acid (RNA) degrading enzymes or ribonucleases (RNases) delayed the discovery of RNA until a decade later. The consequential experiments by Rollin Hotchkiss, Erwin Chargaff, Rosalind Franklin, Maurice Wilkins, Linus Pauling, and eventually, Francis Crick and James Watson, between 1928 and 1966, solved the structure of nucleic acids and the genetic code. In the above historical overview, nucleic acid degrading enzymes played a pivotal role in nucleic acid discovery. Briefly, a DNA unit comprises deoxyribose, phosphoric acid residues, and the organic bases, adenine, guanine, cytosine, and thymine. By contrast, RNA consists of uracil in place of thymine, and the sugar molecule in RNA is ribose instead of deoxyribose, with the ribose $2'$ hydroxyl ($2'$-OH) group in RNA influencing its properties and interaction with various enzymes (Fig. 20.1a). The following chapter discusses the types, diversity, and application of molecular biology enzymes and their applications in recombinant DNA technology and genetic engineering. Biotechnologically relevant enzymes catalyzing reactions involving nucleic acids fall into three broad classes. These enzymes synthesize, cut, or modify nucleic acids.

20.1 Nucleic Acid Synthesis

The synthesis of nucleic acids, i.e., the connection of nucleotides by forming phospho-diester-bonds is catalyzed by polymerases (belonging to the transferases, EC 2.7.7.X). These enzymes are found in all living organisms and are categorized by the type of nucleic acid they synthesize—RNA or DNA polymerases, and by the kind of nucleic acid they use as a template to produce new DNA- or RNA-strands—DNA-dependent-, RNA-dependent-, or independent polymerases.

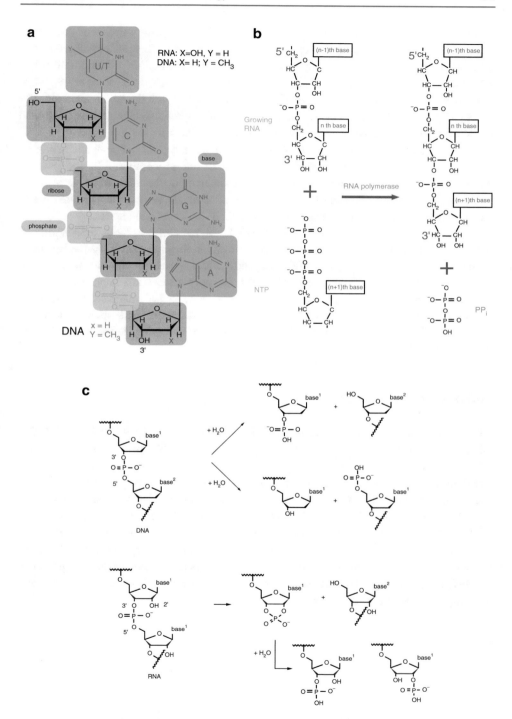

Fig. 20.1 (a) Polynucleotides of DNA and RNA. The bases, phosphates, and ribose are highlighted in blue, green, and red, respectively (https://link.springer.com/referenceworkentry/10.1007/978-3-662-44185-5_1079). The differences between DNA and RNA are highlighted in red letters. (b) Reaction scheme of polymerases using the example of RNA synthesis by RNA polymerase. (https://www.web-books.com/MoBio/Free/Ch4B1.htm). (c) Nucleolytic cleavage of DNA and RNA

During synthesis, phosphodiester bonds between the 3'-hydroxy-group of the former nucleotide and the 5'-α-phosphate of the next one are formed. This nucleophilic attack is strictly direction dependent, and elongation can only happen in 5'-3'-direction, meaning, that at the rear of the nucleic acid chain, a free phosphate group (5' end) can be found, while at the front, a free hydroxy-group is available for the reaction with the next nucleotide. The energy-dependent reaction is driven by the release of a pyrophosphate residue (= β- and γ-phosphates) from the newly attached nucleotide (Fig. 20.1b).

An important characteristic of polymerases, or more generally, of enzymes acting on polymeric substrates, is the synthesis rate, also called processivity. In the case of polymerases, processivity defines the average number of nucleotides which is attached consecutively without releasing the template strand. Processive polymerases are able to attach several nucleotides per second which results in high synthesis rates.

Commonly known is the involvement of DNA polymerase(s) in DNA replication and the role of RNA polymerase(s) in transcription. However, they have several other functions in the cell including DNA repair, RNA processing, and extension of chromosomal ends to ensure the immortality of stem cells. Furthermore, polymerases are important tools for methods in molecular biology and biotechnology, e.g., in polymerase chain reaction (PCR), PCR-based cloning techniques, in vitro transcription, and some sequencing methods. All enzymes which are introduced in the following sections have been summarized in Table 20.1.

20.1.1 DNA Polymerases

DNA polymerases synthesize DNA by linking deoxynucleotidetriphosphates (dNTP). DNA-dependent DNA polymerases which are involved in DNA replication synthesize a complementary daughter-strand from a DNA matrix, pairing always cytosine with guanine and thymine along with adenine, respectively. The correct base pairing is supported by the specific hydrogen bond interactions formed between Watson-Crick base pairs. Generally, fidelity, i.e., accuracy of DNA polymerases, is very high with only one error per 10^7 base pair. Some DNA polymerases contain additionally a 3'–5'-exonuclease domain, also known as "proof-reading activity," allowing to replace an incorrectly incorporated nucleotide with the correct one enhancing the fidelity even further. Mismatches are recognized by the differences in shape and interaction compared to the correct Watson-Crick base pairs resulting in conformational changes.

Table 20.1 Selection of commercially available polymerases applied in molecular biology and biotechnology. All listed enzymes are typically recombinantly expressed and often molecularly engineered for increased performance tailored to the respective application

Enzyme	Description	Application
DNA polymerases		
Taq polymerase	Thermostable DNA polymerase from *Thermus aquaticus*. Does not contain proof-reading activity, creates 3′A overhangs. Tolerates wide range of templates and non-standard nucleotides	Standard PCR
High-fidelity polymerases	Thermostable DNA polymerases from different *Pyrococcus* species. Possess proof-reading activity and have lower error-rate. Create blunt-ends	PCR with higher accuracy
Long-range polymerases	Thermostable DNA polymerases fused to the dsDNA binding protein Sso7d, obtained from *Sulfolobus solfataricus*	PCR amplification of very long fragments
E. coli DNA polymerases I	Contains both 3′–5′ and 5′–3′ exonuclease activity. 5′–3′ exonuclease activity removes nucleotides ahead of the growing DNA strand	DNA-labelling ("nick translation")
Klenow-Fragment	N-terminally truncated fragment of *E. coli* DNA polymerases I without 5′–3′ exonuclease activity	DNA-labelling, probe generation, Sanger sequencing
Phage T7-DNA polymerases	DNA polymerases with high synthesis rate and 3′–5′ exonuclease activity	Sanger sequencing (modified version without exonuclease activity), gap filling, second strand synthesis, site-directed mutagenesis
RNA polymerases		
Phage T7 RNA polymerases	High-level in vitro RNA synthesis from cloned DNA sequences under the T7 promotor	Synthesis of mRNA, labelled RNA-probes, sgRNA (for CrispR/Cas)
Phage T3 RNA polymerases	High-level in vitro RNA synthesis from cloned DNA sequences under the T3 promotor	Synthesis of mRNA and labelled RNA-probes
Phage SP6 RNA polymerase	High-level in vitro RNA synthesis from cloned DNA sequences under the SP6 promotor	Synthesis of mRNA and labelled RNA-probes
E. coli Poly (A) polymerase	Template independent RNA-polymerase, adds adenosine	3′ labeling of RNA with ATP, poly (A) tailing of RNA to enhance

(continued)

Table 20.1 (continued)

Enzyme	Description	Application
	residues from ATP to the 3′ end of RNA	translation in eukaryotic cells or enable affinity purification
Reverse transcriptases		
AMV reverse transcriptase	From Avian Myeloblastosis Virus. Possesses inherently higher processivity and better temperature stability than M-MuLV RT. DNA synthesis is initiated from a primer	cDNA-synthesis from RNA or DNA-synthesis single-stranded DNA, RT-PCR, RNA-sequencing
M-MuLV reverse transcriptase	From Moloney Murine Leukemia Virus. DNA synthesis is initiated from a primer. The most frequently used reverse transcriptase in RNA studies. In combination with Taq polymerase, basis of numerous test systems for detecting SARS-CoV-2	cDNA-synthesis from RNA or DNA-synthesis single-stranded DNA, RT-PCR
Group II intron-encoded reverse transcriptases	Original source not disclosed. Exhibits high processivity, thermostability and inhibitor tolerance according to the manufacturer. Comparable fidelity to retroviral RTs	cDNA synthesis, suitable for synthesis of long cDNA, direct RNA sequencing

Since a free 3′ hydroxy end is needed by replicative DNA polymerases, a so-called primer has to be provided in order to start DNA synthesis. In the cell, this primer is a short RNA molecule synthesized by "Primase," a specialized DNA-dependent RNA polymerase involved in DNA replication. The RNA moieties are later removed and replaced by DNA. For PCR applications, specific DNA-primer molecules are synthetically generated and used to define the DNA section which should be amplified.

The first DNA polymerase, named DNA polymerasw I (Pol I), was isolated from *E. coli* in 1955 by Arthur Kronberg. However, this polymerase plays only a secondary role in replication because of its low processivity. Due to its 5′–3′-exonuclease activity, it removes the RNA primer and replaces it with DNA during replication and is involved in DNA repair. In modern molecular biology, Pol I as well as a truncated version of it, the "Klenow fragment" is used in different cloning techniques as well as for labeling DNA molecules by incorporation of radioactive or fluorescent nucleotides for the generation of probes. The Klenow fragment is an N-terminal truncation that retains polymerase and 3′–5′ exonuclease activity but has lost the proofreading activity. *E. coli* possesses in total five types of DNA polymerases. Pol III is the main replicative DNA polymerase, contains proofreading activity, and is highly processive due to a clamp (β-sliding clamp) binding it tightly to the

DNA template, while Pol II, IV, and V are involved in DNA repair, SOS response, and translesion synthesis.

The mechanisms of replication and the ability of DNA polymerases to synthesize new DNA material complementary to a template are used to rapidly multiply specifically DNA fragments by polymerase chain reaction (PCR). It is a standard method in every molecular biology lab and is used in numerous applications, e.g., cloning techniques, diagnosis and monitoring of genetic diseases, DNA profiling, and fingerprinting. The PCR reaction was developed by Kary Mullis in 1984. The standard method, which has been further developed and diversified over the years, is comprised of three cycles which are typically repeated 25–35 times to exponentially multiply ("amplify") the desired DNA fragment (Fig. 20.2). In the first step, the denaturation phase, the DNA double-strand is separated into single strands. While during the replication process in the cell, this is catalyzed by different enzymes and proteins to break the hydrogen bond between complementary bases, melting is facilitated by heating the reaction mixture to 95 °C in the PCR reaction instead. The dependency of DNA polymerases on free 3' hydroxy groups to start DNA synthesis is used in the PCR reaction to direct the polymerase to a specific part of the DNA molecule and to

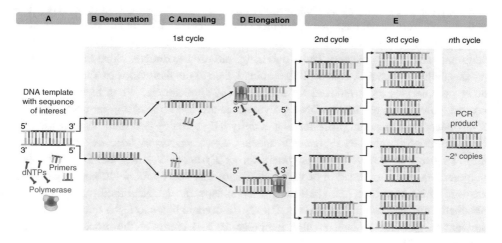

Fig. 20.2 Basic principle of polymerase chain reaction (PCR). (**a**) Double-stranded DNA template. 5' and 3' ends are indicated. (**b**) In the denaturation step at ~95 °C, hydrogen bonds are broken between complementary bases and the DNA double-strand is separated in single-strands. (**c**) By decreasing the temperature in the annealing step to typically 50–65 °C (depending on the primer sequence), primer are able to bind specifically to the DNA template which provide the necessary 3' hydroxy group as initiation point for DNA synthesis by DNA polymerase. (**d**) Elongation step: Depending on the optimal reaction temperature of the used polymerase (typically 72 °C for the standard thermophilic PCR polymerases), DNA is synthesized complementary to the template by linking provided dNTPs. Length of the synthesized fragment is limited in the first elongation step by the duration of the elongation phase. This is the end of the first cycle. (**e**) In the following cycles, the synthesized fragments from previous cycles act now as template and have the correct size. The number of copies increase exponentially with every cycle

define the fragment which shall be amplified. For this purpose, a pair of DNA primers—forward and reverse-, usually 20–25 long nucleotides long, is designed. The forward primer defines the beginning of the fragment to be amplified and binds complementary to the antisense strand, while the reverse primer binds complementary to the sense strand and marks the end of the fragment. The binding of the primers is enabled by lowering the temperature typically to 50–65 °C depending on the primer sequences. This step is called the annealing phase. The annealing temperature has to be as high as possible to ensure specific binding of the primer to the target region and to prevent undesired side-products, but low enough to enable hybridization. If the annealing temperature is chosen too high, the primer will not bind. In the last step, the elongation phase, DNA synthesis takes place. Therefore, a DNA polymerase and dNTPs are part of the reaction mixture. The usage of the thermostable "Taq polymerase" enabled the automatization of the PCR method in a programmable heating block and led to broad application and further improvement of the method. Today, Taq polymerase is still the working horse in standard PCR methods; however, a truncated version of the polymerase deprived of the proofreading activity and recombinantly expressed in *E. coli* is used. It is suitable for the amplification of fragments ≤ 6 kb. The synthesis rate is typically ~1 kb/min which also defines the length of the elongation step. Due to the removed proofreading activity, the fidelity is comparatively low with 8×10^{-6} errors per base pair resulting in the accumulation of mutations. Furthermore, Taq polymerase creates 3′A-overhangs, i.e., at the 3′ end of each DNA fragment an additional adenine is attached. These overhangs are used to directly clone PCR fragments into specialized plasmids, called "T-A-cloning." Due to the limitations of Taq polymerase, other polymerases are employed for specific PCR applications. When high accuracy is needed, so-called high-fidelity DNA polymerases are used possessing an associated 3′–5′-exonuclease-dependent proof-reading activity. High-fidelity polymerases were isolated from thermophilic archaea Pyrococcus species, e.g., *Pyrococcus furiosus* (Pfu polymerase), *Pyrococcus woesei* (Pwo polymerase), or *Pyrococcus* strain GB-D (Deep Vent polymerase). For amplification of very long fragments, i.e., up to 30–40 kb, polymerases are used which have been fused to the sequence-nonspecific dsDNA binding protein Sso7d, obtained from *Sulfolobus solfataricus*. The Sso7d domain binds to DNA double strands at ambient temperature as well as high temperatures and improves the processivity of the polymerase without affecting thermostability or synthesis rate.

DNA polymerases are also employed in different sequencing methods. The first sequencing method was developed by Frederic Sanger in 1977 (Sanger-Sequencing, chain-termination sequencing, dideoxy-DNA sequencing) and is based on the random insertion of dideoxynucleotidetriphosphates (ddNTPs) by DNA polymerase which leads to termination of the fragment elongation. For approximately 40 years, it was widely used as the gold standard in sequencing. Today, even though replaced by next-generation sequencing methods, Sanger sequencing is still routinely used for sequencing smaller fragments and for validation of sequencing results. In the original method, Klenow-Fragment was used, however, it was later replaced with phage T7 DNA polymerase which accepts dNTPs and ddNTPs with the same efficiency.

20.1.2 RNA Polymerases

RNA polymerases synthesize RNA from ribonucleotidetriphosphates (NTPs). In contrast to DNA polymerases, they are able to start the synthesis de novo, i.e., a primer offering a 3′ hydroxyl group is not required. Like DNA polymerases, RNA polymerases are complex proteins and consist of multiple domains; however, they do not possess exonuclease activity and consequently, they do not have proofreading activity resulting in much lower fidelity. The error rate is approximately 1 error per 10,000 base pairs. Furthermore, the synthesis rate is also lower. DNA-dependent RNA polymerases synthesize RNA from a DNA template in the transcription process. The initiation site of transcription is a certain DNA region, the promotor, indicating the beginning of a gene. To bind to the promotor region, transcription factors and mediator complexes are necessary in addition to the RNA polymerase. RNA polymerases synthesize either coding RNA (mRNA), which is further translated into proteins, or non-coding RNAs from RNA-coding genes, which can be further processed but are not translated into a protein. Examples of non-coding RNA species are transfer RNA (tRNA) involved in translation, ribosomal RNA (rRNA), structural component of ribosomes, different small RNAs including several RNAs which form ribonucleoproteins involved in splicing (snRNA), RNA maturation (snoRNA), genome defense (piRNA), maintenance of chromosome structure (telomerase RNA), and micro RNA (miRNA) involved in gene regulation and gene silencing, guide RNA (sgRNA) as part of CRISPR/Cas9-systems and catalytic RNA as part of ribozymes.

In bacteria and archaea, one RNA polymerase transcribes all gene types. Recognition of the different promotors and regulation of transcription frequency and efficiency is mediated by specific sigma factors which bind to the RNA polymerase. In eukaryotes, three to five different nuclear RNA polymerase types exist each responsible for the transcription of a different subset of gene types. Nevertheless, the RNA polymerase of bacteria, archaea, and eukaryotes share a similar core structure and mechanisms, while viral RNA polymerases consist of only one domain. In RNA-viruses, e.g., poliovirus, also RNA-dependent RNA polymerases can be found, synthesizing RNA from an RNA template. Here, they are involved in the replication of the RNA genome and in transcription. The high error rate of the RNA polymerase promotes frequent mutations in the viral genome helping to overcome the host's defense system. In both prokaryotes and eukaryotes, an independent RNA polymerase exists, the Poly(A) polymerase which adds multiple adenosine-residues to mRNA molecules creating the Poly(A)-tail.

In biotechnology, RNA polymerases are used for in vitro transcription to synthesize (m)RNA in large amounts as the template for (in vitro) translation, produce RNA probes, and to investigate RNA stability, promotors, and gene expression. The method was developed by Krieg and Melton in 1984. They showed that by using the RNA polymerase of the bacteriophage SP6 RNA was synthesized from a gene under the SP6 promotor cloned into a plasmid. Subsequently, the system was extended to include promoters and RNA polymerases of bacteriophages T3 and T7. Principally, also other polymerases can be used for in vitro transcription as long as suitable promotors are used and all necessary transcription factors are provided. However, the viral systems have several advantages: RNA polymerases from bacteriophages are comparatively small proteins that can be easily expressed recombinantly in *E. coli* in large amounts. The promotor sequence is short and is therefore standard in many expression vectors in which the promotors are also used as initiation points for sequencing. The synthesis speed of the T7 RNA polymerase is about 8-times faster than the *E. coli* system comprised by *E. coli* RNA polymerase and suitable promotor. In addition, they also accept labeled nucleotides for the generation of probes and modified nucleotides which were especially relevant for the development of mRNA vaccines such as the COVID 19 vaccine Comirnaty from Biontech. Many supplier companies for life science research offer ready-to-use kits for in vitro transcription based on SP6-, T3-, or T7 RNA polymerase, often accompanied by enzymes for polyadenylation and capping to produce mRNA suitable for translation in eukaryotic cells.

20.1.3 Reverse Transcriptase

Reverse transcriptases (RTs) are RNA-dependent DNA polymerases consisting of multiple domains and catalyze the transcription process in reverse order. Starting from a single-stranded RNA template, a DNA strand is complementarily formed by linking dNTPs resulting in an RNA-DNA hybrid. As for DNA-dependent DNA polymerases, a primer is also needed to start the polymerization reaction. In molecular biology, gene-specific primers are added to selectively transcribe a single transcript, or unspecific primers, e.g., Oligo(dT) primer which are able to bind at the poly(A) tail of all eukaryotic mRNA molecules or random primers binding internally of the transcripts, to create libraries. The RNA portion of the hybrid double strand is hydrolyzed by the internal RNaseH-activity of the enzyme. The remaining DNA single strand is finally completed into a DNA double-strand by DNA-dependent DNA polymerase activity of the enzyme. The produced double-stranded DNA is also called complementary or copy DNA (cDNA) to indicate its origin. Retroviral RTs do not contain a proofreading activity, which is why the fidelity is very low with 1 error per 1000–10,000 base pairs resulting in high mutation rates.

RTs were first discovered in retroviruses in 1970 independently and simultaneously by the groups of David Baltimore and Howard Temin. Retroviruses, e.g., human immunode-ficiency virus (HIV), contain two single-stranded 5′ capped RNA molecules with Poly (A) tail mimicking mRNA. Upon infection, these molecules are converted into double-stranded DNA by RT in the cytosol and subsequently integrated into the host genome. Due to the high error rate, mutations occur regularly which may lead to drug resistance in HIV patients.

RTs are also involved in the propagation of retrotransposons, mobile genetic elements which integrate a new copy of themselves in the host genome via an RNA intermediate. Also, the dissemination of self-splicing introns (group II introns), another group of mobile genetic elements, is mediated by RT activity in a process called retrohoming. In most eukaryotes, replication of chromosomal ends, the telomeres, is facilitated by telomerase, a ribonucleoprotein with RT activity.

In molecular biology and biotechnology, recombinantly expressed retroviral RTs, e.g., from Avian Myeloblastosis Virus (AMV) or Moloney Murine Leukemia Virus (M-MuLV), are used to convert mRNA into intron-free cDNA. The obtained cDNAs are amplified by PCR to create cDNA-Libraries in a process called RT-PCR. These libraries can be used to identify splicing sites and modifications which were introduced during RNA processing. By isolating mRNA from cells exposed to different conditions and/or at different time points, differential gene expression can be studied. Also, when attempting to recombinantly express eukaryotic genes in heterologous hosts, cDNA is used by default. Prokaryotic genes do not contain introns; therefore, introns are not removed after transcription, which would lead to erroneous proteins in prokaryotic expression hosts. In heterologous eukaryotic hosts splicing sites might not be recognized correctly, or the splicing process might slow down the expression process; therefore, cDNA is used to ensure optimal expression. Limitations of the native retroviral RTs are their inherently low processivity and fidelity. Also, some RNA templates might contain secondary structures or are even arranged in tertiary structures impeding the reverse transcription process. Therefore, genetically engineered derivatives of retroviral RTs have been created which are characterized by higher fidelity and thermostability to enable higher reaction temperatures beneficial for resolving secondary and tertiary RNA structures. As an alternative, thermostable group II intron-encoded RTs, e.g., from the thermophilic cyanobacterium *Thermosynechococcus elongatus*, have been studied. In contrast to retroviral RTs, they are characterized by high processivity and significantly lower error rate. Group II Intron RTs are also commercially available and can be applied in cDNA generation as well as direct RNA sequencing.

20.2 Nucleic Acid-Cutting Enzymes

Nucleic acid cutting enzymes, known as nucleases, degrade a nucleic acid polymer and depending on the specificity of the reaction, the enzymes serve different biological functions. They are used variously in manipulating DNA or RNA. Nucleases comprise a variety of proteins and catalytic RNAs (ribozymes), which catalyze the breakdown of nucleic acids by attacking the phosphodiester bond (P-O) between nucleosides. Depending on the position of attack, the enzymes form two broad categories, endonucleases, and exonucleases. Exonucleases catalyze the successive cleaving of the nucleotides by hydrolyzing the phosphodiester bonds at a polynucleotide chain's 3' or 5' termini. By contrast, endonucleases hydrolyze the phosphodiester bonds within a polynucleotide chain. Some nucleases possess endo- or exonuclease activity, while enzymes of other classes often carry additional nuclease activities. For example, prokaryotic DNA polymerases also harbor exonuclease (proofreading) activity. Nucleolytic enzymes digest DNA or RNA, but several others break down both nucleic acid types. Classic examples of specific nucleases include DNase I, which digests single and double-stranded DNA and RNase I, which preferentially degrades single-stranded RNA.

DNA hydrolysis involves the cleavage of the phosphodiester bonds (P-O3' or P-O5') through the activity of nucleophiles in acid-base catalysis (Fig. 20.1c). In this reaction, a general base, usually side chains of specific residues (e.g., Histidine) in the catalytic center, deprotonates water converting it into hydroxide. The activated nucleophile (hydroxide) then attacks the scissile phosphate of the substrates to form a pentacovalent intermediate, leading to the subsequent release of a leaving group (the breakaway 5'-linked nucleoside). The leaving group gets protonated by other residues (e.g., Histidine, Arginine and Lysine) acting as general acids. The 2'-OH next to a scissile phosphate in RNA acts as an additional nucleophile involved in the catalytic activity of most RNases. During RNA hydrolysis, the general base (Histidine) deprotonates the 2'-OH group, which then acts as a nucleophile attacking the scissile phosphate to form a transitional 2',3'-cyclic phosphate. A general acid (Histidine) then protonates the leaving group. The next step involves the activation of water by a general base leading to the hydrolysis of the cyclic phosphate to 2'- or 3'-phosphates. The above general mechanism varies with specific enzymes and divalent metal ions such as Mg^{2+}, Z^{2+}, and Ca^{2+}, which act as cofactors. Nevertheless, the nucleases form a complex and diverse group of enzymes varying in catalytic mechanism, including metal ion dependency and substrate range. For their relevance to molecular biology, genetics, and biotechnology, this chapter focuses on broad nuclease grouping, DNases, and RNases.

20.2.1 Deoxyribonucleases

Deoxyribonuclease (DNase) refers to enzymes that selectively catalyze the digestion of DNA over RNA based on the principles outlined above. DNases possess either endo- or exonuclease activity and vary in properties, including biological function and substrate specificity. Prominent applications of DNases involve the removal of DNA from protein or RNA samples using DNase I. Although, Miescher noted nucleolytic activity circa a century earlier, DNase I from beef pancreas represents the first isolated deoxyribonuclease by M. Kunitz in 1948. Owing to modern cloning techniques, DNase I and other nucleases have become commercially available for application in various fields (Table 20.2).

20.2.1.1 DNA-Specific Endonuclease

Deoxyribonucleases I
DNase I is a 31 kDa glycoprotein and exists in at least four isozymes; A, B, C, and D, with isozyme A being the most predominant form. This endonuclease non-specifically cuts DNA into dinucleotides, trinucleotides, or oligonucleotides with 3'-phosphates and 5'--hydroxyl end and acts on isolated or chromatin complexed double and single-stranded

Table 20.2 Selection of nucleases applied in molecular biology and biotechnology.

Enzyme	Description	Application
Endonucleases		
DNase I EC 3.1.21.1	An endonuclease that non-specifically cuts single- and double-stranded DNA into dinucleotides, trinucleotides, or oligonucleotides. Activity depends on divalent cations, e.g., Ca^{2+}, Co^{2+2+}, Mg^{2+2+}, Mn^{2+} and Zn^{2+} and inactivated at 75 °C for 5–10 min or chelating agents, like EDTA	Removal of contaminating DNA from RNA, protein, and primary cell isolation samples. DNase footprinting and isotopic labelling of DNA
APE 1 EC 3.1.25.2 (Transferred to EC 4.2.99.18)	A multifunctional endonuclease that produces a single nucleotide gap with 5′ deoxyribose-phosphate and 3′ deoxyribose-hydroxyl termini by cleaving the DNA immediately 5′ of the AP sites. APE 1 also excises terminal mismatches by 3′–5′ exonuclease, DNA 3′-diesterase and RNase H activities	DNA damage (comet assay) studies and nucleotide incision repair in drug design, target research, and SNP genotyping
Endo IV EC 3.1.21.2	AP endonuclease that produces a similar single nucleotide gap in double-stranded DNA, 3′-diesterase activity, which removes 3′ DNA-blocking groups. The *E. coli* Endo IV depends on Zn^{2+} ions Inactivated by elevated temperature, ~80 °C, but insensitive to EDTA	DNA damage (comet assay) studies and nucleotide incision repair in drug design, target research, and SNP genotyping. Endo V
Endo V EC 3.1.21.7	A deoxyinosine 3′ endonuclease that recognizes single or double-stranded deoxyinosine-containing DNA. It catalyzes the hydrolysis of the second phosphodiester bond 3′ to deaminated base lesions. It also acts on AP sites and other aberrant DNA positions. Depends on Mg^{2+} or other divalent cations	Primarily used in mutation research, DNA shuffling and mismatch repair studies
T7EI EC 3.1.21.10	Selectively binds heteroduplex nucleic acids deformities to introduce two nicks on the non-crossing strands at the first, second or third phosphodiester bond from the 5′ ends. Depends on divalent cations, inactivated at 65 °C or treated with proteinase K	Assessing mutations induced by other enzymes. Resolution of heteroduplex DNA structures and mismatches
Restriction endonucleases	REases recognize and hydrolyze the phosphodiester bonds of double-stranded DNA, and less effectively,	REases are top among the valuable tools of genetic engineering used to create DNA fragments from diverse

(continued)

Table 20.2 (continued)

Enzyme	Description	Application
	single-stranded DNA, at or near restriction sites, generating two incisions on both sides of the double helix DNA. They are classified as type I, II, III, and IV based on their properties. Have an absolute requirement for Mg^{2+} as a cofactor. Types I, III, and IV have limed application in biotechnology because of ATP dependency, low specificity, and distance of cleavage from recognition location. Type II REases (EC 3.1.21.4) occur as monomers or oligomers harboring REase activity distinctively from their methylases. Type II REases cleave DNA at reliably constant positions within or close to the recognition sites to generate either 3′ or 5′ protruding (sticky) and blunt ends	sources, DNA mapping, restriction fragment length polymorphism (RFLP) and mutation studies
Sugar-nonspecific nucleases		
Benzonase EC 3.1.30.2	Sugar is a non-specific endonuclease and degrades all types of DNA and RNA. Requires Mg^{2+} to hydrolyze its substrate to yield 3′-OH and 5′-phosphoryl ends	Removal of nucleic acids from purified proteins
Mung bean nuclease EC 3.1.30.1	Thermostable single-strand specific endonuclease that requires Zn^{2+} and a reducing agent, such as cysteine, for stable activity at pH 5.0. Deactivated using chelating agents like ethylene diamine tetraacetic acid (EDTA) and sodium dodecyl sulfate (SDS)	Blunting 3′ and 5′ single-stranded overhangs in double-stranded DNA, removal of single-stranded nucleic acids, cleavage of DNA hairpin loops and RNA transcript mapping
P1 nuclease EC 3.1.30.1	Zn^{2+}-dependent single-strand-specific endonuclease. Hydrolyzes the phosphodiester bonds in single-stranded nucleic acids yielding 5′ mononucleotides. Active on double-stranded DNA under nonstandard conditions. Inactivated using SDS and reversibly using guanidine hydrochloride and SDS or urea	Removal of nucleic acids during protein purification and DNA damage and modification analysis
S1 nuclease EC 3.1.30.1	Single-strand specific endonuclease. Readily cleaves the opposite side of double-stranded DNA nicked by other	Resolving hairpin loops produced during cDNA synthesis, mutagenesis and DNA/DNA-RNA hybridization

(continued)

Table 20.2 (continued)

Enzyme	Description	Application
	enzymes and single-stranded regions of the molecule. Zn^{2+} dependent endonuclease and inactivated using chelating agents such as EDTA and citrate	studies, and selective digestion and removal of overhangs
MNase EC 3.1.31.1	MNase cleaves all nucleic acid with a preference for single-stranded molecules at A/T-containing dinucleotides. It hydrolyzes nucleic acids generating 3'-phosphate and 3'-P, and 5'-hydroxyl terminated mononucleotides and oligonucleotides. Possesses both endo- and exonuclease properties. MNase requires Ca^{2+} and is inactivated at elevated temperatures or by chelating agents	Nucleosome mapping coupled with sequencing (MNase-seq) and depletion of nucleic acids from protein samples
Exonuclease		
EXOI EC 3.1.11.1	EXOI catalyzes the stepwise hydrolysis of the phosphodiester bonds of linear single-stranded DNA at 3'-hydroxyl ends, yielding deoxyribonucleoside 5'-monophosphates (dNMP) and 5'-terminal dinucleotides. The activity requires Mg^{2+} and high processive. Tolerant to high salt concentrations, only inactivated at elevated temperatures around 80 °C	Removal of single-stranded DNA and fragments from reaction mixtures
ExoIII EC 3.1.11.2	The *E. coli* ExoIII catalyzes multiple nucleolytic activities on double-stranded DNA. As a 3' exonuclease, it cleaves the 3'-hydroxyl termini (a blunt or a 3' recessed ends) of double-stranded DNA. Generates a gap at abasic sites in a double-stranded DNA to generate a gap. EXOIII carries phosphatase and ribonuclease H activities. Requires Mg^{2+} and is inhibited by Zn^{2+}	Mutation studies, synthesis of strand-specific labelled probes and generation of single-stranded DNA
EXOVII EC 3.1.11.6	EXOVII specifically cleaves single-stranded DNA or double-stranded DNA with unpaired termini in both 3'–5' and 5'–3' directions. Requires no divalent cation, and therefore	Removal of single-stranded primers at the end of each cycle of nested PCR reaction. Removal of DNA in a sample. Intron mapping

(continued)

Table 20.2 (continued)

Enzyme	Description	Application
	inactivation requires elevated temperature	
Lambda Exonuclease EC 3.1.11.3	λ exo selectively cleaves on the 5′-phosphorylated strand of double-stranded DNA to yield 5′-mononucleotides and the complementary single-stranded DNA. Highly processive around pH 7.0, requires Mg^{2+} and can be inactivated at elevated temperatures	Conversion of linear double-stranded DNA to single-stranded DNA for various applications, including DNA microarrays, pyrosequencing technology, single-stranded conformation polymorphism technique (SSCP), single-nucleotide polymorphism analysis, among numerous others
Bal 31 nucleases EC 3.1.30.1	Two distinct exonucleases with fast and slow activity rates on duplex DNA. Progressively cleaves 3′ and 5′ termini of duplex DNA without introducing internal scissions yielding a double-stranded molecule and 5′-mononucleotides. It also shows 3′ and 5′ exonuclease and endonuclease activities against single-stranded DNA. Depending on Mg^{2+} and Ca^{2+}, deactivation requires an elevated temperature of 95 °C	Progressive removal of nucleotides from both ends of double-stranded DNA, restriction site mapping and mutation studies
Ribonucleases		
RNase A EC 3.1.13.1	An endoribonuclease that cleaves the 3′,5′-phosphodiester bonds of single-stranded RNA by recognizing uracil and cytosine generating 3′ polycytidylic (Cp) or polyuridylic (Up) acids with 2′,3′-cyclic phosphate intermediates. Cp and Up cleavage distributive but processive for poly(A). Requires Mg^{2+} and is active over a wide temperature range. Ubiquitous among animals and widespread in the environment	Removal of RNA contamination in DNA preparations
RNase H EC 3.1.13.2	RNase specifically hydrolyzes the phosphodiester bonds of RNA in RNA/DNA hybrids. It requires Mg^{2+} or Mn^{2+} and is inactivated by EDTA	High-stringency hybrid selection, cDNA synthesis and antiviral research

DNA. DNase I require activation by divalent cations such as Ca^{2+}, Co^{2+}, Mg^{2+}, Mn^{2+}, and Zn^{2+} with the desired cleavage action modulated by the type of metal. For instance, Mg^{2+} and Ca^{2+} induce maximum DNase I activation while Mg^{2+} and Mn^{2+} activation favor random cleavage (with preference to single-stranded DNA) and same site cleavage of single and double strands, respectively. The enzyme can be inactivated by elevated temperature, usually 75 °C for 5–10 min and different chelating agents depending on the desired application. DNase I remove contaminating DNA from RNA, protein, and primary cell isolation samples. Other uses include its application in DNase footprinting and nicking DNA to incorporate labelled nucleotide bases.

20.2.1.2 Structure-Specific DNA Endonucleases

Apurinic/Apyrimidinic (AP) Endonuclease 1, APE1

APE 1, also known as Ref-1, is a 33–37 kDa multifunctional endonuclease which recognizes positions in a double-stranded DNA known as the apurinic/apyrimidinic (AP) site. As the name implies, the Ap site (also abasic site) is a location that contains a damaged base that occurs spontaneously through the natural or induced activity of DNA glycosylase or hydrolytic depurination or action of free radicals. The human APE1 shares homology with exonuclease III and ExoA of *E. coli* and *Streptococcus pneumoniae*, respectively, and is commonly expressed in the former organism. The principal endonuclease activity of the enzyme produces a single nucleotide gap with 5′ deoxyribose-phosphate and 3′ deoxyribose-hydroxyl termini by cleaving the DNA immediately 5′ of the AP sites. APE 1 also excises terminal mismatches by 3′–5′ exonuclease activity. Less significant activities of the enzyme include DNA 3′-diesterase and RNase H activities. The enzyme is used in DNA damage (comet assay) studies and nucleotide incision repair in drug design, target research, and SNP genotyping.

Endonuclease IV, Endo IV

Like APE 1, Endo IV is an AP endonuclease that produces a similar single nucleotide gap in double-stranded DNA and shares the 3′-diesterase activity, which removes 3′ DNA-blocking groups such as 3′ phosphate, 3′-α, β-unsaturated aldehyde and others resulting from enzymatic or oxidative damage. Both enzymes can be inactivated by elevated temperature, ~80 °C, but Endo IV is insensitive to EDTA inactivation. Unlike APE 1, endonuclease activity of the *E. coli* ~30 kDa depends on nucleophilic and 3′ OH stabilizing Zn^{2+} ions. Endo IV originating and expressed in *E. coli* shares homology with several bacteria and eukaryotic AP endonucleases and has similar applications as APE1.

Endonuclease V, Endo V

Endo V, first discovered in *E. coli*, represents a unique variant of DNA repair protein identified as deoxyinosine 3′ endonuclease. The enzyme recognizes single or double-stranded deoxyinosine-containing DNA. It catalyzes the hydrolysis of the second phosphodiester bond 3′ to the deaminated base lesion, generating a gap with 3′ hydroxyl and 5′ phosphate termini, leaving behind the lesion. Endo V also acts on the signal of AP sites, urea, and other aberrant DNA positions. The catalytic activity of the 25 kDa enzyme depends on Mg^{2+} or other divalent cations such as Co^{2+}, Ni^{2+}, or Mn^{2+} and has an optimum pH of ~7.5. Endo V is overexpressed in *E. coli* and used primarily in mutation research, DNA shuffling, and mismatch repair studies.

Phage T7 Endonuclease I, T7EI

T7EI is about 18 kDa divalent cation-dependent protein encoded by gene 3 of bacteriophage T7 and first isolated from *E. coli* infected by the phage. The enzyme selectively binds heteroduplex nucleic acids deformities to introduce two nicks on the non-crossing strands at the first, second, or third phosphodiester bond from the 5′ end gaps with 3′ hydroxyl and 5′ phosphate termini. Besides metal ions requirement, the T7EI variant expressed in *E. coli* with a 45 kDa maltose-binding protein fusion resulted in increased activity linked to enhanced solubility, stability, and improved folding of the catalytic domain. The activity of T7EI and recombinant derivatives can be inactivated at 65 °C or treated with proteinase K at 37 °C (New England Biolabs). Because of its ability to recognize mismatches and extrahelical loops in double-stranded DNA selective resolving activity, the enzyme is used in assessing mutations generated by CRISPR-Cas9 and other strategies and the resolution of heteroduplex DNA structures and mismatches.

20.2.1.3 Restriction Endonucleases, REases

Restriction endonucleases, popularly known as restriction enzymes, first identified in bacteria between 1952 and 1965, are widespread in prokaryotes, serving as a natural defense or immune system against invading foreign genetic elements. REase recognizes and hydrolyzes the phosphodiester bonds of double-stranded DNA, and less effectively, single-stranded DNA, at or near specific sequences known as restriction sites, generating two incisions on both sides of the double helix DNA. The restriction sites, often palindromic, usually comprise 4–6 nucleotides and occur in the prokaryotic DNA. Prokaryotes deploy methyltransferase (MTase), which adds a methyl group to a specific nucleotide position in the restriction site. Methylation distinguishes "self" from "foreign" DNA, ensuring only the latter gets digested. Together, the combined MTase and REase activity constitute a prokaryotes' restriction-modification (R-M) system. The continued search for REases often results in identifying several enzymes sharing the same recognition patterns and cleaving the DNA at the same or different sites. The new enzymes, while having the features of the classic or prototype enzyme, may provide improved desirable features, including stringent specificity, reduced off-target activity, low purification cost, and high yield. Newly isolated enzymes sharing the same recognition sites as the classic REase are known as isoschizomers or neoschizomers if they cleave the same or different sites, respectively. For example, the isoschizomers pair PaeI and SphI recognize and cut the sequence 5′-GCATG/C-3′ while the neoschizomers pairs AatII and ZraI recognize the 5′-GACGTC-3′ but cleave the respective positions GACGT/C and GAC/GTC.

Naming REases

By convention, naming restriction endonucleases follows the rules proposed by Smith and Nathans with minor modifications. The first part of the name comprises a single upper case and two lower case letters derived from the genus and species that produce the enzyme. The fourth letter, usually in the upper case, represents the strain or serotype designation. Roman numerals indicate the chronology of identification in cases where the host organism harbor multiple REase systems; for example, the first REase (Type IIP) from *Bacillus amyloliquefaciens* H is named BamHI.

Classification of REase

Based on cofactors requirements, recognition sequences and restriction specificity, the position of cleavage relative to the restriction site, and subunit composition, REases form four classes, namely type I, II, III, and IV (Table 20.2). All REases absolute requirement of Mg^{2+} as a cofactor for the endonuclease activity, while those incorporating MTase activity depend on ATP and *S*-adenosylmethionine (AdoMet) as the donor of the methyl group for methylation.

The three REases, types I, III, and IV, find limited application in genetic studies primarily due to the dependency on ATP, low specificity, and location distance of cleavage. Type I comprises a multiunit enzyme complex involving methylation and restriction. The ATP-dependent system cuts DNA at a variable distance from the restriction sites and targets methylated and unmethylated sites. Type III is also composed of two subunits acting MTase or REase, with an enzyme able to act solely as MTase. Cleavage depends on ATP and relies on an inversely oriented non-palindromic recognition site, resulting in cleavage at a specific distance from the recognition site. Type IV is composed of two proteins that cleave DNA at about 30 nucleotide distance from a restriction site, comprising two methylated dinucleotides separated by a distance between 40 and 3000 bases. In some instances, the enzyme complex catalyzes both restriction and methylation.

Type II REases represent the most abundant and common enzymes of interest for recombinant DNA technology and DNA studies. With a few exceptions, Type II REases occur as monomers or oligomers consisting of two or four monomers (dimers or tetramers) and harbor only the REase activity, with the associated MTase activity catalyzed by an independent monomeric enzyme. The ATP-independent Type II REases cleave DNA at reliably constant positions within or close to the recognition sites to generate 3′ hydroxyl, and 5′ phosphate ends. The ends are either 3′ or 5′ protruding (sticky) or blunt. Due to the diversity in this class, all of which share the standard features of the ability to cleave DNA within or close to the restriction sequences and generate discrete restriction fragments, the enzymes are grouped further into subdivisions. For instance, type IIC, which comprises types IIB, IIG, and IIH, harbors restriction and methylation domains within a single enzyme. Like all REases, Type II depends on Mg^{2+}, with some requiring additional ions, including Na^+ and K^+. Most enzymes show optimal activity at 37 °C and are inactivated at elevated temperatures. Incubating REases under suboptimal conditions, for example, using ions other than Mg^{2+} with EcoRI, results in a phenomenon known as star activity. Star activity is restriction enzymes' tendency to cleave sequences like their specific recognition site.

REases are among the most widely applied genetic engineering tools for the above properties. Hundreds of commercially available REases are used to create various fragments from various sources with compatible ends and in combination with other enzymes to generate recombinant DNA. Examples include gene cloning and expression used to produce enzymes and medical products such as insulin. REases are also applied in DNA mapping, restriction fragment length polymorphism (RFLP), and mutation studies.

20.2.1.4 Sugar-Nonspecific Nucleases

Benzonase

Benzonase or Serratia nuclease is a non-specific endonuclease isolated from *Serratia marcescens* that attacks and degrades all types of DNA and RNA. The ~26.7 kDa protein occurs in two main isoforms with nearly identical structures and biochemical properties. The endonuclease requires Mg^{2+} to hydrolyze its substrate to yield 3'-OH and 5'- -phosphoryl ends. The end products of Benzonase degradation are typically 2–5 bp 5'- -phosphorylated oligonucleotides. The enzyme is highly stable within a temperature range of 35–44 °C and could be inactivated at 70 °C. The typical use of the enzyme is to remove nucleic acids from purified proteins.

Mung Bean Nuclease

Mung bean nuclease is a 39 kDa single-strand specific endonuclease purified from mung bean *Vigna radiata*, which degrades both DNA and RNA. The enzyme degrades double-stranded DNA under an extended duration of incubation and high enzyme concentrations. The enzyme requires Zn^{2+} and a reducing agent, such as cysteine, for stable activity at pH 5.0. Mung bean and other single-stranded specific endonucleases are thermostable. The enzyme is deactivated using chelating agents like ethylene diamine tetraacetic acid (EDTA) and sodium dodecyl sulfate (SDS). The enzyme is used in blunting 3' and 5' single-stranded overhangs in double-stranded DNA, removal of single-stranded nucleic acids, cleavage of DNA hairpin loops, and RNA transcript mapping.

P1 Nuclease

P1 nuclease from *Penicillium citrinum* is another Zn^{2+}-dependent and single-strand-specific endonuclease. The 36 kDa protein hydrolyzes the phosphodiester bonds in single-stranded nucleic acids yielding 5' mononucleotides. Like Mung bean nuclease, P1 shows trace activity on double-stranded DNA under nonstandard conditions. P1 nuclease shows optimal activity and stability at temperatures ~37–70 °C and 5–8 ranges. The enzyme can be inactivated using SDS and reversibly using guanidine hydrochloride and SDS or urea. P1 nuclease removes nucleic acids during protein purification and analysis of DNA damage and modification.

S1 Nuclease

S1 nuclease is a 32 kDa metalloprotein purified from Aspergillus oryzae. The endonuclease shows absolute specificity to single-stranded nucleic acids, hydrolyzing them into mono- or oligonucleotide fragments. S1 nuclease and other single-strand specific endonucleases readily cleave the opposite side of double-stranded DNA nicked by other enzymes and single-stranded regions of the molecule. The Zn2+ dependent endonuclease shows optimum activity around pH 4.0 with thermal stability varying with pH. Thus, the enzyme is inactivated using chelating agents such as EDTA and citrate. ATP at 1 mM also completely inhibits the activity of the enzyme. The enzymes are among the most widely used single-strand-specific nucleases. For example, it is used to resolve hairpin loops produced during cDNA synthesis, in mutagenesis and DNA/DNA-RNA hybridization studies, and selective digestion and removal of overhangs, among various applications.

Micrococcal Nuclease, MNase

MNase, first isolated in Staphylococcus aureus, is ~16.85 kDa nuclease that cleaves all nucleic acid with a preference for single-stranded molecules. The enzyme prefers cleavage at A/T-containing dinucleotides and possesses both endo- and exonuclease properties. It hydrolyzes nucleic acids generating 3'-phosphate and 3'-P, and 5'-hydroxyl terminated mononucleotides and oligonucleotides. MNase requires Ca^{2+} and has a pH optimum of 9.2 at 37 °C. The enzyme can be inactivated at elevated temperatures or by chelating agents. The most prominent application of the enzyme was nucleosome mapping coupled with sequencing (MNase-seq) and the depletion of nucleic acids from protein samples.

20.2.1.5 Exonucleases

Exonuclease I (EXOI)

The *E. coli* EXOI catalyzes the stepwise hydrolysis of the phosphodiester bonds of linear single-stranded DNA at 3'-hydroxyl ends, yielding deoxyribonucleoside 5'--monophosphates (dNMP) and 5'-terminal dinucleotides, which the enzyme is incapable of cleaving. The 54.5 kDa Mg^{2+}-dependent enzyme shows high processivity, i.e., continuous activity without dissociating from a substrate before the fully modified polymer, cleaving single-stranded DNA at ~275 nucleotides per second. It also tolerates high salt concentrations. DNA with a terminal 3'-hydroxyl blocked by phosphoryl or acetyl groups resist the enzyme, and those with 3'-phosphoryl terminals inhibit it. EXOI is inactivated at elevated temperatures around 80 °C. The current application of EXOI includes the clean-up of single-stranded DNA and fragments from reaction mixtures. It was a candidate component in the preliminary stages of single molecule sequencing but for its excessive processivity.

Exonuclease III, (ExoIII)

The *E. coli* ExoIII is an ~28 kDa monomeric enzyme capable of catalyzing multiple nucleolytic activities on double-stranded DNA. The enzyme's 3' exonuclease activity cleaves the 3′-hydroxyl termini (a blunt or a 3′ recessed ends) of double-stranded DNA. This activity generates deoxyribonucleoside 5′-monophosphates (dNMP) and a single-strand on the opposite side of the molecule as end products. The generated 5′ ends are not susceptible to EXOIII digestion. On the other hand, the enzyme generates a gap at abasic sites in a double-stranded DNA to generate a gap. EXOIII carries phosphatase and ribonuclease H activities, dephosphorylating DNA chains that terminate with a 3′-phosphate group and cleaves RNA strands in DNA-RNA hybrids, respectively. The enzyme requires Mg^{2+} for maximal activity, and Zn^{2+} inhibits it. The nuclease specificity can be optimized for specific applications by adjusting the reaction conditions, including temperature, salt concentration, and enzyme/substrate ratio. The enzyme is applied in mutation studies, synthesis of strand-specific labelled probes, and generation of single-stranded DNA.

Exonuclease VII (EXOVII)

EXOVII is an 88-kDa DNA-specific exonuclease purified from *E. coli* that specifically cleaves single-stranded DNA or double-stranded DNA with unpaired termini. It catalyzes the cleavage of single-stranded DNA in both 3′–5′ and 5′–3′ directions, yielding short oligonucleotides. EXOVII can remove single-stranded primers at the end of each cycle of PCR reaction involving multiple sets of primers (nested PCR). It is also useful for removing single-stranded DNA in double-stranded DNA in a sample and intron mapping. EXOVII requires no divalent cation and therefore remains active in chelating agents up 8 mM EDTA.

Lambda Exonuclease (λ Exo)

λ exo is an ~85.4 kDa homotrimer purified from an λ-phage infected *E. coli* with exonuclease activity at trimer and monomer levels. It selectively cleaves on the 5′-phosphorylated strand of double-stranded DNA to yield 5′-mononucleotides and the complementary single-stranded DNA. The enzyme is highly processive around pH 7.0, requires Mg^{2+} and can be inactivated at elevated temperatures. λ exo in converting linear double-stranded DNA to single-stranded DNA for various applications, including DNA microarrays, pyrosequencing technology, single-stranded conformation polymorphism technique (SSCP), single-nucleotide polymorphism analysis, among numerous others.

Bal 31 Nucleases

Alteromonas espejiana extracellular nuclease Bal 31 occur in two distinct exonucleases, the fast and slow molecular species, with molecular weights of 109 and 85 kDa, respectively. The designation was meant to distinguish the relative rates at which the two species cleave duplex DNA. Bal31 nuclease progressively cleaves 3′ and 5′ termini of duplex DNA without introducing internal scissions yielding a double-stranded molecule and 5′-mononucleotides. They also harbor 3′ and 5′ exonuclease and endonuclease properties against single-stranded DNA. Deactivation of the enzyme requires an elevated temperature of 95 °C. It is used for the progressive removal of nucleotides from both ends of double-stranded DNA, restriction site mapping and mutation studies.

20.2.2 Ribonucleases

Nucleases that selectively catalyze RNA degradation into smaller components are termed ribonucleases (RNases). Based on their activities, RNases fall into two major categories: possess endoribonucleases, which cleave internal phosphodiesters, or exoribonucleases, which remove nucleotide from the 3′ or 5′ termini of an RNA molecule. RNases are ubiquitous, varying in function, structure, and sources. A prominent application of RNases includes the removal of contaminating RNA from various samples (Table 20.2).

20.2.2.1 Endoribonuclease

RNase A

The bovine pancreatic ribonuclease A (RNase A) is an endoribonuclease that cleaves the 3′,5′-phosphodiester bonds of single-stranded RNA by recognizing uracil and cytosine generating 3′ polycytidylic (Cp) or polyuridylic (Up) acids with 2′,3′-cyclic phosphate intermediates as described. Cleavage of Cp and Up by the enzyme RNase A is distributive, as the enzyme binds its polymeric substrates, catalyze the reaction releasing polymeric products (bind, cleave and release sequence). However, cleavage of poly(A) by the same enzyme is processive. RNAse A belongs to the group of well-characterized ribonucleases with four Nobel prizes for which the enzyme plays a pivotal role in the discoveries. Aside from the descriptive source above, the enzyme is ubiquitous among animals and widespread in the environment due to its hardiness. The enzyme depletes RNA contamination in DNA preparations and is also applied to remove unhybridized RNA in RNA/DNA hybrids. RNase A activity requires Mg^{2+} and is active over a wide temperature range. It remains the most formidable challenge in handling RNA samples for various applications, despite the availability of various commercial RNase inhibitors.

RNase H

RNase H, first purified from the calf thymus, is an endoribonuclease that hydrolyzes specifically the phosphodiester bonds of RNA in RNA/DNA hybrids. Variants of RNase H occur among eukaryotes and prokaryotes. The *E. coli* RNases HI and HII are active as monomeric proteins of molecular weights ~17.6 and 21.5 kDa, respectively. RNase H cleaves RNA at the 3′-phosphodiester bonds of the RNA/DNA hybrid to yield a product with 3′-hydroxyl and 5′-phosphate termini. The enzyme has an absolute requirement of the divalent cation Mg^{2+}, with Mn^{2+} serving as a less effective substitute. RNase H achieved optimal activity pH 7.5–9.1 in the presence of reducing reagents. The enzyme could be inactivated by EDTA and inhibited dextran and sugars. Applications of RNase H include its use in DNA synthesis, poly(A) sequence removal, and antiviral research.

20.3   Programmable Nucleases

Programmable nucleases (Table 20.3) are a class of genetic engineering tools that allow for precise and targeted manipulation of DNA sequences. These enzymes can be programmed to cleave specific sequences of DNA, leading to changes in gene expression or even the deletion or insertion of new genetic material. They typically consist of a nuclease linked to

Table 20.3 Selection of programmable nucleases applied in molecular biology and biotechnology. All listed enzymes are typically recombinantly expressed and often molecularly engineered for increased performance tailored to the respective application

Enzyme	Description	Application
Programmable nucleases		
Argonautes	Consist of a protein component and a short single-strand guide DNA or RNA. Sequenz-specific target binding induces conformational changes which lead to the cleavage of single or double-strand DNA or RNA	Cutting DNA or RNA at a customizable 16–18 nt sequence TtAgo with a guide DNA induces single-strand breaks in DNA
CRISPR/CAS9	In vivo part of the procaryotic adaptive immune system. Consists of a crRNA and a tracrRNA bound to a CAS9 nuclease. Cleavage of foreign dsDNA at a 20-nucleotide complementary sequence followed by a PAM sequence	crRNA and tracr RNA are replaced by a single synthetic guide RNA (sgRNA). Sequence specifically cleavage of ds-DNA inducing non-homologous end joining (NHEJ), or homology-directed repair (HDR), a variant of CAS9 can create site-specific single strand nicks inducing HDR, fusion of a reporter protein, an activator or repressor to the CAS9 nuclease are used to visualize, activate or repress gene transcription
Zinc Finger Nucleases (ZFNs)	Composed of a customizable zinc finger protein that binds to a specific sequence of DNA and is bound to nuclease domain (FokI, EC 3.1.21.4) that hydrolyzes the DNA	They are used for gene therapy, genome editing, and basic research
Transcription activator-like effector nucleases (TALENs)	Composed of a Transcription Activator-Like (TAL) effector protein, which binds to a specific sequence of DNA, and is bound to nuclease domain (FokI, EC 3.1.21.4), which hydrolyzes the DNA	They are used for gene therapy, genome editing, and basic research. TALENs have higher specificity and flexibility and are easier to design and construct compared to ZFNs

a sequence-specific binding subunit. This sequence-specific binding subunit consists of either a protein as in ZFN and TALEN or a nucleic acid like in Argonaut proteins or CRISPR-CAS9.

ZFN (zinc finger nuclease) and TALEN (transcription activator-like effector nuclease) use protein domains called zinc fingers or transcription activator-like effectors to bind to specific DNA sequences, and a nuclease domain to cleave the DNA at that location. FokI, a type II restriction enzyme isolated from *Flavobacterium okeanokoites* is used as a nuclease in ZFN and TALEN.

Each zinc finger consists of a zinc atom bound to a Csy_2His_2 motif in an ~30 amino acid unit and binds 3 bp of DNA. With this it is possible to engineer zinc fingers that bind to almost any specific sequence of DNA, making it possible to target a specific gene for cleavage. As the FokI nuclease has to dimerize for cleavage of DNA, normally two ZFNs are constructed on neighboring sequences.

The TAL effector component is composed of ~34 amino acid units, each binding to a single base pair. With this, TALENs have a simpler structure compared to ZFNs, which makes them easier to design and construct.

Another example of a programmable nuclease is the Argonaute protein, which is a component of the RNA-induced silencing complex (RISC) found in many eukaryotic organisms. Argonaute binds to small RNAs, such as microRNAs, and uses them as a guide to locate and cleave complementary RNA sequences. This process, known as RNA interference (RNAi), can be used to silence specific genes. Prokaryotic Argonaute proteins can use either DNA or RNA as the guide nucleic acid and act on either DNA or RNA. The guide nucleic acid can be designed to bind to a specific sequence making it possible to induce sequence-specific DNA/RNA cleavage. Commercially available is TtAgo, an argonaute protein originated from *Thermus thermophilus*. Here the guide nucleic acid is a short $5'$-phosphorylated single-stranded DNA. TtAgo introduces one break in the phosphodiester backbone of the complementary substrate sequence.

The most famous programmable nuclease is CRISPR-CAS9, a system that originated in bacteria as a defense mechanism against invading viruses. CRISPR stands for Clustered Regularly Interspaced Short Palindromic Repeats, and CAS9 stands for CRISPR-associated protein 9. The CRISPR-CAS9 system is made up of two components: a small synthetic guideRNA that binds to a specific DNA sequence, and the CAS9 enzyme, which cleaves the DNA at that location.

All four programmable nucleases can be used to delete or insert genetic material, or to make precise changes to specific genes. Double-strand breaks introduced by programmable nucleases can be repaired by homology-directed repair, which leads to gene insertion, correction and point mutagenesis, or by faulty non-homologous end-joining, which results in gene disruptions. The repair of two concurrent double-strand breaks can give rise to chromosomal rearrangements such as deletions, inversions, and translocations. They have applications in research, medicine, and agriculture, and have the potential to lead to new treatments and therapies for a wide range of diseases.

20.4   DNA/RNA Modifying Enzymes

The final part of this chapter covers a diverse group of enzymes used in molecular biotechnology acting on DNA and RNA altering the 3′ or 5′ ends of the molecule by adding or removing a phosphate group, adding methyl groups or more nucleotides, facilitating the repair and stabilizing single-strand polynucleotides (Table 20.4).

20.4.1 Ligases

Ligases are used to join DNA double strands, or DNA-RNA hybrids produced by, e.g., restriction digestion or PCR. They are used in cloning experiments as well as in library preparation for NGS sequencing and can close both single-strand and double-strand breaks. The prerequisite is that a 5′ phosphate end meets a 3′ hydroxyl end and ATP and Mg^{++} is added. A new 3′–5′-phosphodiester bond is then formed between these two molecules. The widest use has the T4-DNA Ligase originated from the bacteriophage T4 and produced recombinantly in *E. coli*. The enzyme was first isolated from T4-infected *E. coli*.

20.4.2 Phosphatases

Phosphatases catalyze the cleavage of the 5′- and 3′-phosphate from DNA and RNA and dNTPs. They are used in cloning, preventing self-ligation of cloning vectors, and enhancing the yields of RNA in transcription reactions. To facilitate subsequent steps in cloning, the enzyme must be deactivated. Therefore, temperature sensitive enzymes are used. The Shrimp Alkaline Phosphatase (SAP) can be inactivated after 2 min at 65 °C and is available as a recombinantly produced enzyme (rSAP). The enzyme needs no cofactor.

20.4.3 Kinases

Kinases add a phosphate group to the 5′-end of DNA fragments, a reaction also called phosphorylation. The reaction requires ATP as a cofactor. The T4 Polynucleotide Kinase originally isolated from the bacteriophage T4 has also a 3′-phosphatase activity but is available as a modified recombinant enzyme with no 3′-phosphatase activity. Kinases are used for the phosphorylation of DNA fragments (e.g., PCR Products) prior to ligation to dephosphorylated vectors and for end-labeling of DNA fragments with radioactive phosphates for detection.

Table 20.4 Selection of DNA/RNA-modifying enzymes applied in molecular biology and biotechnology. All listed enzymes are typically recombinantly expressed and often molecularly engineered for increased performance tailored to the respective application

Enzyme	Description	Application
DNA/RNA-modifying enzymes		
Ligase EC 6.5.1.X	Catalyzes phosphodiester bond formation between a 5′-phosphate and a 3′ hydroxyl end of an RNA or DNA, ATP or NAD+ dependent	Ligation of DNA fragments
Phosphatase EC 3.1.3.1	Removal of 5′ phosphate groups from DNA and RNA	Preparation of templates for 5′ end labeling, minimizing recircularization of cloning vectors, optimizing yields of RNA in transcription reactions
Kinase EC 2.7.1.78	Phosphorylation of 5′ prime hydroxyl groups, ATP dependent	Used in the cloning of PCR-generated DNA fragments
Methyltransferase EC 2.1.1.37	Transfers sequence specific a methyl group from SAM to a base in the DNA.	Masking restriction enzyme recognition sites and used in gene expression studies
Capping enzyme EC 2.7.7.50	Creates the Cap Structure (m7Gppp5′ N), combines RNA tri-phosphatase, guanylyl transferase, and guanine methyltransferase activity. SAM and GTP-dependent	In vitro production of translatable eucaryotic mRNA
Glycosylase EC 3.3.3.X EC3.3.3.27	Recognize and remove damaged bases but do not specifically cleave the DNA backbone.	Base editing
ATP Sulfurylase EC 2.7.7.4	Transferring sulfate to the adenine monophosphate moiety of ATP to form adenosine 5′-phosphosulfate (APS) and pyrophosphate	Used in pyrosequencing
Cre-Recombinase EC 6.6.2.1	Cre Recombinase is a Type I topoisomerase.	Excision or inversion of DNA between two loxP sites and fusion of DNA molecules containing loxP sites
RecA EC 3.4.99.37	Are ATP-dependent single-strand binding proteins involved in strand displacement and strand exchange, polymerize along the single-strand DNA, involved in the homologous DNA repair process	Stabilizing and visualization of single-strand DNA
Single-Strand Binding Proteins	Stabilize single-strand DNA	Isothermal DNA amplification, marking and stabilizing of ssDNA
Helicase EC 5.6.2.3	Separates dsDNA into single strands	Isothermal DNA amplification

20.4.4 Terminal Transferases

Terminal transferases, also known as terminal deoxynucleotidyl transferases (TdT), are enzymes that add template-independent nucleotides to the 3′ end of a DNA strand at a rate of up to 1000 nucleotides per minute. They are commonly found in the immune system, where they play a role in the diversification of antibody and T-cell receptor genes.

20.4.5 Methyltransferase

Methyltransferases are a class of enzymes that transfer methyl groups from a methyl donor, such as S-adenosyl methionine (SAM), to a specific site on a substrate. In molecular biology, methyltransferases are particularly important for DNA methylation, a process by which methyl groups are added to the cytosine residues in DNA. This process can have a significant impact on gene expression, as methylated genes are often transcriptionally silent, and on masking restriction enzyme recognition sites. One of the best-known DNA methyltransferases is DNMT3A and DNMT3B, which are responsible for the de novo methylation of CpG dinucleotides. Another DNA methyltransferase is DNMT1, which is responsible for maintaining methylation patterns during DNA replication. DNA methylation is a key mechanism of epigenetic regulation, and alterations in DNA methylation patterns have been linked to the development of cancer and other diseases.

20.4.6 Capping Enzyme

The capping enzyme is a type of enzyme that plays a critical role in the process of mRNA maturation. This enzyme is responsible for adding a 7-methylguanosine cap to the 5′ end of pre-mRNA molecules, a process known as cap addition. The cap structure is an important feature of eukaryotic mRNA, as it plays a crucial role in the regulation of gene expression and the stability of the mRNA molecule.

The cap structure is composed of a 7-methylguanosine residue linked to the 5′ end of the mRNA via a 5′–5′ triphosphate bond. The capping enzyme catalyzes the transfer of the methylguanosine from a guanosine triphosphate (GTP) substrate to the 5′ end of the pre-mRNA. This process is usually coupled with the removal of the 5′ terminal phosphate group by the enzyme 5′-nucleotidase.

The cap structure serves several functions, including protecting the mRNA from degradation by exonucleases, recruiting the ribosome for translation, and providing a binding site for the cap-binding proteins that are involved in the regulation of mRNA transport, translation, and stability. The capping enzyme has been used as a tool for the modification of mRNA molecules, leading to the development of new treatments for diseases and is also essential for the production of mRNA vaccines. The ability to cap and uncap mRNA molecules has also been exploited to create new therapeutics such as siRNA and microRNA and for the synthesis of recombinant protein in vitro.

20.4.7 Glycosylases

Glycosylases recognize and remove damaged bases but do not specifically cleave the DNA backbone. They cleave the glycosidic bond between an abnormal base and the 2-deoxyribose, creating apurinic or apyrimidinic sites (AP-sites).

20.4.8 Cre-Recombinase

Cre-recombinases are enzymes that are responsible for the site-specific recombination of DNA. It was first discovered in the bacteriophage P1, where it plays a role in the process of lysogeny, the integration of the phage's genetic material into the host bacterium's genome. The Cre-recombinase enzyme is a 38 kDa protein that is composed of two domains: the N-terminal recombination directionality factor (RDF) and the C-terminal recombinase core enzyme (RCE). The RDF domain is responsible for determining the direction of recombination, while the RCE domain is responsible for the actual recombination process.

There are two main types of Cre-recombinases: the wild-type Cre-recombinase and the mutant Cre-recombinase. The wild-type Cre-recombinase is able to recognize and recombine the loxP sequence, which is a 34 bp DNA sequence that is present in the P1 bacteriophage genome. The mutant Cre-recombinase, on the other hand, is able to recognize and recombine other specific DNA sequences, such as the FRT and lox221 sequences.

The Cre-recombinase enzyme is commonly used in genetic engineering to target specific genomic regions for manipulation. One of the most popular methods of using Cre-recombinase is called the Cre-lox system, which utilizes the recognition of the loxP sequence by the Cre-recombinase enzyme to target specific genomic regions for deletion, replacement, or activation. In current research, efforts have been directed toward altering the properties of many recombinases by protein engineering to widen the applicability.

20.4.9 Single-Strand Binding Proteins

There are several types of single-strand binding proteins, involved in the replication, regulation, and DNA repair reactions in vivo. The best-known proteins and their application in molecular biotechnology are RecA, single-strand binding proteins, and helicases:

20.4.10 RecA-Type Proteins

RecA-proteins are single-strand binding proteins that stimulate single-strand assimilation, a process by which a single strand of DNA displaces its homolog in a double helix. It is necessary for genetic recombination reactions involving DNA repair, SOS repair, and UV-induced mutagenesis.

20.4.11 Single-Strand Binding Proteins

Single-strand binding proteins (SSB) stabilize single-strand DNA in DNA replication. They are used for isothermal DNA amplification and for marking and stabilizing of single-strand

20.4.12 DNA Helicases

In vivo helicases are involved in the replication of DNA. They have a double-strand DNA unwinding activity. In molecular biotechnology, this activity is applied in isothermal DNA amplification, enabling primer annealing and extension by a strand-displacing DNA polymerase. It is also applied in nanopore sequencing (Fig. 20.3).

A good example of the application of nearly all the enzymes described in this chapter is the manufacturing of mRNA as vaccines or as mRNA drugs. In the first step, the target sequence has to be determined. Therefore, the source mRNA of either the virus protein or a cancer-specific protein has to be transcribed with a reverse transcriptase into DNA and then multiplied by PCR (DNA polymerase). The DNA can then be sequenced, e.g., by Nanopore sequencing where a helicase is applied to produce single-strand DNA. DNA with the target sequence as well as the vector will be prepared for cloning by digestion with restriction endonucleases, phosphatases, and kinases. A ligase will finally stitch the fragments together to form a circular plasmid that can be transformed into the host organism. The plasmid, which is multiplied in vivo, has to be purified from the cells (use of lysozyme, proteases, and RNAses) and linearized (restriction endonucleases). All applied enzymes have to be separated from the linearized DNA template and can be recycled to save cost. For mRNA synthesis, an RNA polymerase, a capping enzyme, and a Poly(A) polymerase is needed. In large-scale applications, the enzymes are recycled, mRNA is capped by cotranscriptional capping, and the final mRNA is purified by chromatography and tangential flow filtration.

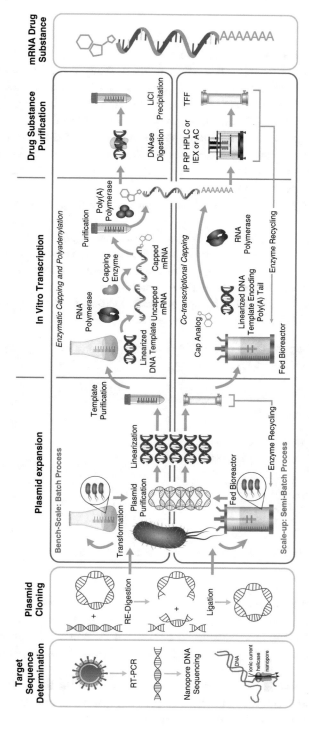

Fig. 20.3 Enzymes involved in large-scale production of mRNA drugs and vaccines including target sequencing, cloning in vitro transcription, and downstream processing. Acronyms: RT-PCR, reverse transcription polymerase chain reaction; IP RP HPLC, ion-pair reverse phased high-performance liquid chromatography; IEX, ion exchange chromatography; AC, affinity chromatography; TFF, tangential flow filtration. Adapted from: Cameron Webb, Shell Ip, Nuthan V. Bathula, Petya Popova, Shekinah K. V. Soriano, Han Han Ly, Burcu Eryilmaz, Viet Anh Nguyen Huu, Richard Broadhead, Martin Rabel, Ian Villamagna, Suraj Abraham, Vahid Raeesi, Anitha Thomas, Samuel Clarke, Euan C. Ramsay, Yvonne Perrie, and Anna K. Blakney (2022). Current Status and Future Perspectives on MRNA Drug Manufacturing; Molecular Pharmaceutics 19 (4), 1047–1058; doi: 10.1021/acs.molpharmaceut.2c00010

20.5 Questions

1. **Q1: Which reaction is catalyzed by polymerases?**
   A1: Polymerases produce nucleic acids, DNA or RNA. They link nucleotides by forming phospho-diester bonds between the 3'-hydroxy-group of the former nucleotide and the 5'-α-phosphate of the next one to form a nucleic acid chain. The sequence is usually determined by another nucleic acid, which is used as a template.

2. **Q2: Which types of polymerases can be defined and how is the distinction made?**
   A2: Polymerases are categorized by the type of template and the product. A distinction can be made between DNA-dependent and RNA-dependent DNA and RNA polymerases as well as independent polymerases.

3. **Q3: Give three examples of biotechnological methods in which polymerases are used.**
   A3: DNA-dependent DNA polymerase, e.g., Taq polymerase, is used in PCR, DNA-dependent RNA polymerase, e.g., from bacteriophage T7, is used in in vitro transcription, and RNA-dependent DNA polymerase/reverse transcriptase from retroviruses is used in reverse transcription/cDNA synthesis.

4. **Q4: Place the following enzymes in the appropriate position in the table below DNase I, EXOVII, EXOIII, and Lambda Exonuclease.**

Enzyme	Polarity	Substrate	Products
1	Endonuclease	All forms of DNA	Di-, tri-, and oligonucleotides, single- and double-stranded DNA
2	3'→ 5'	All forms of DNA except circular DNA	Deoxyribonucleoside 5'-monophosphates, single-stranded DNA
3	both	only linear single-stranded DNA	Short oligos
4	5' → 3'	All forms of DNA except circular DNA	Deoxyribonucleoside 5'-monophosphates, dinucleotide, single-stranded DNA
5	Endonuclease	only linear single-stranded DNA	5' mononucleotides

A4: DNase I, (2) EXOIII, (3) EXOVII, (4) Lambda Exonuclease, and (5) P1 nuclease.

5. **Q5: A scientist interested in generating complementary DNA (cDNA) from messenger RNA (mRNA) encountered the following questions on RNase A and H.**
   (a) **Agarose gel electrophoresis using an EDTA buffer generated a smear of RNA samples, suggesting RNA degradation. What could explain the observation?**
   (b) **Which RNases should the investigator use in removing the RNA template after first-strand complementary DNA (cDNA) synthesis, and is the second enzyme unsuitable for this experiment?**

A5:

(a) Potential contamination with RNases, possibly RNase A, during RNA extraction or handling. RNases are ubiquitous. Unlike RNase A, RNase H is deactivated by EDTA. Potential sources of contamination include the investigator's skin or breath, the air, the electrophoresis equipment and buffers, tubes, pipettes, and tips.

(b) RNase H. It is an endonuclease that specifically cleaves the phosphodiester bonds of RNA in an RNA-DNA hybrid. By contrast, RNase A is an endonuclease that specifically cleaves the 3′ ends of unpaired or single-stranded RNA at cytosine and uracil residues.

6. **Q6: Based on the recognition sites and indicated cleavage position of the following:**
   (a) **Describe the type and sequence of overhangs produced by the REases**
   (b) **Which nuclease could cleave the generated overhangs to yield blunt ends?**
   (c) **Which pair or set are isoschizomers and neoschizomers?**

       (I) AcoI

          5′-Y↓GGCCR-3′
          3′-RCCGG↑Y-5′

       (II) ZraI

          5′-GAC↓GTC-3′
          3′-CTG↑CAG-5′

       (III) AfeI

          5′-AGC↓GCT-3′
          3′-TCG↑CGA-5′

       (IV) EaeI

          5′-Y↓GGCCR-3′
          3′-RCCGG↑Y-5′

       (V) NotI

          5′-GC↓GGCCGC-3′
          3′-CGCCGG↑CG-5′

       (VI) EcoRV

          5′-GAT↓ATC-3′
          3′-CTA↑TAG-5′

       (VII) AatII

          5′-GACGT↓C-3′
          3′-C↑TGCAG-5′

A6:

  (a) Blunt ends (EcoRV, AfeI, and ZraI), sticky ends with 3′ overhangs (AatII:3′--ACGT), and sticky ends with 5′ overhangs (AcoI: 5′-GGCC, EaeI: 5′ GGCC, and VotI: 5′-GGCC).
  (b) Mung bean nuclease is a single-stranded specific endonuclease that removes 3′ and 5′ single-stranded overhangs in double-stranded DNA.
  (c) Isoschizomers (AcoI and EaeI) and neoschizomers (AatII and ZraI).

 7. **Q7: What is the main difference between ZFN and TALEN on the one hand and CRISPR-CAS9 and the Argonaut protein on the other hand?**
    A7: The DNA binding domain in ZFN and TALEN consists of a protein whereas in CRISPR Cas9 and in the Argonaut protein a synthetic guide DNA or RNA is used. The Nuclease is FokI for ZFN and TALEN, whereas CAS9 and the Argonaut protein are endonucleases themself. A sequence-specific guide DNA or RNA is easier to construct than the sequence-specific zinc fingers of TALEN and ZFN.
 8. **Q8: How could the Cre/Lox Recombinase system be used in removing a marker gene in recombinant plants?**
    A8: Marker genes are placed between Lox sequences. Recombination catalyzed by the cre recombinase of these sites lead to deletion of the marker gene.
 9. **Q9: Which enzymes need *S*-adenosylmethionin (SAM) as a cofactor?**
    A9: Methyltransferases transfer the methyl group from SAM to a specific base of the recognition sequence.
10. **Q10: What is the role of the capping enzyme in mRNA vaccine manufacturing?**
    A10: It adds the Cap Structure (m7Gppp5′N) to facilitates the eukaryotic translation of the mRNA.
11. **Q11: Describe the general workflow to produce an mRNA vaccine.**
    A11: The gene of the surface protein is cloned, the plasmid is amplified, purified, and linearized. In vitro transcription is used to produce translatable mRNA which serves as the active compound in the vaccine.

Take Home Message/Conclusion
- Enzymes acting on nucleic acid synthesize, hydrolyze, or modify the molecules.
- Polymerases synthesize nucleic acids by connecting nucleotide triphosphates by forming phospho-diester bonds using a template to produce a new DNA- or RNA-strand.
- Nucleases hydrolyze phosphodiester bonds within or at the ends of nucleic acid, producing various fragment types.

(continued)

- Nucleic acids can be modified at their 3' or 5' ends by phosphatases and kinases, internally by glycosidases and methyltransferases, or stitched together by ligases.

  Neither a complete list of enzymes nor a complete description of all catalyzed reactions in vivo is given in this chapter. In summary, enzymes used in molecular biotechnology are usually applied for basic research to elucidate gene functions or in cloning techniques to produce expression systems in all sectors of biotechnology and pharmaceutical applications. Only a limited number of enzymes are used in the bulk production of nucleic acids, e.g., the production of mRNA vaccines. However, it is to be expected that this field will increase in the future.

References

Further Reading—Books

Desai S, Patil N, Sivaram A. Recent trends and advances. In: Patil N, Sivaram A, editors. A complete guide to gene cloning: from basic to advanced. Techniques in life science and biomedicine for the non-expert. Cham: Springer; 2022. https://doi.org/10.1007/978-3-030-96851-9_8.

Further Reading—Online Resources

Further explanation and reaction conditions can be found for most of the mentioned enzymes in the NEB online catalogue: https://international.neb.com/products

Specific References

To Polymerases in General

Ishino S, Ishino Y. DNA polymerases as useful reagents for biotechnology—the history of developmental research in the field. Front Microbiol. 2014;5:465. https://doi.org/10.3389/fmicb.2014.00465.

To DNA Polymerases for PCR

Cline J, Braman JC, Hogrefe HH. PCR fidelity of pfu DNA polymerase and other thermostable DNA polymerases. Nucleic Acids Res. 1996;24:3546–51. https://doi.org/10.1093/nar/24.18.3546.

Wang Y, Prosen DE, Mei L, Sullivan JC, Finney M, Vander Horn PB. A novel strategy to engineer DNA polymerases for enhanced processivity and improved performance in vitro. Nucleic Acids Res. 2004;32:1197–207. https://doi.org/10.1093/nar/gkh271.

To DNA Polymerases for Sequencing

Chen C-Y. DNA polymerases drive DNA sequencing-by-synthesis technologies: both past and present. Front Microbiol. 2014;5:305. https://doi.org/10.3389/fmicb.2014.00305.

To Non-coding RNA

Mattick JS, Makunin IV. Non-coding RNA. Hum Mol Genet. 2006;15(1):R17–29. https://doi.org/10.1093/hmg/ddl046.

To Reverse Transcription and Retrotransposons

Conlan LH, Stanger MJ, Ichiyanagi K, Belfort M. Localization, mobility and fidelity of retrotransposed Group II introns in rRNA genes. Nucleic Acids Res. 2005;33:5262–70. https://doi.org/10.1093/nar/gki819.

Hughes SH. Reverse transcription of retroviruses and LTR retrotransposons. Microbiol Spectr. 2015;3 https://doi.org/10.1128/microbiolspec.MDNA3-0027-2014.

Pyle AM. The tertiary structure of group II introns: implications for biological function and evolution. Crit Rev Biochem Mol Biol. 2010;45:215–32. https://doi.org/10.3109/10409231003796523.

To Nucleases

Mikkola S, Lönnberg T, Lönnberg H. Phosphodiester models for cleavage of nucleic acids. Beilstein J Org Chem. 2018;14:803–37. https://doi.org/10.3762/bjoc.14.68.

Roberts GC, Dennis EA, Meadows DH, Cohen JS, Jardetzky O. The mechanism of action of ribonuclease. Proc Natl Acad Sci U S A. 1969;62(4):1151–8. https://doi.org/10.1073/pnas.62.4.1151.

Roberts RJ, Belfort M, Bestor T, Bhagwat AS, Bickle TA, Bitinaite J, et al. A nomenclature for restriction enzymes, DNA methyltransferases, homing endonucleases and their genes. Nucleic acids Res. 2003;31(7):1805–12. https://doi.org/10.1093/nar/gkg274.

Valsala G, Sugathan S. Enzymes as molecular tools. In: Sugathan S, Pradeep NS, Abdulhameed S, editors. Bioresources and bioprocess in biotechnology: Volume 2: Exploring potential biomolecules. Singapore: Springer; 2017. p. 99–128.

To Programmable Nucleases

Carroll D. Genome engineering with zinc-finger nucleases. Genetics. 2011;188(4):773–82. https://doi.org/10.1534/genetics.111.131433.

Kim H, Kim JS. A guide to genome engineering with programmable nucleases. Nat Rev Genet. 2014;15:321–34. https://doi.org/10.1038/nrg3686.

To Capping Enzyme

Topisirovic I, Svitkin YV, Sonenberg N, Shatkin AJ. Cap and cap-binding proteins in the control of gene expression. Wiley Interdiscip Rev RNA. 2011;2(2):277–98. https://doi.org/10.1002/wrna.52. Epub 2010 Oct 28. PMID: 21957010

To Cre-Recombinase

Gaj T, Sirk SJ, Barbas III CF. Expanding the scope of site-specific recombinases for genetic and metabolic engineering. Biotechnol Bioeng. 2013; https://doi.org/10.1002/bit.25096.

To Helicase

Vincent M, Xu Y, Kong H. Helicase-dependent isothermal DNA amplification. EMBO Rep. 2004;5 (8):795–800. https://doi.org/10.1038/sj.embor.7400200. Epub 2004 July 9

Index